Advanced Research on Nanotechnology for Civil Engineering Applications

Anwar Khitab
Mirpur University of Science and Technology, Pakistan

Waqas Anwar
Mirpur University of Science and Technology, Pakistan

A volume in the Advances in Civil and Industrial Engineering (ACIE) Book Series

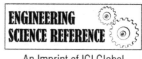

An Imprint of IGI Global

Published in the United States of America by
Engineering Science Reference (an imprint of IGI Global)
701 E. Chocolate Avenue
Hershey PA 17033
Tel: 717-533-8845
Fax: 717-533-8661
E-mail: cust@igi-global.com
Web site: http://www.igi-global.com

Copyright © 2016 by IGI Global. All rights reserved. No part of this publication may be reproduced, stored or distributed in any form or by any means, electronic or mechanical, including photocopying, without written permission from the publisher.
Product or company names used in this set are for identification purposes only. Inclusion of the names of the products or companies does not indicate a claim of ownership by IGI Global of the trademark or registered trademark.

Library of Congress Cataloging-in-Publication Data

Names: Khitab, Anwar, 1972- editor. | Anwar, Waqas, 1988- editor.
Title: Advanced research on nanotechnology for civil engineering applications
  / Anwar Khitab and Waqas Anwar, editors.
Description: Hershey, PA : Engineering Science Reference, [2016] | Includes
  bibliographical references and index.
Identifiers: LCCN 2016006923| ISBN 9781522503446 (hardcover) | ISBN
  9781522503453 (ebook)
Subjects: LCSH: Nanostructured materials--Industrial applications. | Building
  materials. | Nanotechnology. | Materials--Research.
Classification: LCC TA418.9.N35 A32877 2016 | DDC 620.1/15--dc23 LC record available at http://lccn.loc.gov/2016006923

This book is published in the IGI Global book series Advances in Civil and Industrial Engineering (ACIE) (ISSN: 2326-6139; eISSN: 2326-6155)

British Cataloguing in Publication Data
A Cataloguing in Publication record for this book is available from the British Library.

All work contributed to this book is new, previously-unpublished material. The views expressed in this book are those of the authors, but not necessarily of the publisher.

Advances in Civil and Industrial Engineering (ACIE) Book Series

ISSN: 2326-6139
EISSN: 2326-6155

Mission

Private and public sector infrastructures begin to age, or require change in the face of developing technologies, the fields of civil and industrial engineering have become increasingly important as a method to mitigate and manage these changes. As governments and the public at large begin to grapple with climate change and growing populations, civil engineering has become more interdisciplinary and the need for publications that discuss the rapid changes and advancements in the field have become more in-demand. Additionally, private corporations and companies are facing similar changes and challenges, with the pressure for new and innovative methods being placed on those involved in industrial engineering.

The **Advances in Civil and Industrial Engineering (ACIE) Book Series** aims to present research and methodology that will provide solutions and discussions to meet such needs. The latest methodologies, applications, tools, and analysis will be published through the books included in **ACIE** in order to keep the available research in civil and industrial engineering as current and timely as possible.

Coverage

- Engineering Economics
- Earthquake engineering
- Operations Research
- Transportation Engineering
- Materials Management
- Ergonomics
- Urban Engineering
- Construction Engineering
- Production Planning and Control
- Optimization Techniques

IGI Global is currently accepting manuscripts for publication within this series. To submit a proposal for a volume in this series, please contact our Acquisition Editors at Acquisitions@igi-global.com or visit: http://www.igi-global.com/publish/.

The Advances in Civil and Industrial Engineering (ACIE) Book Series (ISSN 2326-6139) is published by IGI Global, 701 E. Chocolate Avenue, Hershey, PA 17033-1240, USA, www.igi-global.com. This series is composed of titles available for purchase individually; each title is edited to be contextually exclusive from any other title within the series. For pricing and ordering information please visit http://www.igi-global.com/book-series/advances-civil-industrial-engineering/73673. Postmaster: Send all address changes to above address. Copyright © 2016 IGI Global. All rights, including translation in other languages reserved by the publisher. No part of this series may be reproduced or used in any form or by any means – graphics, electronic, or mechanical, including photocopying, recording, taping, or information and retrieval systems – without written permission from the publisher, except for non commercial, educational use, including classroom teaching purposes. The views expressed in this series are those of the authors, but not necessarily of IGI Global.

Titles in this Series

For a list of additional titles in this series, please visit: www.igi-global.com

Advanced Manufacturing Techniques Using Laser Material Processing
Esther Titilayo Akinlabi (Univeristy of Johannesburg, South Africa) Rasheedat Modupe Mahamood (University of Johannesburg, South Africa & University of Ilorin, Nigeria) and Stephen Akinwale Akinlabi (University of Johannesburg, South Africa)
Engineering Science Reference • copyright 2016 • 288pp • H/C (ISBN: 9781522503293)
• US $175.00 (our price)

Handbook of Research on Applied E-Learning in Engineering and Architecture Education
David Fonseca (La Salle Campus Barcelona, Universitat Ramon Llull, Spain) and Ernest Redondo (Universitat Politècnica de Catalunya, BarcelonaTech, Spain)
Engineering Science Reference • copyright 2016 • 569pp • H/C (ISBN: 9781466688032)
• US $310.00 (our price)

Emerging Design Solutions in Structural Health Monitoring Systems
Diego Alexander Tibaduiza Burgos (Universidad Santo Tomás, Colombia) Luis Eduardo Mujica (Universitat Politecnica de Catalunya, Spain) and Jose Rodellar (Universitat Politecnica de Catalunya, Spain)
Engineering Science Reference • copyright 2015 • 337pp • H/C (ISBN: 9781466684904)
• US $235.00 (our price)

Robotics, Automation, and Control in Industrial and Service Settings
Zongwei Luo (South University of Science and Technology of China, China)
Engineering Science Reference • copyright 2015 • 337pp • H/C (ISBN: 9781466686939)
• US $215.00 (our price)

Using Decision Support Systems for Transportation Planning Efficiency
Ebru V. Ocalir-Akunal (Gazi University, Turkey)
Engineering Science Reference • copyright 2016 • 475pp • H/C (ISBN: 9781466686489)
• US $215.00 (our price)

Contemporary Ethical Issues in Engineering
Satya Sundar Sethy (Indian Institute of Technology Madras, India)
Engineering Science Reference • copyright 2015 • 343pp • H/C (ISBN: 9781466681309)
• US $215.00 (our price)

Emerging Issues, Challenges, and Opportunities in Urban E-Planning
Carlos Nunes Silva (University of Lisbon, Portugal)
Engineering Science Reference • copyright 2015 • 380pp • H/C (ISBN: 9781466681507)
• US $205.00 (our price)

www.igi-global.com

701 E. Chocolate Ave., Hershey, PA 17033
Order online at www.igi-global.com or call 717-533-8845 x100
To place a standing order for titles released in this series,
contact: cust@igi-global.com
Mon-Fri 8:00 am - 5:00 pm (est) or fax 24 hours a day 717-533-8661

Editorial Advisory Board

N. M. Butt, *Preston University, Pakistan*
Baoguo Han, *Dalian University of Technology, China*
Liaqat Qureshi, *University of Engineering and Technology Taxila, Pakistan*
Mohsin Usman Qureshi, *Sohar University, Oman*
Habib ur Rehman, *Mirpur University of Science and Technology, Pakistan*
Muhammad Kalim Tahir, *Preston University, Pakistan*
George Wardeh, *Université de Cergy-Pontoise, France*
Khawaja Yaldram, *Preston University, Pakistan*

Table of Contents

Foreword xiii

Preface xv

Chapter 1
Classical Building Materials 1
    Anwar Khitab, Mirpur University of Science and Technology, Pakistan
    Waqas Anwar, Mirpur University of Science and Technology, Pakistan

Chapter 2
Nano-Scale Behavior and Nano-Modification of Cement and Concrete
Materials 28
    Liqing Zhang, Dalian University of Technology, China
    Siqi Ding, Dalian University of Technology, China
    Shengwei Sun, Harbin Institute of Technology, China
    Baoguo Han, Dalian University of Technology, China
    Xun Yu, New York Institute of Technology, USA
    Jinping Ou, Dalian University of Technology, China

Chapter 3
Trends in Nanoscopy in Materials Research: Nano-Scale Microscopic
Characterization of Cements 80
    Ahmed Sharif, Bangladesh University of Engineering and Technology
      (BUET), Bangladesh

Chapter 4
Nanotechnology Applications in the Construction Industry 111
    Iman Mansouri, Birjand University of Technology, Iran
    Elaheh Esmaeili, Birjand University of Technology, Iran

Chapter 5
Nanotechnology Future and Present in Construction Industry: Applications in
Geotechnical Engineering 141
> Umair Hasan, Curtin University of Technology, Australia
> Amin Chegenizadeh, Curtin University, Australia
> Hamid Nikraz, Curtin University, Australia

Chapter 6
Applications of Nanotechnology in Transportation Engineering 180
> Imtiaz Ahmed, Mirpur University of Science and Technology, Pakistan
> Naveed Ahmad, University of Engineering and Technology Taxila, Pakistan
> Imran Mehmood, Mirpur University of Science and Technology, Pakistan
> Israr Ul Haq, Mirpur University of Science and Technology, Pakistan
> Muhammad Hassan, Mirpur University of Science and Technology, Pakistan
> Muhammad Umer Arif Khan, Mirpur University of Science and Technology, Pakistan

Chapter 7
Recent Trends and Advancement in Nanotechnology for Water and
Wastewater Treatment: Nanotechnological Approach for Water
Purification 208
> Sushmita Banerjee, University of Allahabad, India
> Ravindra Kumar Gautam, University of Allahabad, India
> Pavan Kumar Gautam, University of Allahabad, India
> Amita Jaiswal, University of Allahabad, India
> Mahesh Chandra Chattopadhyaya, University of Allahabad, India

Chapter 8
Risks and Preventive Measures of Nanotechnology 253
> Waqas Anwar, Mirpur University of Science and Technology, Pakistan
> Anwar Khitab, Mirpur University of Science and Technology, Pakistan

Compilation of References 277

About the Contributors 331

Index 336

Detailed Table of Contents

Foreword xiii

Preface xv

Chapter 1
Classical Building Materials 1
    Anwar Khitab, Mirpur University of Science and Technology, Pakistan
    Waqas Anwar, Mirpur University of Science and Technology, Pakistan

Classical building materials are widely used in Civil Engineering projects. Many ancient generations used fire clay and stone as building blocks. The most common building materials today include stones, concrete, plastics, bitumen, glass, wood, metals, bricks, polymers, tiles and heat resisting materials. Current research work has come up with the considerable improvement in the natural characteristics of these materials. Glasses are available in more variety as they were 50 years ago. Similarly, concrete is now available in ultra-high strengths and even blast resistant form. Recently, use of nano-technology has emerged as a rapidly growing field; success of which is also highly dependent on the basic understanding of classical building materials. Advance research work including implementation of nanotechnology may come up with further improvements in the physical and chemical properties of these materials. This chapter would focus on classical materials in detail covering their physical and chemical characteristics, usages as well as economical suitability.

Chapter 2
Nano-Scale Behavior and Nano-Modification of Cement and Concrete
Materials 28
> *Liqing Zhang, Dalian University of Technology, China*
> *Siqi Ding, Dalian University of Technology, China*
> *Shengwei Sun, Harbin Institute of Technology, China*
> *Baoguo Han, Dalian University of Technology, China*
> *Xun Yu, New York Institute of Technology, USA*
> *Jinping Ou, Dalian University of Technology, China*

Cement and concrete materials are widely used, but the development of them comes cross many problems and challenges, such as high energy consumption, high pollution, poor safety and durability, low smart. Nanotechnology is beneficial to understand the behavior of cement and concrete materials at nano-scale. In addition, nanomaterials have remarkable specific properties and functions which can endow cement and concrete materials high mechanical property and durability and multifunctionality. Therefore, applications and advances of nanotechnology and nanomaterials have injected new vitality into cement and concrete materials. This chapter will give a review about nano-scale behavior of cement and concrete materials, the nano modification methods to cement and concrete materials by using nano-binders and adding nano materials with attention to workability, hydration, mechanical property, durability and other properties of the cement and concrete materials, and the nano modification mechanisms to the cement and concrete materials. Finally, future development and challenge of nano-modificated cement and concrete materials are also discussed.

Chapter 3
Trends in Nanoscopy in Materials Research: Nano-Scale Microscopic
Characterization of Cements... 80
> *Ahmed Sharif, Bangladesh University of Engineering and Technology (BUET), Bangladesh*

Nanotechnology has become one of the most emerging research areas to the researchers of the present world due to the wide application of nanomaterials including structures and buildings. With the rapid advancement of nanotechnology, manipulation and characterization of materials in nano scale have become an obvious part of construction related technology. This chapter will focus on some of the nano characterization techniques that are most frequently used in current research of nano materials. In particular scanning electron microscopy, transmission electron microscopy, atomic force microscopy, scanning tunneling microscopy, tomography, scanning transmission X-ray microscopy and laser scanning confocal microscopy are addressed. The basic principle of these characterization techniques and their limitations were briefly discussed in this chapter. In addition, a number of case studies related to microscopic characterization of nano materials utilizing the aforementioned techniques from the published literature were discussed.

Chapter 4
Nanotechnology Applications in the Construction Industry 111
    Iman Mansouri, Birjand University of Technology, Iran
    Elaheh Esmaeili, Birjand University of Technology, Iran

Nanotechnology refers to the understanding and manipulation of materials on the nanoscale (<100 nm). This can lead to marked changes in material properties and can result in improved performance and new functionality. Nanomaterials with properties such as corrosion resistance, and strength and durability are of particular interests to construction professionals, because, these properties directly affect the selection of construction materials, erection methods, and on-site handling techniques. Applying nanotechnology to construction, in some cases, may result in visionary and paradigm-breaking advances. The incorporation of nanomaterials can improve structural efficiency, durability and strength of cementitious materials and can thereby assist in improving the quality and longevity of structures. This chapter tries to analyze nanotechnology in the context of construction and explores the current scenario of nanotechnology in the construction industry. In order to identify the potential benefits and existing barriers, an extensive literature review is conducted.

Chapter 5
Nanotechnology Future and Present in Construction Industry: Applications in
Geotechnical Engineering 141
    Umair Hasan, Curtin University of Technology, Australia
    Amin Chegenizadeh, Curtin University, Australia
    Hamid Nikraz, Curtin University, Australia

After the introduction of nanotechnology, it has been widely researched in geotechnical engineering field. This chapter aims to study these advancements with specific focus on geotechnical applications. In-situ probing of soil and rock masses through nanomaterials may help in providing better safeguards against natural hazards. The molecular dynamics and finite element methods may also be used for the modelling of the nanostructures to better understand the material behavior, causing a bottom-up approach from nano to macroscopic simulations. Nanoclays, nano-metallic oxides and fibers (carbon nanotubes) can enhance the mechanical characteristics of weak, reactive and soft soils. Nanomaterials may also be used for improving the performance of reinforced concrete pavements by enhancing the thermal, mechanical and electrical characteristics of the concrete mixes. The chapter presents a review of the current researches and practices in the nano-probing, nanoscale modelling and application of nanomaterials for soil, pavement concrete mortar and subgrade stabilization.

Chapter 6
Applications of Nanotechnology in Transportation Engineering 180
    Imtiaz Ahmed, Mirpur University of Science and Technology, Pakistan
    Naveed Ahmad, University of Engineering and Technology Taxila, Pakistan
    Imran Mehmood, Mirpur University of Science and Technology, Pakistan
    Israr Ul Haq, Mirpur University of Science and Technology, Pakistan
    Muhammad Hassan, Mirpur University of Science and Technology, Pakistan
    Muhammad Umer Arif Khan, Mirpur University of Science and Technology, Pakistan

Nanotechnology is the latest development in science, where design, construction and applications of various particles involve at least one dimension in nanometers. The nanotechnology has been utilized in many of the scientific and societal disciplines including electronics, medicine, materials science and many more. It has also influenced the broader fields like civil engineering as well as the sub-disciplines including transportation, structural, geotechnical, water resources and environmental engineering. The current focus of the researchers in transportation field is to develop the materials for sustainable transportation facilities, by using the concepts of nanotechnology. The chapter is concerned with the literature review of potential applications of the nanotechnology in transportation engineering including safety, durability, sustainability and economy. The practical applications of the nanotechnology and nanomaterials shall prove to be an asset in transportation engineering.

Chapter 7
Recent Trends and Advancement in Nanotechnology for Water and Wastewater Treatment: Nanotechnological Approach for Water Purification 208
    Sushmita Banerjee, University of Allahabad, India
    Ravindra Kumar Gautam, University of Allahabad, India
    Pavan Kumar Gautam, University of Allahabad, India
    Amita Jaiswal, University of Allahabad, India
    Mahesh Chandra Chattopadhyaya, University of Allahabad, India

Fast growing demand of fresh water due to increasing population and industrialization dictated research interests towards development of techniques that offers highly efficient and affordable methods of wastewater treatment. In recent decades water treatment using nano-technological based expertise have gained significant attention. Varieties of nanoparticles were synthesized and proficiently used in treatment of wide range of organic and inorganic contaminants from waste streams. This chapter encompasses recent development in nano-technological approach towards water and wastewater treatment. The authors tried to compile up to-date development, properties, application, and mechanisms of the nanoparticles used for decontamination purpose. This piece of work offer a well organized comprehensive assessment of the technology that delineates opportunities as well as its limitation in water management practices moreover few recommendations for future research are also proposed.

Chapter 8
Risks and Preventive Measures of Nanotechnology 253
    Waqas Anwar, Mirpur University of Science and Technology, Pakistan
    Anwar Khitab, Mirpur University of Science and Technology, Pakistan

Application of Nanotechnology in Civil Engineering is a rapidly growing field. It has brought improvements in construction materials as well as practices. Moreover, further developments are foreseeable in this field based on the positive outcomes of the current research works. Utilization of nanoparticles in Civil Engineering has been proved advantageous from several aspects of strength, durability and sustainability. Unfortunately, there are not only benefits associated with the application of nanotechnology. According to various studies, nanoparticles are supposed to damage human organs through physical contact and inhalation. Considering the environmental impacts, atmospheric transport, as well as transport in saturated and unsaturated regions in the subsurface are possible. Nowadays, nanoparticles are progressively produced and they could easily be released in air, water, and eventually contaminate the soil which is harmful for the environment and its habitats. The following chapter would address these issues as well as preventive measures in order to improve benefit-risk ratio.

Compilation of References 277

About the Contributors 331

Index 336

Foreword

This book, *Advanced Research on Nanotechnology for Civil Engineering Applications*, is composed of eight chapters which cover applications of this newly emerging technology in various disciplines of Civil engineering. Civil engineering has always faced a number of challenges; though with the improvement in knowledge and technology over decades, Civil engineers have kept on tackling them in a good manner, however, construction industry still has considerable gaps and significant number of structures fails every year all across the world causing loss of human lives as well as property. Now, it is time to get over the traditional ways for improving properties of construction materials and come up with something new. Material characterization techniques are now moving from just 'Physical' to 'Physicochemical' and 'Chemical'. Similarly, technology has moved from 'Macrotechnology' to 'Microtechnology' and 'Microtechnology' to 'Nanotechnology'. However, almost all the industries, may it be medicine and pharmaceutical, space industry, electronics, environment, energy, oil and gas, defence or whatever, including of course Civil construction across the world are now moving towards the application of nanotechnology to the concerned industrial products, referring Nanotechnology as an industrial revolution. On research side, the work has been done on application of nanotechnology in Civil engineering and it has also been applied in structural engineering, transportation engineering, environmental engineering as well as materials engineering; all of which are major disciplines of Civil engineering.

This book has described the nano-scale properties and various classes of nanomaterials. Furthermore, the book contains how the application of this technology has brought tremendous improvements in properties of construction materials and has increased the output to the marvelous levels. For example; steel when replaced by specific nano-materials gives many times higher tensile strength. Nano silica concrete incorporates nano silica instead of micro silica particles or well-known silica fumes. This concrete results in higher initial and final compressive strengths, higher workability, and lower permeability. Additionally, higher tensile strength and segregation resistance are also achieved.

However, as in case of any other technology, the nanotechnology does come with risks for which precautionary guidelines are necessary and this book has also covered them so that risks can be minimized. This book has covered a large number of applications in a very extensive way and has discussed the research work in this field very comprehensively. This book is important for undergraduate students to know about this new field and also for research students so that they may do research work in this specific field and take application of nanotechnology in the construction industry one step higher. For all those persons who are working in the civil construction industry, this book tells them how this technology can accurately be applied in the industry. The book also describes the limitation aspects of the nano engineered materials including the necessary precautions which field workers should take while dealing with such small size particles to keep their health and the environment safe.

The authors have already published research work in this specific field in international refereed journals and have taken contributions of researchers from different organizations and countries in this edited book.

I congratulate the authors Dr. Anwar Khitab and Engr. Waqas Anwar on the production of this book at the appropriate time.

N. M. Butt
Preston University, Pakistan
22 February 2016

Preface

Nanotechnology is a subject of material science that encompasses nano particles. A nano particle has at least one of its dimensions in nanometers. Its significance was revealed when it was observed that the materials drastically change their characteristics when they are ground to nano size. The importance of nanotechnology related research and development is now well-recognized worldwide. A huge public and private investment is now involved into research and development in a number of industrial sectors, where nanotechnology has become established and has led to new commercial products.

During the last few years, nanotechnology has offered vast opportunities from microelectronics to aerospace and civil engineering is not an exception. Nanomaterials are gaining very rapid recognition in all engineering fields due to some unique advantages, they offer over conventional materials. Its utilization in civil engineering is twofold: Direct incorporation in existing materials like concretes and paints or grinding of materials itself like cement to nano-scale. During the past few years, it has been well established that nano materials owing to their higher strength and lower density are very useful for construction industry.

A DESCRIPTION OF WHERE THE TOPIC FITS IN THE WORLD TODAY

Construction industry faces broad range of challenges ranging from performance of materials to the health & environmental issues. Recent improvements in various areas of nanotechnology have shown noteworthy promise in addressing many of these challenges. Application of nanotechnology can improve the performance of traditional construction materials; such as concrete, glass, steel, paint, insulating materials, etc. Significant improvements in concrete strength, durability and sustainability are being achieved with considered use of nanoparticles nanoparticles. A range of nanomaterials are also being used to add new functionalities, such as self-cleaning properties, to traditional construction industry products, for example paint, glass, mortar, cement and concrete. Nanomaterials such as nano-alumina and nano-silica may be used for the stabilization of the concrete mixes in reinforced concrete pavement. Nano-silica has been observed to improve the rheological characteristics and microstructural properties of cement paste. The fabrication of nanoclays composites has also been conducted by researchers to produce engineered nanoclays which have mineralogical materials such as montmorillonite. It has been used by researchers along with other sedimentary clays to produce the nanosoils. These nanoclays can then be used as an additive that may enhance the asphalt binders' mechanical characteristics. Researchers have observed that the addition of nanoclays may increase the complex shear modulus and the asphalt binders' viscosities and may also produce significant effects on the lower temperature resistance against cracking depending upon the percentage, type and mixing procedure of the nanoclays and the asphalt binders. Our ability to design new materials from the bottom up is influencing the building industry. New materials and products based on nanotechnology can also be found in building insulation, coatings, and solar technologies. Work now underway in nanotech labs will soon result in new products for lighting, structures, and energy. In the building industry, nanotechnology has by now brought to market self-cleaning windows, smog-eating concrete, and many other progresses. But, these developments and currently available products are slightly compared to those incubating in the world's nanotech labs these days. There, work is happening on illuminating walls that change color with the flip of a switch, nanocomposites as thin as glass yet capable of supporting entire buildings, and photosynthetic surfaces making any building frontage a source of free energy. Furthermore, treatment procedure of wastewater, which is one of the major issues in the current era, can also be enhanced by the use of nanoparticles. The TiO_2 nanoparticles can also attach themselves to the chitinous exoskeleton of the animals leading to obstruct molting, which is necessary for the growth in juveniles. This phenomenon may kill such animals thus having serious negative effects on the environmental balance in long term. Nano scale adsorbents exhibit remarkable adsorption performance which owes to its enhanced characteristic features such as extremely high specific surface area, small size, availability of good number of active sites for different kind of pollutants, short intraparticle diffusion distance tunable porosity and easily recyclable without significant loss in adsorption capacity. In short, the application of nanotecnological innovations deals with a highly multidisciplinary field of engineering. Nanotechnology is expected to bring massive changes in robotics, chemical, mechanical, biological, civil as well as electrical engineering. This book deals especially with the application of nanotechnology in Civil Engineering/construction industry.

A DESCRIPTION OF THE TARGET AUDIENCE

Nanotechnology in Construction will attract the researchers already working in this field as well as those deciding to make carrier in it. It will also inform governmental and other endowment agencies of the potential future implications. Practical applications are considered and explanations of the underlying basics are given, raising awareness and understanding of what nanotechnology can offer to construction professionals in general. Furthermore applications of nanotechnology in Civil Engineering can also be taught at undergraduate and postgraduate level in universities and this book will be of great help to all the students and teachers. Lastly, workers who are dealing with nanotechnology are also very important targeted audience of this book as it covers health and safety aspects of technology in detail.

A DESCRIPTION OF THE IMPORTANCE OF EACH OF THE CHAPTER SUBMISSIONS

This book is organised into many broad sections; the first chapter covers an overview of the classical construction materials and the significance of nanotechnology for the construction industry. Concrete, the backbone of civil engineering is conventionally a mixture of cement, sand, gravel and water. Along with its conventional components, some ancillary materials are also sometimes added into concrete, to gain some additional advantages like strength, workability and crack control. One such addition is silica fume, which apart from some other useful effects, predominantly makes concrete resistant against chemicals. Steel is an iron-carbon alloy having carbon content up to 1.8% along with some other elements like manganese, copper, silicon, nickel and molybdenum present in minor quantities. Apart from iron, the other alloying elements are intended to increase strength, hardness and corrosion resistance of steel. Paints or coatings are intended to protect the surface from harmful weathering effects as well as to provide beauty to the surface. Paints are composed of many constituent materials for example, lead or aluminium are used as base; resins are used as vehicle or binder; oil or water are used to adjust the viscosity; lead or cobalt are used as drier. In the presence of so many useful materials, the question arises, as what the nanotechnology is going to offer in addition?

Chapter 2 gives a comprehensive review about nano cement and concrete. Firstly, the morphology of C-S-H, the nano characteristics of C-S-H at elevated temperatures, and the effects of C-S-H on the elasticity, durability, creep and shrinkage of cement and concrete materials are introduced for comprehensively understanding nano-scale gene and behaviors of cement and concrete materials in nature and describing the nano-scale blueprint to control properties in nano-scale of cement and concrete materials. Secondly, the effects of nano-binders and other nano materials on the rheological behavior and workability, the hydration, the mechanical properties, the durability, the functional properties and the smart/intelligent properties of the cement and concrete materials are presented. Thirdly, the underlying modification mechanisms of nano materials to cement and concrete materials are summarized. Finally, the challenge and future development and deployment of nano-modificated cement and concrete materials are discussed.

It is essential to adopt new tactics of making things through understanding and control over the fundamental building blocks (i.e. atoms, molecules and nanostructures) of all physical things. It is fact that, potential properties of nanomaterials have been brought out in the construction industries but still there is a long way to go. There have been many limitations mainly due to the insufficient control of interfaces and edges of the nanomaterials with the conventional building materials. Considering the recent rapid advances in technology, the control of interfaces and edges is expected to be possible to provide a sustainable solution for nanoscale engineering of cementitious materials. Characterization of nano materials poses enormous challenges to researchers. Thus nano characterization of materials has become a significant field of research now-a-days. In this book chapter 3, author has chosen some frequently used characterization techniques of nano materials, briefly mentioned their basic principles and current research trends with these methods. Furthermore, emphasis has been given on various case studies published in the literature on cement based materials.

Usually, nanotechnology has been related with developments in the areas of microelectronics, medicine and materials sciences. However, the potential for application of many of the developments in the nanotechnology field in the area of construction engineering is growing. Though different chapters cover application of nanotechnology to specific civil engineering fields. However, in chapter 4 a wide overview of the potential application of various nanotechnology developments in the construction engineering field is considered, and the potential for further basic research that may lead to improved systems is also evaluated.

It is true to say that the potential effects of nanotechnology on construction are mostly unidentified to the construction profession in general, although specific research is being done in universities and other institutes all over the world. These provide indicators to what would soon be available to industry. Many of these developments are in line for arriving within the next five years. In order to fully benefit from this new industrial revolution, a concerted effort is needed to overcome the vital barriers of lack of knowledge and conservatism in construction as regards nanotechnology. Nanotechnology is a multifaceted and deep subject and it is nearly impossible to grasp for those who are not actively involved; thus, awareness of research done can simply be increased by educating students and professionals through easily digestible information made available by universities, relevant institutions, journals and other sources. Focused research into the timeous and directed research into nanotechnology for construction infrastructure should be pursued to ensure that the potential benefits of this technology can be harnessed to provide longer life and more economical infrastructure. This chapter 4 concludes with a roadmap and strategic action plan on how nanotechnology can have its biggest impact on the field of civil engineering.

Chapter 5 is an attempt to cover the recent developments in the field of nanotechnology for the construction industry, with an emphasis over the sector of geotechnical engineering. The global initiatives on the R&D and practical use of nanotechnology has garnered a certain buzz around nanomaterials, nano-probing and nano-sensing and nanoscale modelling is the new interest of researchers and innovators alike. As such, the chapter provide a brief background and introduction to nanotechnology, including nanoscale properties and classes of nanomaterials, and role of carbon nanotubes in concrete for enhanced durability, safety, strength and performance results. The nature of underlying soil plays a vital role in geotechnical and pavement structures as the durability of overlying structures is highly dependent upon the soil characteristics, behavior and composition. Clay particles have a nanostructure, as 2D sheets of SiO_4 tetrahedrons linked by sharing of corners, forming hexagonal meshes, where particle interactions and factors like van der Waals forces are more dominant than gravity and ratio of surface area to particle mass is important. To broaden study on nanotechnology applications, like engineered clay particles, nanoindentation, MEMS, FTIR and AFM, self-assembling clay particles, CNT-reinforced clay matrixes and nano, nanoscale Monte Carlo techniques and molecular dynamic simulations, this chapter covers three primary aspects of nanosensors for soil characterization, numerical modelling and empirical relations for soil and rock behavior and soil.

In summary, the applications of nanotechnology in the field of geotechnical and pavement engineering hold great potential. Nano-modification of ground under foundations or subgrades may increase the mechanical and chemical characteristics and produce economical, greener and safer structures that are more durable. All these mentioned applications and future research potential are briefly described in this chapter.

Chapter 6 deals with the application of nanotechnology in the field of transportation engineering. Transportation Engineering is one of those fields that got benefitted from this tremendous growth in engineering sciences. Not very far back into the history, the engineers used to chew bitumen in order to have an idea of its consistency (hardness or softness). The aggregates were just hammered or crushed to determine their suitability for a certain project or application. Similar practices are still being used but they are now complimented by more refined tools and sophisticated techniques. 'State of Practice' is now approaching the 'State of the Art'. The pavement design techniques are now progressing from the 'Empirical' to the 'Mechanistic-Empirical' and purely 'Analytical'. Material characterization techniques are now moving from just 'Physical' to 'Physico-chemical' and 'Chemical'. Similarly, science is moving from 'Macro' to 'Micro', to 'Nanotechnology'. Now we are even using Atomic Force Microscopy (AFM) for research on materials. Nanotechnology being one of the most emergent fields of the current age affects all the disciplines of the science, technology and engineering. The chapter gathers the literature unfolding the use and importance of this emerging field in Transportation Engineering. It organizes all the elementary information in such a manner that it becomes very easy for the readers to understand the technology under discussion. The chapter summarizes all the basic nanomaterials that can and have been used in the Transportation Engineering for enhancing the function and life of the pavements. It also discusses the aspects in which the nanotechnology can benefit Pavement Engineering including economy, sustainability and safety etc. The potential hazardous effects of nano-materials in Pavement Engineering have also been discussed. Overall, the chapter is a very good attempt to summarize the advantages and disadvantages of using nano-materials in a literature perspective. It will also be very beneficial for the researchers as it has been tried to collect and condense the literature related to the topic in a single chapter.

Chapter 7 titled "Recent Trends and Advancement in Nanotechnology for Water and Wastewater Treatment: Nano-Technological Approach for Water Purification" helps in providing a brief overview to the readers regarding various possible synthesis techniques of nano scale materials. Moreover, the chapter also entails a concise outline about some of the basic instrumentation techniques that helps in ascertaining the size as well as various physical and chemical properties of the nano materials. Besides, the chapter also focused on the application of the some of the recently synthesized nano materials in resolving many problems related to water purification and water quality. The chapter also encompasses the role of various process parameters influencing the adsorption behaviors of nanomaterials. Furthermore, a general idea on the limitation aspect of the nano engineered materials was also briefly represented.

Lastly, Chapter 8 deals with the risks associated with the dealing and usage of nanotechnology. The use of nanotechnology in any field requires great care and any sort of negligence is likely to bring negative effects for the environment and its habitats. Usually, nanomaterials become threat when they are discharged in undesirable quantities into the wrong destinations. . Exposure to nanomaterials may occur unintentionally in the environment or through the use of nanotechnology based products in our daily lives. Some examples include physical contact at manufacturing units, inhalation of nanomaterials released to the atmosphere and use of drinking water or food having accumulated nano particles. Moreover, absorption by soil and then transportation in saturated and unsaturated regions in the subsurface leading to the final destination at vegetation is also possible. Nanoparticles are so small when absorbed; they may reach the inner bio molecules in the body. Studies have shown that inhaling nanoparticles can even affect the central nervous system. Moreover, their extremely small size is very tangible to affect the skin and eyes of people exposed to them. Some nanoparticles may even cause mesothelioma and lungs cancer. So, there is need to adopt wide precautions. The chapter covers risks of nanotechnology to the environment and its habitats in detail. Furthermore, it also discusses remedial measures from public awareness to the engineering control. Important precautionary aspects including hazard identification, dose response assessment, exposure assessment as well as risk characterization are also discussed in this chapter. Lastly considering above mentioned disastrous effects, waste monitoring and recycling have also been included in this chapter. In short, this chapter brings forth both the bright and dark side of nanotechnology and discusses important aspects to get benefits from it by keeping the harmful impacts at minimum level.

CONCLUSION

Nanotechnology has tremendous potentials in construction industry. The examples are germ-free laboratories and hospitals, waterproof buildings, urban environmental protection. The important developments made in concrete technology are ultra high strength concrete, photo-catalytic concrete, self-heating concrete, bendable concrete and concrete containing CNTs. Nano Silica Concrete incorporates nano silica instead of micro silica particles or well known silica fumes. This concrete results in higher initial and final compressive strengths, higher workability, and lower permeability. Additionally, higher tensile strength and segregation resistance are also achieved. The new concrete is named as Ultra High Strength Concrete. The advantages of this concrete are numerous: the column sections in buildings can be reduced. The amount of steel reinforcement in concrete can also be reduced. And in highways and railway tunnels, thinner tunnel segments can be constructed leading to a great saving in excavations. It is a well-known fact that nano TiO_2 on UV irradiation can be used as an effective way to reduce the contaminants and enhance environmental safety. Photocatalytic concrete is a green material. With this concrete, structures looking new for decades can be constructed. Inside hospitals and laboratories, the spread of germs can be minimized and urban air quality can be improved. Various contaminants like algae and barnacles cannot cling to CNT-containing nano paints, which can increase the time saving in repainting massive marine structures. Serious health issues related to the use of nano materials must be well understood and remedies are mandatory; these risks are remedial measures are also well discussed in this book. The investigation for various applications of nanotechnology to build up novel building materials continues. It is by now obvious that the science of the very small is creating big changes, with various economic benefits to the construction industry.

Chapter 1
Classical Building Materials

Anwar Khitab
Mirpur University of Science and Technology, Pakistan

Waqas Anwar
Mirpur University of Science and Technology, Pakistan

ABSTRACT

Classical building materials are widely used in Civil Engineering projects. Many ancient generations used fire clay and stone as building blocks. The most common building materials today include stones, concrete, plastics, bitumen, glass, wood, metals, bricks, polymers, tiles and heat resisting materials. Current research work has come up with the considerable improvement in the natural characteristics of these materials. Glasses are available in more variety as they were 50 years ago. Similarly, concrete is now available in ultra-high strengths and even blast resistant form. Recently, use of nano-technology has emerged as a rapidly growing field; success of which is also highly dependent on the basic understanding of classical building materials. Advance research work including implementation of nanotechnology may come up with further improvements in the physical and chemical properties of these materials. This chapter would focus on classical materials in detail covering their physical and chemical characteristics, usages as well as economical suitability.

INTRODUCTION

Construction materials are the backbone of all civil engineering structures. They are being used since prehistoric times. Many ancient generations used fire clay and stone as building blocks. The most common building materials today include stones, concrete, plastics, bitumen, glass, wood, metals, bricks, polymers, tiles and heat resisting materials. Among these materials, stones, various metals and wood are the naturally occurring materials whereas the materials like plastics, cement concrete, bricks and glass etc. are among the famous man-made materials. The modern structures are amalgam of different construction materials; for example, the foundations, beams and slab are constructed using concrete and steel, whereas doors and windows are comprised of wood and various other metals. Similarly, glass is used in doors and windows. Some of the common construction materials are briefly discussed in this chapter.

Stone

Stone is one of the oldest and basic construction materials. Stones are obtained from ground and rocks through various processes. Site from where stones are extracted is known as 'quarry site' and the process of extraction is termed as 'quarrying'. Initially, the stones have irregular shapes and dressing of stones is carried out to bring them in the requisite size and shape. Extraction and dressing of stones requires intense man power, hand tools and machinery. Millions of tones of stone are required every year for the building purposes. It is extensively used in the construction of walls, beams, columns, paving slabs and several other structural components. Many mega structures including dams require huge amount of stones in their construction. Stone is also required to prepare several other construction materials; for example, crushed stone is used in the preparation of concrete. Use of stone in construction is also considered good from sustainability point of view. A house made up of stones catches sunlight in winter but not in summer making it an ideal living place for the whole year. Furthermore, beautiful texture of several stones is accountable for their extensive use as an aesthetic material in buildings.

History

Stone has been used in construction works since Paleolithic times (Smith, 1999). From early days, even before the existence of bricks, stones were used in dwellings, roads and ornamental works (Varghese, 2010). Most of the prehistoric monuments and ancient temples were built using stone (Singh, 2005; Varghese, 2010). Another important example for the substantial use of stone is the construction of the pyramids

Figure 1. Pyramids of Giza

in Egypt (Khitab, 2012). In current era, despite the development of many strong and durable construction materials, stone is still chosen over other materials for a number of reasons including its extreme durability, availability and appealing aesthetics.

CLASSIFICATION OF STONES

Based on the origin of stones, they are geologically classified into sedimentary, igneous and metamorphic rock stones. Rocks from where stones are obtained are classified into the following three types: (Varghese, 2010)

- **Igneous Rocks:** Rocks which are formed by the solidification of molten lava are known as igneous rocks. These are further classified into two types:
  ○ Intrusive igneous rocks which are formed below the surface of earth, and
  ○ Extrusive igneous rocks which are formed when erupted molten matter solidifies above the surface of earth.
- **Sedimentary Rocks:** Rocks, which are formed by the deposition of materials, are termed as sedimentary rocks. Basically particles from decayed rocks are deposited by streams of water or wind and their consolidation results in the formation of sedimentary rocks. Sedimentary rocks are formed at the earth surface or even within the water bodies. Some common sedimentary rocks are limestone, sandstone, conglomerate and shale.
- **Metamorphic Rocks:** Metamorphic rocks are formed by the physical or chemical alteration of the existing sedimentary or igneous rocks when subjected to immense heat and pressure. Examples include the transformations of sandstone into quartzite, shale into slate and limestone into marble.

In addition to above mentioned categories, stones from rocks can also be classified into the following types based on their physical and chemical characteristics, (Syed, 1970)

- **Siliceous Stone:** Stones having silica as its principal constituent are known as siliceous stones. Examples include granite, chert and sandstone. Siliceous stones are durable and relatively easy to clean with mild acidic solutions; therefore they are preferred in the construction of common houses and commercial buildings.
- **Argillaceous Stone:** Stones containing clay as its principal component are known as argillaceous stones. Argillaceous stones are the most abundant sedimentary rock stones, varying according to different estimates from 44 to 56% of the entire sedimentary rock column (Siever, 1974). Slate, mudstones and shale are some common argillaceous stones.
- **Calcareous Stone:** Calcareous stones also belong to the sedimentary rocks and are mainly composed of calcium carbonate. Limestone and marble are the most common calcareous stones used in construction industry.

Industrial Stones

Today stone are available in various types and each type has its own physical and chemical characteristics. Some important types of industrial stones are briefly discussed below:

- **Granite:** Granite is a crystalline rock composed of a mixture of quartz and feldspar with very small amounts of mica or hornblende. It belongs to the igneous family and slow cooling process of molten magma is accountable

Figure 2. Igneous-rock at Rushikonda beach in Visakhapatnam
© *[2011], [Adityamadhav83]. Used with permission.*

Figure 3. Middle Triassic marginal marine sequence of siltstone (below) and limestone (above), southwestern Utah, USA

Figure 4. White marble rocks at Bhedaghat
© *[2012], [Sandyadav080]. Used with permission.*

for giving it a crystalline structure. Granite is either grey or pink in color; moreover concentration of a color also varies from light to very dark. Due to the high strength of granite, it is used in retaining walls, foundations and paving blocks.

- **Shale:** Shale is a fine grained sedimentary rock, which results from the compaction of silt and clay particles. Shale is fissile and laminated. It is usually gray in color but alteration of various minor components transforms its color to a high degree. Red, green and black shales are the most commonly available stones.
- **Slate:** Slate is a metamorphic argillaceous rocks. It is characterized by high density, strength and water resistance. It is commonly used for damp proofing

purposes in flooring and roofing. Color of slate is usually dark grey or dark brown.
- **Gneiss:** Gneiss is a type of metamorphic rock with medium to coarse grained texture. It consists of alternate layers of different materials and may contain abundant quantities of quartz and feldspar. It has a rough surface and is considered one of the most suitable cladding materials for construction purposes. Gneiss also possesses very high tolerance to various atmospheric effects including temperature variations.
- **Marble:** Marble is formed by the metamorphism of sedimentary limestone rocks. Marble is normally white but presence of various mineral impurities such as clay, silt, sand, iron oxides or chert produces a variety of colors. Marble is vastly used for ornamental and architectural purposes all over the world. Marble also possess good resistance to heat and fire. Due to its firm crystalline structure, marble can be polished to improve its shine and is thus a common and attractive stone for building applications.

Characteristics of Stones

Selection of the most appropriate stone is made on the basis of its appearance, strength and chemical behaviors. Several tests are carried out for this purpose including crushing strength, specific gravity and absorption. The important properties of stone which are generally considered for the construction purposes are as follows,

- **Crushing Strength:** Crushing strength of stone is one of the most important properties to be considered. Stone is often used as a load bearing material in walls and foundations, which requires high crushing strength. Crushing strength of majority of stones is higher than that of bricks which is a common reason for their extensive use in hilly areas, where stone is easily available as compared to bricks.
- **Appearance:** Appearance plays a very important role when stone is to be used for architectural purposes. Granite, basalt and sandstone are some common stones used for aesthetic purposes due to their attractive colors.
- **Durability:** Though stone is a highly durable material but certain stones when exposed to harsh environment such as high temperature, sun exposure and humidity undergo considerable deterioration. So, proper consideration should be made before selection of stones subjected to be used in aggressive environment. Moreover, there are certain other properties including hardness, density, resistance to impact and fire which should be well considered before the selection of stones.

Khitab, 2012; Varghese, 2010.

Classical Building Materials

Table 1. Crushing strength of stones

Type of Stone	Specific Gravity	Water Absorption (kg/m^3)	Crushing Strength (tons/ft^2)
Sandstone	2.3	57.7	400-800
Limestone	2.5	112	100-400
Granite	2.7	9.61	900-1000
Marble	2.7	16	300
Basalt	2.8	11.2	1000

Artificial Stone

Artificial stone is manufactured of small size crushed stone mixed with cement. Its use in construction works is very popular nowadays. The name `Artificial stone' is used for a variety of synthetic products, which are widely used for the last two centuries. Its use in cladding and paneling is very popular. Artificial stones have many advantages over natural available stones. There are fewer defects in artificial stones as compared to natural stones. They can easily be manufactured at places where natural stone is scarce. They can be given different shapes as well as colors. Different grooves can be made depending upon the architectural requirements and off course they are economical as cost on quarrying and transportation is saved.

BRICKS

Brick, a rectangular block of clay made up of fired or sun- dried clay is also one of the oldest construction materials. Bricks are used in buildings, pavements and linings. Bricks are subjected to compressive and transverse stresses and are not designed to take the tension. According to Syed (1970), bricks can still take tensile stresses up to 1.6 tons/ ft2 for 1:4 cement masonry. Bricks are available in different qualities and various tests are conducted to analyze the quality of bricks. Size of a brick varies to some extent from one country to the other. They are usually laid in layers in the construction of buildings and a variety of bonds are provided to keep them intact and stable. Selection of a bond depends upon various factors. Moreover, an indention known as frog is provided in the brick which is filled with mortar thus making stronger bonds between different layers of a wall.

Apart from brick blocks, brick ballast or broken bricks are also used in several ways; as an aggregate when natural aggregate is expensive or not available, under flooring of buildings and also as a bedding of sewer lines.

Figure 5. Common clay brick

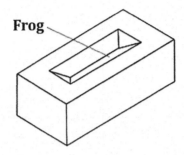

History

Bricks are also one of the oldest construction materials. Archaeologists have found bricks in the Middle East dating 10,000 years ago. The ancient city of Ur (modern Iraq) was built with mud bricks around 4,000 B.C. (Science Jrank). Romans, Jews, Arabs and Christians constructed the ancient city of Toledo with bricks as the major construction material (López et al., 2003). Bricks also remained a very popular construction material among Egyptians (Varghese, 2010) and the Indus Valley Civilization (Khitab, 2012). Moreover, the Great Wall in China was also built using bricks. With the passage of time, quality of bricks improved as several brick casting machines have been invented which provide sufficient pressure thus producing highly dense and uniform bricks.

Figure 6. Great Bath at Mohenjo-Daro
BBC UK, 2014.

Classification of Bricks

- **First Class Bricks:** These are the high quality bricks, having uniform shape, color and texture. They have sharp edges. They are sound and well burnt. If scratched with a nail, they leave no mark. Moreover, if struck against each other, a clear metallic sound is produced. They are free from efflorescence, cracks and other flaws. The crushing strength of these bricks is approximately 105 kg/cm^2. The water absorption after 24 hours in water is no more than 10-15 percent by mass of the brick. Owing to highest quality, they are used in load bearing walls and pavements.
- **Second Class Bricks:** Although sound and well burnt, the second class bricks have certain irregularities in shape and texture. Like first class bricks, they are also free from cracks, considerable efflorescence and flaws. The water absorption after 24 hours in water is about 20 percent by mass of the brick. They are also used in load bearing walls but of moderate capacity usually in single storey buildings. However, they are widely used in partition walls. Moreover, they are preferred as brick ballast in foundations and floorings.
- **Third Class Bricks:** Slightly under burnt bricks are termed as third class bricks. They are not uniform in shape and size. Their crushing strength is about 30 kg/cm^2 far below than that of first class bricks. The water absorption is about 25 percent by mass of the brick after 24 hours of submersion in water. They produce a dull sound when struck against each other. Though they have poor strength and texture, yet they are highly economical and can be used in the construction of temporary and rural structures.
- **Fourth Class Bricks:** They are slightly over burnt thus have very low strength and are not recommended for application in major construction works. However, they can be used in flooring and inferior construction works.

In addition to the above classification, bricks have also been classified on the basis of their method of manufacture as follows:

- **Ground-Molded Bricks:** In ground molding, bricks are cast on the ground surface. It requires huge land area and sufficient manpower. Moreover, the casting ground should be properly leveled and cleaned before molding of bricks is carried out. This methodology is adopted when manpower is cheap.
- **Table-Molded Bricks:** Molding is done on a table of size 1m x 2m with the help of hands, moulds and various tools.
- **Machine-Molded Bricks:** Molding of bricks is carried out with the help of a machine. This method results in better shape and high production rate. This approach is used when a larger number of bricks are required in a very limited time span.

Besides solid rectangular form of brick, there are many special forms of bricks available. Few of them are discussed below:

- **Perforated Bricks:** These bricks have hollow holes, produced by pushing iron bars within the bricks. Purpose to produce these holes is to reduce the overall weight of the brick thus minimizing the self weight of the structure. However, these bricks do transmit sound and are also not suitable to be used in the hydraulic structures.
- **Hollow Bricks:** These bricks are used for insulation purposes. They are strong against distributed loads however they may easily fail against concentrated loads. They are different from perforated brick as number of holes in the hollow brick is less but size of holes is bigger as compared to those in perforated bricks.
- **Bull-Nose Bricks:** These bricks have rounded corners and are usually preferred in buildings from architectural point of view. They are mostly used in the construction of steps, sills and capping walls.

Drying of Bricks

Bricks obtained after molding are initially wet and are termed as green bricks. These bricks are then dried to remove the excess moisture. This process reduces the chances of cracking and distortion in bricks. Moreover, these dried bricks also take less time for burning in kiln thus reducing the fuel costs extensively. Drying of bricks requires great care and can be achieved naturally or artificially. However, following is recommended: Bricks should be dried at a slow rate under shade. The molded bricks should be loaded and unloaded at the same time. The drying ground should be raised from ground level, so that any rain water does not stay therein. However, in the first hand the drying area should be facilitated with temporary shades to be used in case of any occasional rain.

Properties of Bricks

Bricks are usually preferred as they are strong, economic, require low maintenance and have good thermal resistance. Moreover, variety of colors and textures is also a key factor for their extensive use in the construction industry. Several tests are conducted to check the physical and chemical properties of brick, which include compressive strength, water absorption and efflorescence tests. Quality of bricks can be enhanced by the addition of fly-ash, nano-coatings and several other additives etc. Thermal insulation and fire resistance of bricks can be increased tremendously with the application of nano-coatings. Moreover, bricks are also resistant to the noxious insects. Bricks can easily be recycled to make aggregates for use as a

Classical Building Materials

Figure 7. Brick types

general fill, in landscaping and as a plant substrate. Damaged bricks are also used in the manufacturing of new bricks and to make athletic tracks.

CEMENTITIOUS MATERIALS

Mortar and concrete are the two most important cementitious materials used in the construction industry: Thanks to the invention of Ordinary Portland Cement (OPC) by the Scottish mason Joseph Aspdin in the year 1824 (Linch, 2002).

Mortar

Mortar is a composition of cement, fine aggregate and water. It is used as a binding material. As time passes it becomes hard, and binds firmly the bricks and the blocks together.

On the basis of binder, mortar is generally classified into Lime, Cement, composite and mud mortars:

- **Lime Mortar:** Lime mortar consists of lime, sand and Surkhi. However sometimes, lime is also mixed with cinder and the resultant product is known as black mortar. It provides moderate strength and is not preferred to be used in structures subjected to heavy loads. If coarse and well grained sand is used, sufficient strength can be achieved. Surkhi should be well ground and it should be prepared by grinding the well burnt bricks. One of the advantages of Lime mortar is that it can be used within 24 hours of its preparation, which is considerably more than that of its competitor cement and composite mortars.
- **Cement Mortar:** Cement mortars comprise of cement and sand in different proportions. It is usually preferred for bearing heavy loads. Cement mortars should be used within 30 minutes of adding water in it. Cement mortars can be prepared by hand mixing or machine mixing. Hand mixing is preferred when mortar is required in smaller quantities.

- **Composite Mortar:** It is also known as lime cement mortar. It is composed of cement, lime and sand in different proportions and has more setting time as compared to the lime mortar (Singh, 2005).
- **Mud Mortar:** Mud mortar consists of a plastic mix of clay and sand. It offers poor resistance against seepage and is only recommended for inferior and temporary structures being economical as compared to other mortars. In order to deal with the seepage issue, proper drainage must be provided in the structure. Depending upon the type of soil used and the addition of bitumen with lime, mud mortar may become a water proof material.

Moreover latex based mortar, fire resistant mortar and x ray shielding mortar are also prepared based on the requirement. Fire resistant mortar is usually used in kitchens whereas x ray shielding mortar is preferred in hospitals and diagnostic centers.

Properties of a Good Mortar

Mortar should have the following properties to be safely used in the construction works. It must have good strength and durability. However the strength of mortar is directly related to the strength of materials to be joined with mortar. Providing an over strong mortar for a moderate strength blocks or bricks is of no use and only leads to the increased cost of the structure. It must provide a good bond with the blocks or bricks. Mortar should provide no hindrance to the painting works. Workability of mortar should be sufficient for easy application. Addition of lime in mortar enhances the workability. It should provide as minimum permeability as possible to not allow seepage of rainwater.

Concrete

Concrete is a composition of cement, fine aggregate, coarse aggregate and water. According to Kevin Hunt (2000), concrete as a building material has been around for a very long time. In Yugoslavia, some 7000 years old concrete has been identified.

Concrete is strong in compression but weak in tension. It is reinforced with steel for use against tensile loads and the combined concrete is known as reinforced concrete owing to the fact that steel has high very tensile strength. Cement added in concrete acts as a binder; it forms a paste with water which hardens and binds the aggregates together. Broken stone or brick ballast can be used as coarse aggregates in concrete. Moreover, certain admixtures may be added in concrete to further improve its properties. According to Abdullahi (2006), almost 60 to 80 percent of the volume of concrete is occupied by the aggregates.

Classical Building Materials

Good water is essential for the production of quality concrete. Mixing water should be good enough to drink, free of trash, organic matter and excessive chemicals and minerals. The strength and other properties of concrete are highly dependent on the water-cement ratio. According to Domone et al. (2010), the strength of concrete increases with the passage of time; in early stages, rate of strength gain is significantly higher and concrete gains almost 50-60% of its ultimate strength in just 7 days. It takes almost 28 days to gain its strength up to 80-85%. This process continues for many years, even very small improvements in strength have been found in concrete after 30 years.

Concrete types have become very vast over the decades. By the addition of different admixtures, many new forms of concrete have been developed. Successful approaches towards reducing the self weight of concrete and increasing the tensile strength have been made. Some of the important types of concrete are discussed below,

- **Lightweight Concrete:** Lightweight concrete is a versatile material in the modern construction. It reduces the self weight of the structure, improves heat and fire resistance and reduces the usage of heavy formworks. According to Mukherjee (2014), it offers the same strength as ordinary concrete and is yet 35% lighter in weight. Moreover, it is also observed that lightweight concrete reduces the high expenditure on transportation. Lightweight concrete is produced from a variety of lightweight aggregates which may be obtained from volcanic pumice, clay, shale, slate, fly ash or slag.
- **High Strength Concrete:** It is often required to use concrete having strength more than 10,000 psi. Conventionally, it is not possible to achieve such a high strength by using typical ingredients. However, over last two decades engineers and researchers have succeeded in manufacturing even ultra high strength concrete. High strength concrete is widely used in the multi-storey buildings, bridges and highways where concrete components are subjected to very heavy loads. Moreover, sometimes architectural requirements limit the size of different structural components and it becomes necessary to achieve high strengths via smaller beams, this is only possible by the use of high strength concrete. Fly ash and silica fumes are the most commonly used admixtures used to achieve high strength concrete. According to The Concrete Society, it is important to limit water to cement ratio to 0.35 in high strength concrete; generally a value between 0.3 to 0.35 is considered good.
- **Air Entrained Concrete:** Air entrained concrete is produced either by using air entrained Portland cement or by the addition of several air entraining agents. There is a variety of air entraining agents commonly used now-a-days including synthetic detergents, petroleum acid salts, alkylbenzenesulfonic

acid and fatty acid salts (Gutmann, 1987). This type of concrete contains billions of tiny air pockets in one cubic foot of concrete. These tiny pockets prevent the development of high pressure and thus preferred to be used in structures which are subjected to freeze and thaw actions. Air entrained concrete also possesses strong resistance against bleeding and segregation.

- **Fiber Reinforced Concrete:** Fiber reinforced concrete is composed of cement, aggregates and fibers. These fibers serve as reinforcement in the concrete and are produced using steel, glass, organic fibers and synthetic fibers. The purpose of adding these fibers is to improve the tensile strength of concrete. Fiber reinforced concrete has so far been successfully used in slabs, shotcrete, offshore structures, structures in seismic regions, foundations, hydraulic structures and many other applications.
- **Bacteriological Concrete:** Development of cracks in concrete is a common phenomenon. These cracks prolong over time and may cause the ultimate failure of the structure (Edvardsen, 1999). As far as micro cracks are concerned, they can heal themselves. For this purpose, bacterial spores, calcium lactate and nutrients are introduced in concrete by embedding them in capsules. When cracks develop, these capsules break and bacteria become active on interaction with water thus making limestone out of calcium lactate and nutrients. This limestone fills up the cracks and prevents the further seepage of water (Jonkers et al., 2010).
- **Stamped Concrete:** Concrete is imprinted to obtain texture of different other construction materials including bricks, wood, tile etc. Purpose is to get high strength of concrete giving aesthetics of different other materials. An example of stamped concrete is shown in Figure 9.

Figure 8. Fiber reinforced concrete

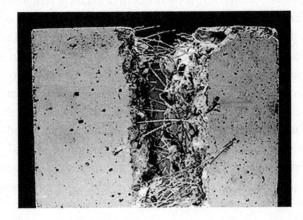

Figure 9. Stamped concrete in Nawab cafe (M.U.S.T) Mirpur Azad Kashmir

Stamped concrete preparation is different from the conventional concrete in several ways. Base color and accent color are added during the preparation and later on stamping of a pattern into the concrete is done.

Shotcrete

Shotcrete is a term used for mortar and concrete which is conveyed via a hose pipe. Shotcrete is ideal for application in very wet, frozen, elevated and confined places. Moreover, shotcrete can easily be applied on very thin sections where application of ordinary concrete is not possible. Shotcrete is produced by wet and dry processes.

- **Dry Mix Shotcrete:** Dry components are combined in a pot and then transferred to a nozzle where water is added and the whole mixture is sprayed on the surface under the action of compressed air. This method is employed for high strength shotcretes and is economical as compared to the wet mix methodology.
- **Wet Mix Shotcrete:** Unlike dry mix shotcrete, water is also added in the pot with other components and the readymade concrete is pumped to the nozzle and then compressed air injects this mixture onto the receiving surface. The method is suitable for low to moderate strength shotcretes. However, addition of several admixtures is well achieved in a wet mix as compared to the dry mix shotcrete.

Properties of Good Concrete

Concrete is suited in construction industry due to its unique properties. It has high compressive strength and durability. It can be molded into any shape. Beams, columns and shells of different shapes are easily cast with concrete. It is economical and can be reused in many forms. With proper composition, it offers great resistance towards acids, water and fire. Manufacturing of concrete does not require highly skilled labor and the materials required for the preparation of concrete are easily in the local markets.

Though concrete has several advantages and is still used in the world but there are some drawbacks which must be catered. Concrete may undergo shrinkage due to evaporation of water. Excessive shrinkage results in cracks which may initiate a failure in the structure. With proper care, excessive shrinkage is easily tackled. Concrete structures have very high self weights and require strong foundations. In case of collapse, more human loss is expected as compared to the steel structures.

PLASTICS

Plastic is a synthetic material, which is derived from a variety of organic polymers and can be molded into any desired shape. Acrylic, amino resin, cellulosic, phenolic, polyamides, polyester, polyolefin, polyurethane, styrene and vinyl are the important types of plastics. The first plastic was invented by Alexander Perkes in the late 19th century. In construction industry, plastics are used in roofing, walls, windows and pipes etc. Use of plastics in construction can help save energy and maintenance costs. Plastics are today's and tomorrow's materials of choice because they fulfill our needs with least negative impact on the environment. Different additives are also added into plastics to improve their properties; to change their color and hardness

Figure 10. Plastic elbows for pipes

and make them long lasting. Some appropriate tests to check physical properties of plastics are flexural strength, tensile strength and izod impact tests.

A strong link between nanotechnology and plastics has already been established. The plastic industry currently uses nanotechnology in a variety of ways. Thermoplastics are comprised of materials reinforced through nanotechnology because of their heat resistant, dimensional stability and electricity conduction properties. Plastic nano-tubes are flexible, durable and lightweight, which are currently used in different industries.

Types of Plastics

Plastic is commercially available in so many types, some of the important plastics commonly used in construction works are briefly discussed below:

- **Acrylic:** It is a thermoplastic having methyl methacrylate as its major constituent (Syed, 1970). Acrylic based plastics are more transparent than glass and provide very high strength, about 10 to 15 times more than that of ordinary glass (Varghese, 2010). They are therefore preferred in building constructions. Acrylics are very light in weight and possess high tolerance against harsh detergents.
- **Epoxy Resin:** Epoxy resins are capable of forming cross linked structures, which are tough, strong and offer low shrinkage. They are mostly used for flooring purposes. Moreover, they can also be injected in concrete or masonry cracks, where they harden and seal the cracks.
- **Fiberglass:** Fiber glass is a type of fiber reinforced plastic. It can be made into a variety of forms including wool, strands and cloth etc. It is strong, lightweight and very economical. It has good thermal insulation and is resistant towards harsh environments. Fiberglass is widely used in the construction of domes and sheds.
- **Polyvinyl Chloride:** Polyvinyl chloride is obtained from vinyl chloride and acetates. It is one of the cheapest plastic materials available in the market. It has very high resistance to heat, termites, acids, alkalis and various other chemicals. Therefore it is preferred to be used as a conduit for water. However, it has high expansion coefficients and may fail under very high loads, so proper precautions should be made for their usage in the buildings.
- **Acrylonitrile-Butadiene-Styrene (ABS):** ABS is a combination of acrylonitrile, butadiene and styrene. Impact resistance and the toughness are the key properties responsible for its usage in the construction industry. Depending upon the requirements, its impact resistance can be enhanced by increasing the ratio of poly-butadiene as compared to acrylonitrile and styrene.

Properties of Plastics

Just like other materials, plastics also offer some unique characteristics. Most of the plastic materials are very economical and easily available. They are easy to install and dismantle. They provide moderate to good insulation against heat and sound. They are aesthetically appealing. PVC paneling is preferred all across the globe. It is economical and gives fine and attractive texture. Plastic pipes are very popular being light and flexible. Moreover, they are rust free and long lasting. Polyethylene, polypropylene, polyvinyl chloride (PVC), or acrylonitrile butadiene styrene (ABS) are the most commonly used plastic materials. Plastics also offer very good stiffness. However, adding wood to plastics can double and sometimes triple the stiffness of the plastic matrix.

GLASS

Glass is an inorganic non-crystalline fusion product, solidified by cooling. Glass achieves its maximum strength under compression and fails suddenly when subjected to tensile stresses. Glass has become an integral part of many construction materials. It can be converted into any desired shape by blowing, rolling and pressing. Glass is a transparent, chemically inert and biologically inactive material. Glass, which was considered fragile years ago has now come out as one of the strongest construction materials with the proper selection of dimensions and strength enhancing additives. There are many types of glass, varying from foam glass to bullet proof type. Use of glass in construction improves aesthetic looks. Glass partitions and claddings are very common in modern structures. Some of the disadvantages of glass including poor tensile strength may be well tackled by the addition of nano particles, however future research work is needed in this field.

Figure 11. The Moor House Building in London, England

Types of Glass

Likewise concrete and steel, glass can be manufactured in a variety of ways, giving it some unique characteristics. Some important types are described as follows:

- **Translucent Glass:** This glass has a texture embedded in it to hinder the visibility. Therefore, it is used in doors and windows of residential and commercial buildings. The textured side faces inside of room and the plane side faces the outer side.
- **Wired Glass:** This type of glass has steel wire mesh embedded in it. It provides higher safety in case of accidents as wire mesh keeps the broken pieces with in the glass structure and avoids any harm to human beings, which may have caused by the flying glass particles. This type of glass also offers good fire-resistance (Singh, 2005)
- **Tinted Glass:** This type of glass has color and provides better resistance against solar radiations as compared to the plane glass. It is used for decorations works in buildings.
- **Ground Glass:** In this type of glass, one face of the glass sheet is rough while other side is smooth. Like translucent glass, it allows the transmission of light but obstructs the vision, therefore keeping the privacy of structure undisturbed.
- **Laminated Glass:** It consists of two or more glass plates adjoined under the action of heat and pressure. It provides better strength and insulation as compared to a single sheet glass. This type of glass also remains intact as a single body when broken and its particles does not fly off thus giving higher safety to the people in case of accidents.
- **Bulletproof Glass:** This type of glass does not allow common bullets to pass through it. It is widely used in buildings demanding higher safety including banks, jewellery shops and finance departments. The thickness of bulletproof glass varies from 15mm to 75mm or even more (Khitab, 2012).
- **Foam Glass:** Foam glass is prepared from powdered glass. It is a lightweight glass with millions of completely sealed pores (Khitab, 2012). The material is watertight and achieves an efficient barrier against humidity. Foam glass can be manufactured from waste glass thus use of foam glass in buildings is a good approach towards sustainability.

Advantages of Glass Use in Construction

Glass offers many advantages in particular situations. Its use in buildings improves the aesthetics. Glass claddings in buildings fulfill the purpose of heat retention and energy saving. Glass is light in weight and adds up very little to self weight of the structure. Glass is easy to dismantle and requires less maintenance during its life cycle. Glass is a recycled material.

METALS

Metals and their alloys are extensively used in engineering projects. They can be classified into two categories: Ferrous metals and Non ferrous metals. Ferrous metals are the alloys containing iron as its main constituent and non-ferrous are those one, which either contain zero or small quantity of iron.

Types of Metals

About 91 of the 118 elements in the periodic table are metals but not all of them are used in construction works. Some important ferrous metals which are commonly used in construction industry are described below:

- **Pig Iron:** The crudest form of iron available in nature is known as pig iron. It is extracted from several iron ores. It has very high compressive strength but low to moderate tensile strength (Singh,2005). It is used in door brackets, base plates and columns.
- **Cast Iron:** Cast iron is an advanced form of pig iron having lime stone and coke added to it. Like pig iron, it is strong in compression but weak in tension. According to Varghese, (2010), its compressive strength is 3 to 5 times more than its tensile strength.

Both pig iron and cast iron have high resistance to rusting. Cast iron is used for railings, gutter inlets and man hole covers etc.

- **Wrought Iron:** It is obtained by oxidizing carbon and other elements from the pig iron and is left with only 0.25 percent of carbon. It is a purest form of iron as impurities in it do not exceed 0.5 percent (Singh, 2005). Unlike pig iron and cast iron, it is ductile and can be welded. However, its resistance to corrosion is far less than that of pig and cast irons. It is used in the construction of corrugated sheets, grills, railings, gratings etc.
- **Steel:** Steel is an alloy of iron, carbon and other metals. The main difference between steel and cast iron is the amount of carbon content. As carbon contents in steel are increased, it becomes harder. Based on the carbon quantity, steel is further classified into the following types: Low carbon steel having carbon content not more than 0.25%, mild carbon steel, containing 0.25% to

Classical Building Materials

0.7% of carbon and high carbon steel, which contains 0.7 to 0.15% of carbon (Syed, 1970).

- **Stainless Steel:** Steel having at least 10% of chromium in it. Chromium is a blue white metal having high resistance towards corrosion. It makes a passive film of chromium oxide, which prevents the surface corrosion thus blocks the rust spreading into the internal surface of metal (Khitab, 2012).

Besides these ferrous metals, many non-ferrous metals are also used in the construction industry; few of them are described below:

- **Aluminum:** Aluminum and its alloys are widely used in the construction works. Aluminum has many advantages over several other metals. Pure aluminum is ductile, light, air tight, economic and durable. Its strength to weight ratio is the major reason for its use in the construction works. Characteristics of aluminum can be increased by the addition of several other metals like silicon, manganese, and magnesium (Varghese, 2010). Unlike steel, aluminum has good scrap value. Aluminum is also used as an additive in paints for insulation of roofs. Moreover, in modern buildings, frames for office partitions are made up of aluminum and glass.
- **Zinc:** Zinc is a bluish white crystalline metal and is used for coating steel or iron to increase their resistance against corrosion. Zinc is also used in the manufacturing of paints. According to Khitab (2012), the large deposits of Zinc are found in Australia, Asia and the United States.

Figure 12. The Blackpool Tower (steel structure) in Lancashire, England

- **Tin:** Tin is one of the most extensively used protective materials in the construction works. It is highly ductile and offers good plasticity. Moreover, it has good resistance against acid attacks (Singh, 2005).
- **Lead:** Lead is also used in construction works as it is highly ductile and offers good resistance to corrosion. According to Hodge, (1981) it is not recommended to use lead pipes as lead is poisonous and may cause serious health risks if taken in excess quantities. However, use of lead for manufacturing waste water conduits is commonly practiced.

Advantages of Metal Use in Construction

Metals are considered as the backbone of all Civil Engineering projects based on the numerous benefits they offer to the construction industry (Singh, 2005). For example, use of steel in construction provides reliability, uniformity, durability, high strength as well as lesser weight. Furthermore, it offers good scrap value and there exists good possibility for its reuse (Siddiqi, 2009). As far as environmental impact is considered, composite steel floors are given the highest A+ environmental rating in the (BRE's Green Guide to Specification). Furthermore, pre-fabricated steel frame structures allow off-site manufacture resulting in less site activity which leads to enhanced safety during construction.

PAINTS

Paints are coatings which are applied on a variety of surfaces including wood, plaster and metals for protection as well as good appearance purposes (Varghese, 2010).

Paints are applied in a variety of ways. Besides manual application; distribution of paints, solvents and varnishes can be carried out using several distributing devices including spray guns (Tchebinyayeff, 1981). Moreover, paints can be applied in solid, gaseous suspension or in liquid form and are usually built in layers as priming coat, first coat and second coat.

Types of Paints

Enough advancement has been achieved in the manufacturing of paints over the last few decades and paints are now available in a variety of types. Besides typical types, there are some special forms of paints developed for specific purposes e.g. heat resisting paints for insulation purposes, chlorinated rubber paints for protection against acid fumes and luminous paints to make appearance more visible in the dark environment (Singh, 2005). Some important types of paints are discussed below,

- **Oil Paints:** These are one of the earliest types of paints and consist a base, binder and a thinner. Zinc oxide, iron oxide, white lead and various metallic powders are the commonly used bases. Binder used is mostly linseed oil however other oils and waxes are also used for this purpose. Purpose of thinner is to increase the fluidity of the paint and mostly oil of turpentine is used as a thinner. Oils paints are used on walls when a glossy surface is required. They are also easily washable and when applied to metals, they provide good corrosion resistance (Khitab, 2012; Singh, 2005).
- **Emulsion Paints:** Emulsion is a mixture of one liquid with another liquid having water as a base along with vinyl or acrylic resins added to make them hard wearing. One of the advantages for using such paints is quick drying time (Khitab, 2012). Thus they should be preferred in places where weather remains unfriendly most of the time and only few limited hours are available for carrying out the painting activity. Furthermore, they are washable many times and give a very appealing finish to the plastered walls.
- **Distempers:** Distempers contain powdered chalk as base along with some other binders. Distempers are categorized as low grade wall paints and are preferred only due to their cheap prices (Varghese, 2010). Distempers are easily affected by weathers so their use in restricted only to the interior works (Singh, 2005).
- **Fire Proof Paints:** Paints having sodium tungstate and asbestos provide good resistance towards fire. These paints are usually applied on wooden structures to make them fire resistant to some degree (Singh, 2005).

Paint Additives

Paint additives are generally added in paints to further improve their properties. There are several paint additives such as leveling agents, film formation promoters etc. available these days. Leveling agents give a smooth and uniform textured finish once the paint is dried whereas film formation promoters reduce the film forming temperature. Similarly, wetting additives are also used to prevent the flocculation of pigment particles thus giving a uniform and haze free color (Freitag et al., 2008).

Characteristics of a Good Paint

Color of paint should not fade over time. However some fadedness over decades is common in paints. Paint should also offer good resistance towards weathering effects. Furthermore, paint should not crack once dried. It should be easy to apply and being able to be laid in thin coats. Lastly, a good appealing appearance is also an important characteristic of paint (Singh, 2005).

TIMBER

Timber is a natural polymer which is composed of cells in the shape of long thin tubes with tapered ends. The cell wall consists of crystalline cellulose aligned parallel to the cell axis. These cellulose crystals are bonded by a complex amorphous lignin composed of carbohydrate compounds. Generally, wood is 50 to 60% cellulose and 20 to 35% lignin, the remaining percentage is occupied by other carbohydrates and mineral matter (Merritt, F. S et al., 2001)

Timber was used in the construction industry for roofs, rails, sleepers, signal poles and water wheels etc in the 19th century. However, in 20th century an extension was seen in use of timber in some specific areas and a decline in others. Timber is used in the construction industry due to its high strength as well as appearance. Rays are an important structural feature that adds to the attractive appearance especially when the rays are both deep and wide as in the case of oak. When the surface of the plank coincides with the longitudinal–radial plane, these rays can be seen as sinuous light colored ribbons running across the grain. Furthermore, knots which are considered troublesome from the mechanical aspects can also be regarded as a decorative feature; the fashion for knotty pine furniture and wall paneling in the early seventies is a very good example of this (Domone, P., 2010).

Types of Timber

Following are some important timbers which are commonly used in the construction industry,

- **Chir:** It is pale grayish to reddish or yellowish brown in color. Furthermore it is light in weight and seasons very well. Durability of Chir is low to moderate and is preferred for interior works only however if creosoted, it may be used for railway sleepers (Khitab, 2012; Singh, 2005).
- **Deodar:** Deodar is generally found in hilly areas. It is more durable as compared to Chir and is also easily workable. However, its strength is not as satisfactory to use it for bearing high intensity loads therefore it is used only in moderate structural works (Khitab, 2012).
- **Mulbery:** Mulbery is a strong and elastic wood having brown color. It is easily workable and is mostly used where curved sections are required. However, when polished it does not give the appealing texture. Its density is 6750 N/m3, which is quite more than Chir and Deodar having 5700 N/m3 and 5450 N/m3 respectively (Khitab, 2012; Singh, 2005).

Classical Building Materials

- **Bamboo:** Babboo is also one of the strongest and durable woods. Bamboos are used for scaffolding, construction of temporary structures including bridges as well as in construction of cheap and big sheds (Singh, 2005).
- **Shisham:** Shisham is a very tough, strong and hard wood (Syed, 1970). It is used in the construction of houses and due to its high weight density of 7000 N/m^3, its use has been extended to the bridge piles (Khitab, 2012).

Though timber use in construction industry offers a number of advantages, however there are some issues which should be addressed to avoid any problems. Timber is affected by light, rain and wind which results in a complex degrading mechanism that renders the timber silvery grey in appearance. Moreover, surface integrity is also reduced subjected to this weathering process (Derbyshire and Miller, 1981; Derbyshire et al., 1995). Prolonged exposure of timber to elevated temperatures also results in a reduction in strength and a very marked loss in toughness (Shafizadeh and Chin, 1977). But still, if timber is of good quality and processed well, it definitely offers very good service life economically. All that matters is to use the most suitable type of timber for the specified purpose. A timber suitable for interior use may deteriorate very quickly when used outside. Furthermore, according to (Domone, P., 2010), though timber is an organic product it is surprising at first to find that it can withstand attack from fungi and insects for very long periods of time, certainly much longer than its herbaceous counterparts.

REFERENCES

Abdullahi, M. (2006). Properties of Some Natural Fine Aggregates in Minna, Nigeria and Environs. *Leonardo. Journal of Science*, (8), 1–6.

Adityamadhav. (2011). Retrieved August 12, 2015, from http://commons.wikimedia.org/wiki/File%3AIgneousrock_(Tuff_type)_at_Rushikonda%2C_Visakhapatnam.jpg

Anderson, J., & Shiers, D. (2009). *Green guide to specification*. John Wiley & Sons.

Anwar Khitab. (2012). *Materials of Construction (Classical and Novel)*. Allied Book Company Brick-History. Retrieved July 13, 2015 from http://science.jrank.org/pages/1022/Brick-History.html

Derbyshire, H., & Miller, E. R. (1981). The photodegradation of wood during solar irradiation. *Holz als Roh- und Werkstoff*, *39*(8), 341–350. doi:10.1007/BF02608404

Derbyshire, H., Miller, E. R., Sell, J., & Turkulin, H. (1995). Assessment of wood photodegradation by microtensile testing. *Drvna Industrija, 46*(3), 123–132.

Dipanjan Mukherjee. (2014). *Low Cost Light Weight Concrete Making By Using Waste Materials*. Global Journal of Engineering Science and Research Management.

Domone, P., & Illston, J. (Eds.). (2010). *Construction materials: their nature and behaviour*. CRC Press.

Edvardsen, C. (1999). Water permeability and autogenous healing of cracks in concrete. *ACI Materials Journal, 96*(4).

Freitag, W., & Stoye, D. (Eds.). (2008). *Paints, coatings and solvents*. John Wiley & Sons.

Gutmann, P. F. (1987). Bubble characteristics as they pertain to compressive strength and freeze-thaw durability. In *MRS Proceedings* (Vol. 114, p. 271). Cambridge University Press.

High Strength Concrete. (n.d.). Retrieved May 04, 2015 from, http://www.concrete.org.uk/fingertips_nuggets.asp?cmd=display&id=528

Hodge, A. T. (1981). Vitruvius, lead pipes and lead poisoning. *American Journal of Archaeology,* 486-491. Retrieved June 02, 2015 from, http://www.bbc.co.uk/schools/primaryhistory/indus_valley/technology_and_jobs/

Hunt. (2000). Material Evidence Conserving historic building fabric. *New Heritage Materials*.

Jonkers, H. M., Thijssen, A., Muyzer, G., Copuroglu, O., & Schlangen, E. (2010). Application of bacteria as self-healing agent for the development of sustainable concrete. *Ecological Engineering, 36*(2), 230–235. doi:10.1016/j.ecoleng.2008.12.036

Linch, K. D. (2002). Respirable concrete dust--silicosis hazard in the construction industry. *Applied Occupational and Environmental Hygiene, 17*(3), 209–221. doi:10.1080/104732202753438298 PMID:11871757

López-Arce, P., Garcia-Guinea, J., Gracia, M., & Obis, J. (2003). Bricks in historical buildings of Toledo City: Characterisation and restoration. *Materials Characterization, 50*(1), 59–68. doi:10.1016/S1044-5803(03)00101-3

Merritt, F. S., & Ricketts, J. T. (2001). *Building design and construction handbook* (Vol. 13). New York, NY: McGraw-Hill.

Sandyadav. (2012). Retrieved September 02, 2015 from http://commons.wikimedia.org/wiki/File%3AWhite_Marble_Rocks_at_Bhedaghat.jpg

Siddiqi, Z. A., & Ashraf, M. (2009). *Steel Structures* (2nd ed.). Lahore: Help Civil Engineering Publisher.

Siever, R. (1979). Plate-tectonic controls on diagenesis. *The Journal of Geology, 87*(2), 127–155. doi:10.1086/628405

Singh. (2005). Engineering Materials (5th ed.). Academic Press.

Smith, M. R. (Ed.). (1999). *Stone: Building stone, rock fill and armourstone in construction*. Geological Society of London.

Syed. (1970). *Materials of Construction*. Academic Press.

Tchebinyayeff, M. R. (1981). *U.S. Patent No. 4,306,587*. Washington, DC: U.S. Patent and Trademark Office.

Varghese, P. C. (2010). *Building Materials*. PHI Learning Private Limited.

Chapter 2
Nano-Scale Behavior and Nano-Modification of Cement and Concrete Materials

Liqing Zhang
Dalian University of Technology, China

Baoguo Han
Dalian University of Technology, China

Siqi Ding
Dalian University of Technology, China

Xun Yu
New York Institute of Technology, USA

Shengwei Sun
Harbin Institute of Technology, China

Jinping Ou
Dalian University of Technology, China

ABSTRACT

Cement and concrete materials are widely used, but the development of them comes cross many problems and challenges, such as high energy consumption, high pollution, poor safety and durability, low smart. Nanotechnology is beneficial to understand the behavior of cement and concrete materials at nano-scale. In addition, nanomaterials have remarkable specific properties and functions which can endow cement and concrete materials high mechanical property and durability and multifunctionality. Therefore, applications and advances of nanotechnology and nanomaterials have injected new vitality into cement and concrete materials. This chapter will give a review about nano-scale behavior of cement and concrete materials, the nano modification methods to cement and concrete materials by using

DOI: 10.4018/978-1-5225-0344-6.ch002

nano-binders and adding nano materials with attention to workability, hydration, mechanical property, durability and other properties of the cement and concrete materials, and the nano modification mechanisms to the cement and concrete materials. Finally, future development and challenge of nano-modificated cement and concrete materials are also discussed.

INTRODUCTION

Cement and concrete are the most widely used construction materials because they are resistant to water, easily formed into various shapes and sizes, the cheapest and readily available everywhere. In the foreseeable future, cement and concrete materials will continue to play an important role in construction materials. However, the development of cement and concrete materials is encountering enormous problems and challenges. Firstly, cement manufacturing consumes high energy and spits out a large amount of carbon dioxide. Additionally, most raw materials of cement are non-renewable. These disadvantages mentioned above are likely to pose great pressure on environment. Secondly, increasing attention has been paid on the security of cement and concrete structures since cement and concrete are brittle materials and usually work with cracks. Thirdly, the durability of cement and concrete structures is a very important issue, in particular during the process of their design and application. Fourthly, cement and concrete materials are complex composites in nature. There are difficulties in how to bring about huge revolution in the cement and concrete material field unless it is available to have a good understanding of cement hydration process, main hydrate phases and so on. Fifthly, the multifunctional and smart cement and concrete materials are required since traditional cement and concrete that just serve as structural materials can not meet the requirement of construction of advanced engineering structures.

Nanotechnology is an emerging field related to the understanding and control of matters at nanoscale. Some remarkably specific properties and functions are exhibited when materials reach nano-size. Recent developments of nanotechnology show significant promise in addressing many of the challenges in various areas. To date, applications and advances of nanotechnology have injected new vitality into cement and concrete materials (Bartos, 2009). The vitality mainly reflects in understanding more about behaviors of cement and concrete materials at nano-scale, fabricating nano-cement and modifying cement and concrete materials by addition of nano-materials. The main advance is that we can learn about basic phenomena in cement at nano-scale (Bartos, 2009; Scrivener & Kirkpatrick, 2008). Jo et al. (2014) used the bead milling to produce ultrafine cement (~220 nm) without changing the chemical phase. The application of nanotechnology makes cement and concrete materials have high mechanical properties, high durability, and even multifunctionility. Li et al. (2004) studied the impact of addition of 3 wt. %, 5 wt. % and 10 wt. % of nano-silica (NS) to cement mortars, and achieved an increase in compressive and flexural strengths. Haruehansapong et al. (2014) fabricated cement mortars containing different contents (3%, 6%, 9% and 12% of cement weight) of NS with various sizes of 12 nm, 20 nm and 40 nm. They observed that 40 nm NS is more effective for improving compressive strengths compared with 12 nm and 40 nm NS. The optimum content of 9% NS can improve compressive strength to 1.5 times at 28 d against the plain cement mortars. Li (2004) incorporated 4% NS into concrete with 50 wt. % fly ash and obtained an increase of 81% in compressive strengths compared with concrete with only 50 wt. % fly ash at 3 d. Using of NS shows a new approach to save cement so that the pressure on environment can be reduced. Oltulu et al. (2014) found that NS can refine pore structure of hardened cement and concrete materials. Apart from NS, nano-TiO_2 (NT) and nano carbon materials (NCMs) were also used to modify cement and concrete materials. Cement and concrete materials filled with NT have the function of decomposing environment pollution, self-cleaning and self-disinfecting (Chen & Poon, 2009). Sáez De Ibarra et al. (2006) reinforced cement paste by using carbon nanotubes (CNTs). They found addition of only 0.1 wt. % of CNTs can increase hardness by 177% and Young's modules by 227%. Besides mechanical properties, NCMs can endow cement and concrete with thermal property, sensing property, electromagnetic interference property and other properties (Han et al., 2015). Some researchers have also studied the properties of cement and concrete materials with other nano-materials, such as nano-fly ash, nano-Al_2O_3 (NA), nano-Fe_2O_3, nano-ZnO_2, nano-CuO, nano-Cu_2O_3, nano-ZrO_2, nano-$CaCO_3$ (NC) and nano-clay.

This chapter is predicted to review on nano-scale behavior of cement and concrete materials, nano modification methods to cement and concrete materials by using nano-binders and adding nano materials, and nano modification mechanisms of cement and concrete materials. The future development and challenge of nanotechnology in the field of cement and concrete materials will also be discussed.

NANO-SCALE BEHAVIOR

As the rapid development of cement manufacturing since the initial production of Portland cement in 1824, preparation techniques of cement have become considerably mature at present. However, the hydration process and productions, being central to cement and concrete materials, still cannot be absolutely understood, especially in nano-scale. With the development of nanotechnology, we can better understand the behavior of cement and concrete materials in nano-scale.

The particle size of cement and admixture is mostly in the microscale level. However, the size of hydration productions of cement and admixture, like Calcium Silicate Hydrate (C-S-H), is in the nano-scale level. To some extent, the hydration can be seen as a top-down process to produce nanomaterials. In addition, C-S-H (about 65 vol. % of solid hydration production) is the main binding phase in all Portland cement-based systems (Richardson, 1999). Their structure and chemical composition have strong effect on the mechanical properties, durability and other properties of cement and concrete materials.

Morphology of C-S-H

According to the original boundaries of clinker particles, Taplin (1959) termed those products within the boundaries as 'inner' products (Ip) and those products outside the boundaries as 'outer' products (Op), i.e. two types of C-S-H are Ip C-S-H and Op C-S-H which are also called high-density C-S-H and low-density C-S-H, respectively. This scheme has been adopted widely (Constantinides & Ulm, 2007; Groves, Sueur, & Sinclair, 1986; Kjellsen, Jennings, & Lagerblad, 1996; Williamson, 1972). The Ip C-S-H presenting in larger cement grains typically has a fine-scale and homogeneous morphology which can be observed through TEM *(Richardson, 1999)*. The Op C-S-H in cement paste looks like a fine fibrillar. With the increasing of slag in cement paste, the morphology of Op C-S-H changes from fine fibrillar to foil-like *(Richardson, 1992)*.

Nano Behavior of C-S-H at Elevated Temperatures

De Jong et al. (2007) studied the behavior of C-S-H when the temperature increased to 700°C. They drew a conclusion that the loosening of packing density resulted from the shrinkage of C-S-H nanoparticles which occurred at high temperature and was probably responsible for the thermal degradation of cement paste.

Lim et al. (2014) reported that there were two main factors causing the thermal degradation of cement and concrete materials. The first one was the different thermal expansion rate between unhydrate cement particles and the paste matrix. In addition, reversible expansion of unhydrated cement grain and irreversible shrinkgae of C-S-H which probably resulted in a gap interface at 300°C. The gaps would become wider and develop to micro-crack when the temperature was over 300°C. The second one was that the nano-structure of C-S-H continuously rearranged which made the micro-structure more porous at elevated tempreture. These changes would cause loss of strength and pore-coarsening ultimately.

Effect of C-S-H on the Elasticity

Constantinides et al (2004) studied the effect of two types of C-S-H on the elastic properties of cement and concrete materials by means of nanoindentation. They found that the decalcification of the C-S-H phase was the primary source of the macroscopic elastic modulus degradation. In addition, the high-density C-S-H was less affected by calcium leaching compared with low density C-S-H. This suggests that a cement past with higher volumetric proportion of high-density C-S-H would be less affected by calcium leaching during age.

Effect of C-S-H on Durability

In cement-slag blends, the fibrillar morphology was gradually replaced by the foil-like morphology with increasing of slag (Richardson & Groves, 1992). The foil-like morphology appears to be more efficient at filling space without leaving large interconnected capillary pores. Therefore, this change in morphology of Op C-S-H from fibrillar to foil-like may be largely responsible for the improved durability.

Effect of C-S-H on Creep and Shrinkage

Creep and shrinkage result from aging of C-S-H, which depends on time, the temperature, relative humidity, as well as the history of the C-S-H. Aging is a process of increasing the number of bonds between globules of C-S-H, causing C-S-H to become stiffer, stronger and denser (Jennings, 2003; Thomas & Jennings, 2006). With increasing age, the changes of creep and shrinkage become lower. In another investigation, Hou et al. (2014) thought the displacement of water layer under tension loading leaded to shrinkage and creep of cement and concrete materials. There are also other standpoints of C-S-H behavior on creep and shrinkage, but it is no doubt that C-S-H has strong impact on creep and shrinkage.

Although many experimental methods such as nanoindentation, atomic force microscopy, and nuclear magnetic resonance have been used to explore behavior of C-S-H, effect of C-S-H on the properties of cement and concrete materials is still unclear.

MODIFICATION METHODS

Modification to Cement and Concrete Materials by using Nano-Scale Binders

Nano Cement

Nanotechnology provides new methods to modify cement. One of the methods is production of environmental friendly nano cement. Jo et al. (2014) synthesized nano cement (the fineness was 167 nm) by using silica source and alumina source. They fabricated two types of nano cement based concrete: type one is named as NC-CS group containing various weights of fine and coarse aggregate, type two is named as NC-CN group with varying weights of alkali activator. They studied the workability and mechanical properties of these concretes.

1. **Workability:** Jo et al. (2014) found that the slump value increased with the increase of fine aggregate/coarse aggregate ratio and the slump value also increased with the increase of activator content at a fixed aggregate/coarse aggregate ratio.
2. **Mechanical Properties:** Jo et al. (2014) studied the compressive strengths of concrete samples at 3, 7, 14 and 28 d and splitting tensile strengths at 28 d. They observed that the compressive strength of concrete mixed with chemically synthesized cement was able to reach 43 MPa after 14 d of curing at an optimized amount of aggregates and alkali activator content.

The chemically synthesized nano cement may provide a new way to produce cement. However, it is a long way to widely use nano cement in cement and concrete materials. The properties of concrete mixture with nano cement should be studied more about the mechanical properties, durability and etc..

Nano-Fly Ash

Fly ash has been considered as a cementitious component of concrete in British stands since 1980s (McCarthy, 1999). It has been used to replace a part of cement in order to produce concrete with higher durability (Sumer, 2012; Thomas, 1999) and better economy. However, many studies (Arulraj & Carmichael, 2011; Hannesson, Kuder, Shogren, & Lehman, 2012; Sumer, 2012; Swamy, Ali, & Theodorakopoulos, 1983) observed that the hydration process of concrete was slow down and the early strengths were lower than that of normal concrete due to addition of fly ash. The nano-fly ash may enhance the durability and early strength at the same time. Arulraj et al. (2011) produced nano-fly ash by high intensity ball milling and fabricated different grades of concrete, namely M20, M30, M40 and M50, with nano-fly ash at levels of 0%, 10%, 20% and 30% by replacing of coarse aggregate. Their research results about the workability and compressive strengths were showed as follow.

1. **Workability:** Arulraj et al. (2011) found the slump value of four grades of fresh concrete all gradually increased with the addition of nano-fly ash. For M20 Grade of concrete, the slump value of mixture increased by 92%, 112% and 132% as the levels nano-fly ash are 10%, 20% and 30%, respectively. For M30 grade of concrete, the slump value of mixture increased by 85.71%, 107.14% and 135.71% as levels of nano-fly ash are 10%, 20% and 30%, respectively. For M40 grade of concrete, the slump value of mixture increased by 86.66%, 113.33% and 140% when levels of nano-fly ash are 10%, 20% and 30%, respectively. For M50 grade of concrete, the slump value of mixture increased by 70.58%, 111.76% and 132.35% as nano-fly ash levels are 10%, 20% and 30%, respectively.
2. **Mechanical Properties:** Arulraj et al. (2011) studied the compressive strengths of nano-fly ash filled concrete after curing 28 d and found that the compressive strengths increased gradually as the addition of nano-fly ash increased. When the contents of nano-fly ash are 10%, 20% and 30%, the enhancements of 28 d compressive strengths of concrete were range from 17% to 46%, 19% to 56% and 21% to 60% according to concrete grade, respectively.

Modification to Cement and Concrete by Adding Nano Materials

Nano-Scale Nonmetallic Oxides

Up to date, many kinds of nanomaterials, including nano-scale nonmetallic oxides, nano-scale metallic oxides, nano-scale carbon materials and other nano-scale materials, have been added to cement and concrete materials. What's more, small dosages of nanomaterials always bring a significant enhancement in the properties of cement and concrete materials.

Nano-Silica
NS, seen the development of silica fume, is applied more early into cement and concrete. NS has smaller particle size and higher activity (Ghafari, Costa, Julio, Portugal, & Duraes, 2014; Ye, Zhang, Kong, & Chen, 2007) than silica fume. NS can reduce the workability, increase hydration heat, short setting time, enhance mechanical properties and durability.

Nano-Scale Behavior and Nano-Modification of Cement and Concrete Materials

1. **Workability:** Many researchers have studied the workability of cement and concrete materials containing NS. Although the workability of cement and concrete with NS is test by using different methods including flow table test (Senff, Labrincha, Ferreira, Hotza, & Repette, 2009), slump (Hosseini, Booshehrian, &Madari, 2011), Viskomat pa viscometer (Ye, Zhang, Kong, & Chen, 2007) and so on, the results showed consistently that addition of NS reduced the flowability of fresh mixture. For example, Collepard et al. (2002) studied the workability of self-compacting concrete with NS. They found that the addition of NS made the mixture more cohesive and reduce tendency of bleeding water and segregation. A higher amount of water reduce agent was needed with increasing dosage of NS. Ye et al. (2007) tested the consistency of fresh cement paste according to ISO 9597:1989. They observed that the fresh cement paste with NS is thicker than that of plain cement paste. The test results of consistency were 34mm, 33mm and 32mm at replacement levels of 0%, 2% and 5%, respectively. Senff et al. (2009) found that adding NS to mixture decreased the spread on the flow table due to the increase of cohesion in the mixture. The spread diameter of mortar with 2.5% NS showed a 32.8% reduction compared with plain mortar. The spread diameter decreased 19.6% and plastic viscosity only increased 3.6% when the content of NS changed from 1.0% to 2.5%.

   Hosseini et al. (2011) reported that the workability (tested by slump) decreased with the increase of NS dosage. The slumps of concrete with 1.5% and 3.0% NS decreased about 47% and 70.5% respectively compared with the concrete without NS. Berra et al. (2012) studied the workability of cement paste with NS at different ratios of water/binder by using Mini-slump device. They found that the addition of 0.8% and 3.8% NS by weight of cement can always reduce the workability of cement paste at different water/binder ratios. Ghafariet al. (2014) reported that addition of NS to ultra high performance concrete reduced the flowability which was measured by flow table test according to ASTM: C1437. They found incorporation 2% and 3% NS showed a reduction of 9mm and 12mm, respectively, in comparison to the mixture without NS.

2. **Hydration:** Addition of NS can increase the hydration heat, accelerate the hydration process and reduce the setting time of cement. Li et al. (2004) reported that the maximum hydration heat of the 50% fly ash filled concrete incorporating 4% NS, the concrete with 50% fly ash and Portland cement concrete are 61°C, 51°C and 65°C, respectively. Jo et al. (2007) studied the heat evolved from the mixtures which include the cement paste with 10 wt. % NS, cement paste with 10 wt. % silica fume and plain cement paste. They found that the amounts of heat evolved in 72 h from the three different mixtures are

245.5 J/g, 235.7 J/g and 231.1 J/g, respectively. In addition, NS can increases the amount of heat evolved during the process of setting and hardening of the cement. Senff et al. (2009) found that the time to reach maximum temperature decreased 51% compared with cement without NS. Lin et al. (2008) observed the initial and final setting time of pure cement paste, cement paste with 10% sludge/fly ash, cement paste with 10% sludge/fly ash and 1% NS were 140min and 230min, 200min and 270min, and 140min and 170min, respectively. Senff et al. (2009) fabricated cement paste with water/binder ratio of 0.35 and NS at levels of 0%, 1.0%, 1.5%, 2.0% and 2.5%. With increasing amount of NS, the setting time including initial and final setting time reduced. The setting time of cement paste with 2.5% NS decreased 60% in comparison to the cement paste without NS. Zhang et al. (Zhang, Islam, & Peethamparan, 2012; Zhang & Islam, 2012) and Hou et al. (2013) studied the setting time of fly ash cement system with NS and got the similar conclusion that addition NS could significantly shorten the initial and final setting time of fly ash cement system.

3. **Mechanical Properties:** NS can enhance the mechanical properties of cement and concrete materials which can make ultra high performance concrete. In addition, NS can also improve the early age mechanical properties of fly ash concrete, concrete with ground ceramic powder, concrete with recycled aggregate and other green concrete, thus broadening the road of green concrete.

Shih et al. (2006) studied the compressive strength of cement paste containing NS at levels of 0.2%, 0.4%, 0.6% and 0.8% by weight of cement. They observed that the compressive strengths at 7d, 14d, 28d and 56d increased before NS level reaches optimal amount of 0.6% and then has a drop at NS level of 0.8%. Jo et al. (2007) found that the compressive strength of cement mortars with NS is higher than that of cement mortars with silica fume at 7 d and 28d. The compressive strengths at 28 d were improved by 112.11%, 141.79%, 166.41% and 168.75% compared with plain cement mortars at replacement levels of 3%, 6%, 10% and 12%, respectively. Li et al. (2004) reported that the compressive strength increased by 5.7%, 20.1% and 20.1% at the age of 7d, 13.8%, 17.0% and 26.0% at age of 28 d at the replacement levels of NS at 3%, 5% and 10%, respectively. Lin et al. (2008) found that NS can increase the compressive strength of sludge/fly ash cement mortars. Hou et al. (2013) also studied the compressive strength of fly ash cement mortars with NS. They observed that the higher dosage of NS, the greater of improvement of early-age compressive strength. 50% fly ash filled concrete incorporating 4% NS has an increase of 81% in compressive strength at 3 d compared with the concrete with 50% fly ash (Li, 2004). Nazari et al. (2011a) found that the addition of NS significantly increased the compressive, split tensile and flexural strengths. Especially, addition of 5% NS increased the 28 d compressive, split tensile and flexural strengths by

74.4%, 118.8% and 64.3%, respectively. Ghafari et al. (2014) prepared ultra high performance concrete with NS. The amount of NS partially replaced cement were 0%, 1%, 2%, 3% and 4% by weight. They found that NS significantly increased the early compressive strength. However, it had a modest effect on the compressive strength at 28 d and 90 d.

Table 1. Effect of NS on mechanical properties of cement and concrete materials

Types	Mechanical Properties	Enhancement of the Properties(%)		Content of NS (%)	Water/Binder	Reference
		7d	28d			
Cement paste	Compressive strength	/	24.8	5	0.22	(Ye et al., 2007)
	Bond strength	43.1	87.9	5	0.22	(Ye et al., 2007)
Cement mortars	Compressive strength	115.9	112.1	3	0.50	(Jo et al., 2007)
		151.9	141.8	6		
		169.4	166.4	10		
		177.1	168.8	12		
		5.7	13.8	3	0.50	(Li et al., 2004)
		20.1	17.0	5		
		20.1	26.0	10		
	Flexural strength	28.0	27.0	5	0.50	(Li et al., 2004)
Concrete	Compressive strength	68.6	39.3	4	0.28	(Li, 2004)
		22.1	56.4	2	0.38	(Jalal, Mansouri, Sharifipour, & Pouladkhan, 2012)
		21.2	18.5	1	0.40	(Givi, Rashid, Aziz, & Salleh, 2010)
		16.5	12.0	1.5	0.40	(Givi et al., 2010)
		101.0	74.4	4	0.44	(Nazari & Riahi, 2011a)
		25.5	43.9	1.5	0.40	(Najigivi, Khaloo, Iraji Zad, & Abdul Rashid, 2013)
	Flexural strength	31.0	31.8	1	0.40	(Givi et al., 2010)
		23.8	22.7	1.5	0.40	(Givi et al., 2010)
		43.2	64.3	4	0.44	(Nazari & Riahi, 2011a)
	Split tensile strength	100.0	83.3	1	0.40	(Givi et al., 2010)
		93.3	72.2	1.5	0.40	(Givi et al., 2010)
		141.7	118.8	4	0.44	(Nazari & Riahi, 2011a)

Table 1 summarizes the previous research results about the mechanical properties of cement and concrete materials with NS. From the table, the conclusion can be made as follow:

- Addition of NS at suitable amount can significantly improve the compressive strength, flexural strength, split tensile strength and other mechanical properties;
- The enhancement degree depend many factors which include the water/binder ratio, the particle of NS, curing age, etc..

4. **Durability:**
   a. **Freeze-Thaw Resistance:** Wang et al. (2008) fabricated C60, C70 and C80 high performance concrete containing NS (about 3.3% of binder, by weight) and used fast-freezing methods to investigate the freeze-thaw resistance. They found that the weight loss of C60, C70 and C80 of high performance concrete reduced by 1.0%, 0.6% and 0.3%, respectively. The relative dynamic modulus increased 7.0%, 4.0% and 6.0% after 300 times freeze-thaw cycle, respectively. Quercia et al. (2012) found that addition of NS can increase the freeze-thaw resistance. In addition, the self-compacting concrete (SCC) with colloidal NS had better freeze-thaw resistance than that with powder NS.
   b. **Chloride Penetration and Permeability:** He et al. (2008) added nano-materials (nanoparticles (Fe_2O_3, Al_2O_3, TiO_2 and SiO_2) and nano-clay) by 1% weight of cement into cement mortars. Fixed water/cement ratio of 0.5 was used. The chloride permeability property was measured by electromigration test after curing 28 d. They found that addition of nano-materials especially NS and nano-clay improved the chloride penetration resistance. Zhang et al. (2011) prepared concrete with different NS' content levels. They found that the use of 1% NS by weight of binder (the sum of cement and NS) enhanced the resistance to chloride penetration by 18.04%. Said et al. (2012) studied the resistance to chloride penetration of six types of concrete through the rapid chloride ion permeability test. Group A includes three mixtures only using cement as binder. Group B also includes three mixture containing 30% of cement replaced by fly ash. In group A and B, the amounts of NS were 0%, 3% and 6%. They found that addition of NS can significantly improve the resistance to chloride penetration. In addition, Kong et al. (2012) and Quercia et al (2012) also reported that NS could improve the resistance to chloride penetration of concrete and SCC. Although addition of NS into cement and concrete

materials can improve the resistance to chloride penetration, the enhance extent is different because the different mix proportions, different raw materials, different test methods and so on.

c. **Leaching:** Calcium leaching is a degradation process that calcium ions migrate from cement hydration production to the aggressive solution. Gaitero et al. (2008) firstly used NS to reduce the calcium leaching of cement past. They prepared cement past with 6% NS by weight of cement by using a fixed water/binder ratio of 0.4. They used a 6M ammonium nitrate solution as aggressive solution to accelerated calcium leaching process. The samples after curing 28 d were immersed in the aggressive solution. They found that the four types of NS can reduce the calcium leaching on basis of compressive strength and bonding strength at 9, 21, 42 and 63 d after immersing in the solution. Kong et al. (2012) studied the calcium leaching of cement mortars containing two types of NS, namely precipitated silica (PS) and fumed silica (FS). The samples were immersed in 6 M NH_4Cl solution after curing 28 d in water. The neutralization depth was measured to characterize the degree of calcium leaching of the samples. For the samples with 1.0% PS, the neutralization depth was decreased 17.6% and 20.0%, respectively after immersing in the aggressive solution for 28 and 180 d, compared with plain mortar. For the samples with 1.0% FS, the reduction were 34.4% and 26.1%, respectively after 28 and 180 d of immersion, relative to the reference samples. In short, the research results show that addition of NS to cement and concrete materials can improve the calcium leaching resistance.

d. **Abrasion Resistance:** Li et al. (2006) fabricated concrete with 0%, 1% and 3% of NS and, fixed water/binder ratio of 0.42, and studied the abrasion resistance property at 28 d. They found that surface index of abrasion resistance increased by 157.0% and 100.8% as NS replacement levels are 1% and 3%, respectively. The side index of abrasion resistance of concrete with 1% and 3% NS enhanced by 139.4% and 89.0%, respectively. Riahi et al. (2011a) and Nazari et al. (2011b, 2011c) studied the abrasion resistance of concrete with NS cured in different media. They found that the abrasion resistance of concrete increased with increasing of NS. In addition, the abrasion resistance of concrete with NS cured in saturated limewater was higher than that of concrete with NS cured in water.

e. **Fire Resistance:** Ibrahim et al. (2012) studied the fire resistance property of mortar with high-volume of fly ash combined with NS. Mortars were prepared with water/binder ratio of 0.4, and about 0.18% polypropylene fiber by weight of binder. They found that mortar with high-volume fly ash and NS had high residual strength after exposure to high temperature of 400°C and 700°C. Shah et al. (2013) reported the spalling behavior of NS high strength concrete exposed to temperatures ranging from room temperatures to 800°C. They found the maximums palling appeared in specimens containing NS and micro silica. However, the maximum strength loss and temperature caused cracks appeared in specimens containing NS.

Nano-Scale Metallic Oxides

Nano-scale metallic oxides have been shown to improve properties of cement and concrete materials without corrosion and degradation by ions in the environment (e.g. chloride and sulfate ions). The addition of nano-scale metallic oxides to cement and concrete materials can reduce the permeability to ions and increase the strength, thereby improving durability of cement and concrete materials. Many research works are focused on cement and concrete materials with nano-Al_2O_3, nano-TiO_2 and nano-Fe_2O_3, while there are a few studies which are focused on other nano-scale metallic oxides, such as nano-Fe_3O_4, nano-ZnO_2, nano-CuO, and nano-ZrO.

Nano-Al_2O_3

1. **Workability:** Nazar et al. (2010) observed that the workability of fresh mixture decreased with the increase of NA. In addition, Nazar et al. (2010a; 2011d; 2011e) partially replaced cement with NA at levels of 0%, 0.5%, 1.0%, 1.5% and 2.0% by weight in concrete. They found that the slump value decreased with the addition of NA.
2. **Hydration:**
   a. **Hydration Heat:** Nazar et al. (2011d, 2011e, 2011f) carried out conduction calorimetry test at 22°C for a maximum of 70 h. The cement contained 0%, 0.5%, 1.0%, 1.5% and 2.0% NA, by weight, mixed with water or saturated limewater. They found that the decrease of the percentage of nanoparticle in mixture retarded peak time and the heat release rate values were raised. What's more, the specimens mixed with saturated limewater showed a decreased peak time and heat release rate values with respect to the corresponding specimens mixed water.
   b. **Setting Time:** Nazar et al. (2010, 2010a) studied the initial and final setting time of cement paste with 0%, 0.5%, 1.0%, 1.5% and 2.0% NA with average particle size of 15 nm. The fixed water/binder was 0.40. They reported that the initial time and final setting time decreased as the addition of NA increased.
3. **Mechanical Properties:** Barbhuiya et al. (2014) studied compressive strength of NA filled cement paste at early age (1d, 3d and 7d). They found the NA had slight effect on early age compressive strength. Li et al. observed that addition

of NA to mortar can significantly improve the elastic modulus. However, the improvement to compressive strength was not obvious. The elastic modulus of mortar with 5% NA are enhanced by 154%, 241% and 243% after curing 3, 7 and 28 d, respectively (Li, Wang, He, Lu, & Wang, 2006).

Campillo et al. (2007) studied the compressive strength of belite cement mortars with two types of NA, namely agglomerated dry alumina (ADA) with average grain size ranging from 100 nm to 1000 nm and colloidal alumina (CA) with average grain size 50 nm. They found that NA is very effective in improving compressive strength of belite cement mortars. The CA was more effective than ADA. The compressive strength of mortar with 9% ADA increased by 142% and 119% at age of 7 and 28 d, respectively, compared with the plain mortar. The compressive strengths of mortar with 1.8% CA enhanced by 84% and 118% after curing 7 and 28 d, respectively.

Oltulu et al. (2011) observed that the compressive strength of mortar containing 5% silica fume and 1.25% NA enhanced by 11.0%, 13.4%, 20.2%, 21.7% and 21.4% at age of 3, 7, 28, 56 and 180 d, respectively. Oltulu et al. (2013) found that the compressive of mortar containing 15% fly ash and 1.25% NA increased by 2.7%, 3.9%, 4.3%, 5.6% and 5.0% after curing for 3, 7, 28, 56 and 180 d, respectively.

Arefi et al. (2011) reported that the compressive, tensile and flexural strength of mortar with 3% NA reached their maximum and enhanced by 63.38%, 81.46% and 70% relative to plain mortar at age of 7 d, respectively.

Shekari et al. (2011) fabricated concrete by adding 1.5% NA into cementitious materials consisting cement, metakaolin and nanoparticles. They found the compressive strength increased by 55% at the age of 28 d.

In a word, the mechanical properties of cement and concrete materials with appropriate NA can be enhanced (as shown in Table 2).

4. **Durability:**
   a. **Abrasion Resistance:** Nazar et al. (2011f) studied the abrasion resistance of concrete with NA and NS at levels of 0%, 0.5%, 1.0%, 1.5% and 2.0% by replacing cement in weight. The samples were curing in water and saturated limewater. They found that the nanoparticles could significantly improve the abrasion resistance.
   b. **Water Absorption:** Shekari et al. (2011) prepared concrete with 1.5% NA and found that the nanoparticle could reduce the water absorption property of concrete.
   c. **Chloride Penetration:** He et al. (2008) reported that cement mortars containing 1 wt. % NA with average particle size of 10 nm could reduce apparent diffusion coefficients of chloride anion. Shekari et al. (2011) observed that the addition of 1.5% NA to concrete significantly improve the chloride penetration resistance.

Table 2. Effect of NA on mechanical properties of cement and concrete materials

Types	Mechanical Properties	Enhancement (%) 7d	Enhancement (%) 28d	Content of NA (%)	Particle Size (nm)	Water/Binder	Reference
Cement mortars	Compressive strength	142	119	3	100-1000	0.80	(Campillo et al., 2007)
		84	113	1.8	50	0.80	(Campillo et al., 2007)
		13.4	20.2	1.25	13	0.40	(Oltulu & Şahin, 2011)
		63.38	-	3	20	0.42	(Arefi et al., 2011)
	Flexural strength	21.4	18.2	1	15	0.40	(Nazari et al., 2010)
		70	-	3	20	0.42	(Arefi et al., 2011)
	Tensile strength	81.46	-	3	20	0.42	(Arefi et al., 2011)
	Split tensile strength	73.3	55.6	1	15	0.40	(Nazari et al., 2010)
	Elastic modulus	241	243	5	<150	0.40	(Li et al., 2006)
Concrete	Compressive strength	-	55	1.5	10-25	0.25	(Shekari & Razzaghi, 2011a)
		16.1	14.9	1	15	0.40	(Nazari & Riahi, 2011e)

d. **Capillary Permeability:** Oltulu et al. (2011) fabricated cement mortars with NA at levels of 0.5%, 1.25% and 2.5% by weight of cement and studied the capillary permeability of samples at age of 180 d. They found the capillary permeability decreased by 12%, 29% and 15% as the additions of NA to mortar are 0.5%, 1.25% and 2.5%, respectively. However, Oltulu et al. (2013) fabricated mortar containing 15% fly ash and NA. They found the capillary permeability of mortar with 0.5%, 1.25% and 2.5% NA by weight of binder after curing 180 d increased by 3%, 4% and 10%, respectively.

e. **Freeze-Thaw Resistance:** Behfarnia et al. (2013) studied the frost resistance of concrete with NA at levels of 0%, 1%, 2% and 3% by weight of the sum of cement and nanoparticles after 28 d of curing. The average particle size of NA was 8 nm and the water/binder ratio was 0.48. They found that addition of NA to concrete can significantly improve the frost resistance. The reduction of compressive strengths of concrete containing 1%, 2% and 3% of NA after 300 freezing and thawing cycles was 23.84%, 21.31% and 18.19%, respectively, while the compressive strength loss of plain concrete after 300 cycles was 100%.

Nano-TiO$_2$

1. **Workability:** Due to the huger specific surface area, addition of NT to cement and concrete materials would reduce the workability. For example, Chen et al. (2012) found that the amount of super plasticizer added must be adjusted with the increase of the amount of TiO$_2$ to maintain the workability. Nariza et al (2011) added 1wt. % polycarboxylate admixture in to SCC to ensure workability. Meng et al. (2012) reported that the fluidity of cement mortars decreased with increasing of NT. The fluidity of cement mortars with 5% and 10% NT decreased by 21% and 40%, respectively. Noorvand et al. (2013) also found that the flowability decreased with addition of NT. Lucas et al. (2013) increased the content of mixing water in order to maintain workability of the mortar with NT. On the other hand, addition of NT improved the consistency of SCC and reduced the probability of bleeding and segregation (Jalal, Ramezanianpour, & Pool, 2013).
2. **Hydration:**
    a. **Hydration Heat:** Chen et al. (2012) studied the rate of heat evolution for the cement paste with 0%, 5% and 10% addition of two kinds of NT. The particle size of P25 and Anatase were 21 nm and 350 nm, respectively. The main peak occurred earlier for all the samples containing NT compared with the reference sample. In addition, the exothermic hydration process accelerated by P25 was more than that of Anatase, at the same content.
    b. **Setting Time:** Chen et al. (2012) also studied the setting time of cement paste with two types of NT, i.e. P25 and Anatase. The setting time was shorter for the samples with 5% and 10% NT. Moreover, the addition of P25 particles having larger surface area resulted in a shorter setting time compared with addition of Anatase particles at the same NT content.
3. **Mechanical Properties:** Many researches (Li et al., 2006; Nazari & Riahi, 2010, 2011, 2011g; Shekari & Razzaghi, 2011) showed that addition of NT to cement and concrete materials could improve the mechanical properties. Some researchers (Chen et al., 2012; Meng et al., 2012) found that the enhancement was more obvious at early age than that of the late age.

Li et al. (2006) partially replaced cement with NT at levels of 0%, 1%, 3% and 5%, by weight in concrete. When the content of NT was 1%, the compressive strength and flexural strength of mortar reached their maximum and enhanced by 18.03% and 10.28% compared with plain concrete, respectively.

Nazari et al. (2011g) studied the compressive strength, split tensile strength and flexural strength of concrete containing 45% ground granulated blast furnace slag modified with NT up to 90 d. The compressive strength of concrete with 3% NT increased by 71.3%, 36.4% and 20.1% at age of 7, 28 and 90 d, respectively, relative to plain concrete. The split strength of concrete with 3% NT enhanced by 80.0%, 33.3% and 27.5% after 7, 28 and 90 d of curing, respectively. The concrete with 3% NT showed an increase of 71.4%, 27.8% and 16.4% in flexural strength at age of 7, 28 and 90 d, respectively. Shekari et al. (2011) observed that an addition of 1.5 wt.% NT can improve the compressive strength by 22.8% at age of 28 d. Chen et al. (2012) prepared cement mortars with two types of NT, namely P25 (75% anatase and 25% rutile, Degussa, 21 nm) and Anatase (99% anatase, Sigma–Aldrich, 350 nm). They found the NT can obviously increase the compressive strength, especially at the early age. Meng et al. (2012) prepared NT filled cement mortars and observed that the early strength increased and the late strength decreased with increasing content of NT. For example, the compressive strength of cement mortars with 5% and 10% NT increased by 46% and 47% at age of 1 d respectively, while the compressive strength of cement mortars with 5% and 10% NT decreased by 6% and 9% after 28 d curing, respectively. Jalal et al. (2013) fabricated SCC with 0%, 1%, 2%, 3%, 4% and 5% of NT. The compressive strength of concrete containing 4% NT enhanced by 22.5%, 22.0% and 26.9% at age of 7, 28 and 90 d, respectively. The splitting tensile strength of concrete with 4% NT increased by 34.5%, 36.1% and 46.2% after 7, 28 and 90 d of curing, respectively. Noorvand et al. (2013) reported addition of NT to cement mortars containing black rice husk ash (BRHS) could improve the compressive strength. They found that the compressive strength of cement mortars containing 10% BRHS and 1.5% NT increased by 10%, 18% and 19% compared with cement mortars containing 10% BRHS at age of 7, 28 and 90 d, respectively.

Table 3 summarizes the previous research results about the mechanical properties of cement and concrete materials.

4. **Durability:**
   a. **Chloride Penetration and Permeability:** Shekari et al. (2011) reported that addition of 1.5 wt. % NT can improve the resistance of chloride penetration. Jalal et al. (2013) fabricated SCC with NT at levels of 0%, 1%, 2%, 3%, 4% and 5%. They found that addition of NT to the concrete is effective in decreasing the chloride penetration. When the content of NT was 4%, the chloride percentage was the least at different depths among

Table 3. Effect of NT on mechanical properties of cement and concrete materials

Types	Mechanical Properties	Enhancement (%) 7d	Enhancement (%) 28d	Content (%)	Particle Size (nm)	Water/Binder	Reference
Concrete	Compressive strength	-	18.03	1	15	0.42	(Li et al., 2006)
		82.5	58.5	4	15	0.40	(Nazari & Riahi, 2010)
		22.5	22.0	4	20	0.38	(Jalal et al., 2013)
	Flexural strength	-	10.28	1	15	0.42	(Li et al., 2006)
		29.7	50.0	4	15	0.40	(Nazari & Riahi, 2010)
	Split tensile strength	100	81.3	4	15	0.40	(Nazari & Riahi, 2010)
		34.5	36.1	4	20	0.38	(Jalal et al., 2013)

all the concrete samples. Li et al. observed that concrete filled with 1% NT in the weight of cement showed a lower diffusion coefficient with respect to common concrete (Li, Xiao, Guan, Wang, & Yu, 2014).

b. **Abrasion Resistance:** Li et al. found that abrasion resistance improved notably through adding NT into concrete. They reported that the abrasion resistance of concrete with 1% NT by weight of binder enhanced by 180.7% for the surface index and 173.3% for the side index. Even for the concrete with 5% NT by weight of binder, the abrasion resistance enhanced by 90.4% and 86% for the surface index and the side index, respectively (Li et al., 2006).

c. **Photocatalytic Abatement of NOx:** It has been 40 years since the revelation of photocatalytic activity of NT (Shon et al., 2008). Adding photocatalytic abatement of NO_x to cement and concrete materials has been studied. The photocatalytic activity of NT in cement and concrete materials depends on many factors such as the phases of NT, the content, the dispersion of NT, and the age of cement and concrete materials.

Cárdenas et al. (2012) fabricated cement paste with 0.0%, 0.5%, 1.0%, 3.0% and 5.0% of NT to evaluate the degradation of NO_x pollutants. The titania phases of anatasa: rutile ratios were 100:0, 85:15 and 50:50 for each addition. They found that the samples with the best photocatalytic activity were the ones with a ratio of

85:15 and 5.0% of NT at early age of 65 hours, while the samples with a ratio of 100:0 and 5.0% of NT at late age of 28 d. Moreover, photocatalytic activity increased with the content of NT. However, it decreased with the aging of the cement paste.

Lucas et al. (2013) prepared mortars using aerial lime, cement and gypsum as binders with NT at levels of 0%, 0.5%, 1%, 2.5% and 5% by weight of cement. They found that all the specimens with NT exhibited high photocatalytic activity even in lower additions (0.5 and 1 wt. %). It should be noted that low porosity together with a strong presence of micro pores (>1 µm) was not an obstacle for the degradation process while mortars with a strong presence of nano pores (<0.1 µm) exhibited an efficient decrease.

Yousefi et al. (2013) found that effective dispersion of NT along with the formation of $CaTiO_3$, obviously enhanced photocatalytic properties of cement paste. The agglomeration of NT particles in cement mixes could be effectively avoided by suitable dispersion of the particles in lime-saturated water, following by an ultrasonication treatment before mixing with cement.

Nano-Fe_2O_3

1. **Workability:** Nazari et al. (2010b) studied the workability of concretes by partial replacement of cement with nano-Fe_2O_3 particles. Nano-Fe_2O_3 with the average diameter of 15 nm was used with four different contents of 0.5%, 0.1%, 1.5% and 2.0% by weight of cement. The results showed a reduction in the workability with increasing contents of nano-Fe_2O_3.

2. **Hydration:** Khoshakhlagh et al. (2012) studied the heat of hydration of pastes modified with nano-Fe_2O_3 up to 70 h. Cement was partially replaced with nano-Fe_2O_3 at levels of 0%, 1%, 2%, 3%, 4% and 5%, by weight, respectively. The results showed that the addition of nano-Fe_2O_3 up to 4 wt. % could accelerate the appearance of the first peak in conduction calorimetry tests, which is related to the acceleration in formation of hydrated cement products. Nazari et al. (2010c) studied the initial and final setting time of concretes modified with nano-Fe_2O_3. The results showed that both initial and final setting time decreased with the addition of nano-Fe_2O_3. With increasing the volume fraction of nanoparticles, the decrease of setting time indicated that nano-Fe_2O_3 has a faster hydration reaction speed than the cement due to its unique surface effects, smaller particle sizes, and higher surface energy.

3. **Mechanical Properties:** Li et al. firstly studied the mechanical properties of nano-Fe_2O_3 filled cement mortars. They found that the compressive and flexural strengths of the cement mortars mixed with the nanoparticles at 7 d and 28 d were higher than that of a plain cement mortars with the same water/cement (Li et al., 2004). Khoshakhlagh et al. (2012) studied the compressive, flexural and splitting tensile strengths of high performance self-compacting

concrete (HPSCC) containing different amount of nano-Fe_2O_3 at age of 2, 7 and 28 d. The strengths of the composites at all age were improved by adding nano-Fe_2O_3 in the cement paste up to 4.0 wt. %. The enhancement in the compressive and flexural strengths at 28 d can go up to 76.19% and 93.75% respectively as the addition of nano-Fe_2O_3 is 4.0 wt. %.

Nazariet al. (2011h) investigated the optimal nano-Fe_2O_3 content that gave the highest splitting tensile strength of concrete by using two computer-aided models based on artificial neural networks and genetic programming. Two models were built by using experimental results from 144 specimens produced with 16 different mixture proportions. The simulation results in both models fitted well with the testing results, indicating that two models are of strong potential for predicting the split tensile strength of concrete containing nano-Fe_2O_3. The optimal nano-Fe_2O_3 content was within the range of 2.0 wt. %-4.0 wt. %. The deficiency in dispersion of nanoparticles more than the mentioned values causes the reduction of improvement effects of nanoparticles on the split tensile strength of concrete.

4. **Durability:** The mixture of nano-Fe_2O_3 has an obvious influence on durability of concrete, such as permeability and water absorption. He et al. (2008) studied the permeability of cement mortars modified with nano-Fe_2O_3 mixed 1% by weight of cement. They reported that the incorporation of nano-Fe_2O_3 improved the chloride penetration resistance of mortars. Oltulu et al. (2011) studied the capillary permeability of 5% silica fume filled cement mortars modified with 0%, 0.5%, 1.25% and 2.5% nano-Fe_2O_3 at age of 180 d. The results showed that the addition of 0.5%, 1.25% and 2.5% nano-Fe_2O_3 reduced the capillary permeability by 26%, 13% and 6%, respectively. They also replaced silica fume by fly ash as the mineral additive. The results showed that the addition of 0.5% and 1.25% nano-Fe_2O_3 reduced the capillary permeability by 5% and 1%, respectively (Oltulu & Şahin, 2013). Khoshakhlagh et al.(2012) investigated the coefficient of water absorption of HPSCC mixed with nano-Fe_2O_3. The results showed the percentage of water absorption of concrete with nano-Fe_2O_3 at age of 2 d is higher than that at age of 7 and 28 d. This may be due to the formation of more hydration products of nano-Fe_2O_3 at early age of curing. Nano-Fe_2O_3 accelerated the formation of cement hydrates and hence the concrete needed more water to produce these products. However, at age of 7 and 28 d, the pore structure is improved and water absorption decreases along with the hydration.

5. **Other Properties:** Nano-Fe_2O_3 has been found to improve the compressive and flexural strengths as well as to provide concrete with self-sensing capabilities. Li et al. found that the volume electric resistance of cement mortars

with nano-Fe_2O_3 changed with the applied load, demonstrating that mortar with nano-Fe_2O_3 could sense its own compressive stress (Li et al., 2004). Such sensing capabilities are invaluable for real-time, structural health monitoring and for the construction of smart structures.

Other Nano-Scale Metallic Oxides

The addition of other metal oxides nanoparticles such as nano-Fe_3O_4, nano-ZnO_2, nano-CuO and nano-ZrO to cement and concrete materials have also been reported to improve properties of cement and concrete materials. For example, Shekari et al. (2011) found that the addition of 1.5% (by weight of cement) of nano-Fe_3O_4 increased both compressive strength and indirect tensile strength by about 28.9% and 28.6%, respectively. Moreover, the percentage of water absorption and chloride penetration of concrete decreased by about 81.7% and 21.3%, which indicated that nano-Fe_3O_4 had noticeable influence on improvement of durability. Nazari and Riahi (2011b) investigated the flexural strength of SCC with different amounts of nano-ZnO_2. The flexural strength at 28 d of curing increased by 46.2% as content of nano-ZnO_2 goes up to 4.0 wt. %. In addition, they studied the percentage of water absorption (Research, 2011) and porosity (Nazari & Riahi, 2011i) of concretes modified with nano-ZnO_2. They found that the percentage of water absorption decreased with the addition of nano-ZnO_2 as well as the porosity of concretes. Fan et al. (2004) studied the effect of synthetic nano-ZrO_2 addition in cement on the strength development of Portland cement paste. Reduction in porosity and permeability, enhancement in compressive strength, and improvement in microstructure of cement paste were observed due to the addition of nano-ZrO_2 powder in cement. Nazari et al. conducted a series of studies on the mechanical properties such as compressive strength, splitting tensile strength and flexural strength and durability of SCC modified with nano-CuO (Nazari, Rafieipour, & Riahi, 2011;Nazari & Riahi, 2011j). The effects were identical with the addition of nano ZrO_2, TiO_2, Al_2O_3 or Fe_3O_4.

Nano-Scale Carbon Materials

NCMs include CNTs, carbon nanofibers (CNFs), nano graphite platelets (NGPs) and nano carbon black (NCB). They all have excellent mechanical, electrical, chemical and thermal properties. The representative properties of NCMs are listed in Table 4. The addition of NCMs to cement and concrete materials can endow the composites high performance at mechanical properties and multifunctional properties. A lot of researches have been done and gotten many achievements.

Table 4. Properties of NCMs

Properties	CNTs		CNFs	NGPs	NCB
	MWCNTs	SWCNTs			
Elastic modules/TPa	0.3-1	1	0.4-0.6	1 (in-plane)	/
Strength/GPa	10-60	50-500	2.7-7.0	10-20	/
Electrical resistivity/ μΩ•cm	5-50		55	50 (in-plane)	0.22 Ω•cm*
Dimensions	Diameter: 2-30 nm	Diameter: 0.75-3 nm	Diameter: 50-200 nm	Diameter: 1-20 μm	Diameter: 10-400 nm
	Length: 0.1-50 μm	Length: 1-50 μm	Length: 50-100 μm	Thinckness:~30 nm	
Surface area/m2/g	>400		~200	~2630	~1056*
Aspect ratio	~1000		100-500	50-300	/

Han et al., 2015; Dai, Sun, Liu, & Li, 2010; Han, Yu, & Ou, 2014.
* Reference from (Dai, Sun, Liu, & Li, 2010).

Carbon Nanotubes

CNTs have a hollowly cylindrical nanostructure which can be visualized as rolling of a graphite sheet (Sindu, Sasmal, & Gopinath, 2014). According to the number of rolled layers of graphite, CNTs can be categorized into single-walled carbon nanotubes (SWCNTs) and multi-walled carbon nanotubes (MWCNTs).

1. **Workability:** The workability of cement and concrete materials is greatly affected by the presence of the CNTs, especially the untreated nanofilaments due to the difficulty in the dispersion of CNTs. The methods of covalent and noncovalent modification of CNTs are commonly utilized to improve the workability of cement and concrete materials incorporated with CNTs. For example, Al-Rub et al. (2012) found that the addition of treated-CNTs produced an increase in workability of cement paste compared with the untreated CNTs due to the increased wettability caused by the functional groups (-COOH and -OH).
2. **Mechanical Properties:** Since early investigations showed that CNTs had strong effect on the hydration process and hardness of cement, a great deal of research has been done to ascertain mechanical properties of CNTs filled cementitious composites, such as strength, deformation and toughness.

Cwirzen et al. (2008) found that the addition of only 0.045% of the polyacrylic acidpolymer-treated MWCNTs to cement paste can mostly increase the compressive strength by 50% compared with the pure cement paste. Musso et al. (2009) compared the mechanical behavior of cement paste obtained by mixing different kinds of MWCNTs with different structures (pristine MWCNTs, annealed MWCNTs and carboxy-group functionalized MWCNTs), and found that the annealed MWCNTs and pristine MWCNTs produced an improvement on the compressive strength of 10–20% with respect to plain cement (Musso, Tulliani, Ferro, & Tagliaferro, 2009). Nasibulin et al. (2009) prepared cement pastes by using the CNTs grown cement and achieved as high as a 200% increase in the compressive strength. Ludvig et al. (2011) prepared cement pastes and cement mortars with 0.3% CNTs which grew on clinker and found that the tensile strength achieved a 34.28% of increase with a simple physical mixture. Hunashyal et al. (2011) investigates the behavior of cement beams reinforced with the addition of 0.75 wt. % of MWCNTs and found the flexural strength of the beam increased by 43.75% compared with the plain cement control beams.

Makar et al. (2005) reported the addition of 2% SWCNTs to cement paste can increase hardness by up to 600% at early age. Sáez et al. (2006) observed that the Young's modulus of cement-based material with 0.1 wt. % SWCNTs reveals as high as a 2.27 times improvement compared with that of the plain cement paste. Luo et al. (2009) achieved a 149.32% increase in the fracture toughness of cement-based materials by using 0.5 wt. % of MWCNTs. Moreover, Luo et al. found the critical opening displacement of the cement-based material with 0.5 wt. % MWCNT has an increase of 34.96% (Luo, Duan & Li, 2009).

3. **Durability:** Han et al. first found that even a very small dosage of MWCNTs can effectively decrease the water sorptivity coefficient, water permeability coefficient, and gas permeability coefficient of cement and concrete materials, which indicates that CNTs can effectively decrease the transport properties of cement and concrete materials (Han, Yang, Shi, & Yu, 2013).
4. **Other Properties:**
    a. **Electrical Properties:** Singh et al. studied the electrically conductive characteristic of composites for various composition of MWCNTs admixed cement matrix. They found that conductivity increases as the MWCNTs increases in the matrix and 10^{-7} order difference in the conductivity values of pristine cement as compared to composites with the addition of 15 wt. % MWCNTs (Singh et al., 2013). Han et al. investigated the effect of CNTs on the capacitance of cement-based materials at different test frequencies during the direct current (DC) measurement. They observed that the capacitance of cement-based composites with 0.5% of CNTs is higher than that of cement-based composites with 0.1% of CNTs. This indicates that the addition of CNTs can increase the capacitance of cement-based materials (Han, Zhang, Yu, Kwon, & Ou, 2012).

Wansom et al. investigated the impedance response of cement-based material with different contents of MWNTs at different stages by using alternating current (AC)-impedance spectroscopy technique. The results show a noticeable different behavior of impedance between CNTs filled cement-based composites and the plain cement paste. The addition of CNTs leads to a decrease in DC resistance of cement-based materials due to the forming of extensive percolating paths in the equivalent circuit in the matrix (Wansom, Kidner, Woo, & Mason, 2006). They also studied the effect of CNTs on the dielectric properties of cement-based materials and observed that the presence of CNTs clearly enhances the dielectric constant in the 10^3-10^6 Hz range above the baseline dielectric constant of the cement paste (the baseline dielectric constant of the plain paste is ~300). This enhancement may be attributed to polarization phenomena within the C–S–H gel phase (Wansom et al., 2006). In addition, Gong et al. studied the influence of CNTs on the piezoelectric and dielectric properties of CNTs/PZT cement-based composites. They found that the CNTs additives improve the electrical conductivity of the composites. The dielectric constant and the dielectric loss increase with increasing CNTs content, indicating an increase in poling efficiency of the composites. The piezoelectric strain factors and the piezoelectric voltage factors all increased with increasing CNTs content upto 0.3 vol. %. Above this value, they decreased with further increase of CNTs content (Gong, Zhang, Quan, & Che, 2011). It is believed that the CNTs-modified cementitious composites with excellent electrical properties can be utilized as sensing materials in the field of civil engineering.

b. **Thermal Properties:** Yakovlevet al. observed that the addition of CNTs (0.05% by mass) decreased the heat conductivity of foam concretes up to 12%-20% (from 0.07 W/m K to 0.056 W/m K) (Yakovlev, Kerienė, Gailius, & Girnienė, 2006). Kerienė et al. observed that the non-autoclaved aerated concrete and the autoclaved aerated concrete with MWCNT showed better resistance thermal distortion than those cement-based composites without these additions. Within creasing of addition of MWCNTs, the temperature of maximal and thermal effect is increased (Kerienė et al., 2013). In addition, Yao et al. employed 0.5 wt. % CNTs to improve the positive thermoelectric power of the carbon fibers (CFs)/cementitious composites, which leads to an increase of 260% in the positive thermoelectric power (from 8.8 µV/°C to 22.6 µV/°C). Besides, the addition of CNTs can effectively improve the linearity and reversibility of the relationship between the Seebeck voltage and the temperature differential (Yao, Zuo, & Wu, 2013).

c. **Sensing Properties:** Shukla et al. studied the smoke sensing property of cement-based composites with varying MWCNTs contents, and found that the DC transient studies depict an increase in electrical conductivity when the composites are exposed to smoke (Shukla et al., 2012). Materazzi et al. investigated the electrical response of prismatic specimens made of CNTs filled cement-based composite subjected to sinusoidal stress–strains in the typical frequency range of large civil structures. The results demonstrate that the composites' output retains all dynamic features of the input thus providing useful information for structural health monitoring, and the composites appeared to be suitable for performing dynamic measurements in civil engineer concrete structures (Materazzi, Ubertini, & D'Alessandro, 2013). Saafi embedded cement-based composites with SWCNTs into concrete beams and subjected to monotonic and cyclic loading to evaluate the effect of damage on their response. The results show that the wireless response of the embedded nanotube sensors changes due to the formation of cracks during loading. In addition, the nanotube sensors were able to detect the initiation of damage at an early stage of loading (Saafi, 2009). Han et al. developed sensing cement-based composite containing CNTs and investigated its feasibility for traffic monitoring. The developed composite shows remarkable responses to vehicular loadings and can accurately detect the passing of different vehicles under different vehicular speeds and test environments. It indicates that the self-sensing CNTs/cement composite has great potential for traffic monitoring use, such as in traffic flow detection, weigh-in-motion measurement and vehicle speed detection(Han, Yu, & Ou, 2014; Han, Yu, & Kwon, 2009; Han, Yu, & Ou, 2011; Han, Zhang, Burnham, Kwon, & Yu, 2013; Han et al., 2012).

Carbon Nanofibers

CNFs are nanoscale carbon fibers whose diameter is from 50-200 nm between that of CNTs and CFs (Han et al., 2015). In structure, CNFs can be thought of formed as regularly stacked truncated conical or planar layers along the filament length. The researches about cement and concrete materials with CNFs are mainly focused on the mechanical properties, sensing properties and durability.

1. **Mechanical Properties:**
    a. **Strength:** Gao et al. (2009) fabricated normal concrete with CNFs at levels of 0%, 0.16%, 0.31%, 0.78% and 1.55% by volume of binder. As the content increased, the compressive strength decreased at age of 28 d. They found that the compressive strength of concrete with 0.16 wt. %

CNFs was 42.7% higher than that of plain concrete. Gay et al. reported that addition of 0.2 wt. % of CNFs can reach a 22% increase in the tensile strength (Gay & Sanchez, 2010). Nasibulina et al. (2012) synthesized CNF/clinker hybrid materials, namely direct synthesis of CNFs on the surface of particles, and prepared mortar with CNF-clinker hybrid materials. The amount of CNF-clinker hybrid materials were 7%, 30% and 100%, which correspond to 0.4%, 1.5% and 5% of CNF. They got a more than 2.5 times increase in compressive strength through adding 0.4 wt. % of CNFs.

b. **Deformation:** Howser et al. fabricated CNFs filled concrete. They observed that addition of 1.0 vol. % CNFs can improve the ductility, ultimate normalized capacity, and deflection by 35.1%, 30.7% and 34.9% compared with the plain concrete column, respectively (Howser, Dhonde, & Mo, 2011). Al-Rub et al. studied the mechanical properties of cement paste with untreated CNFs and acid treated CNFs. They found that the mechanical properties was better than cement paste with treated CNFs. Addition of 0.1 wt.% untreated CNFs achieved a 170% increase in the modulus of toughness (Al-Rub, Tyson, Yazdanbakhsh, & Grasley, 2012). Shah et al. studied the Young's modulus of cement paste containing CNFs and found that the addition of 0.048 wt. % CNFs increased 50% in the Young's modulus relative to cement paste (Shah, Konsta-Gdoutos, & Metaxa, 2010).

c. **Other Mechanical Properties:** Peyvandi et al. (2013) fabricated ultra high performance concrete with 0.145 vol. % of CNFs and 3.55 vol. % of steel fibers. They found the energy sorption and impact resistance have great increase by 3400% and 280%, respectively.

2. **Durability:**
   a. **Abrasion Resistance:** Peyvandi et al. studied the resistance abrasion weight loss of ultra high performance concrete and found that addition 0.145 vol. % CNFs and 3.55 vol. % steel fiber to the mixture can improve the resistance abrasion weight loss by 1700% (Peyvandi, Sbia, Soroushian, & Sobolev, 2013).
   b. **Leaching:** Brown et al. (2013) fabricated Portland cement paste with and without 0.2% CNFs and put those samples into ammonium nitrate solutions after curing for 28 d. They found that the mass loss and flexural strength loss were less than those of plain cement paste after exposure to ammonium nitrate solutions for 125 d.

3. **Other Properties:** Galao et al. (2013) fabricated cement paste with CNFs at levels of 0%, 0.5%, 1.0% and 2.0% by weight of cement. The water/cement ratio was fixed at 0.5. They found that the samples with CNFs all have good

strain-sensing properties after curing 28 d. Furthermore, addition of 2.0% CNFs to cement paste made the material sensitive to its own structural damage. Howser et al. (2011) built shear-critical columns using self-consolidating concrete containing CNFs. The columns containing CNFs can used as a structure health monitoring device and had strain self-sensing capability under a reversed cyclic load and damage self-assessment capability. Portland cement pastes with 2% CNFs (by cement mass) were fabricated as sensors and were attached to the reinforced concrete (RC) beam in different service locations after curing 28 d. Baeza et al. found that these sensors were stain sensitive during the applied load progressively up to the RC beam failure (Baeza, Galao, Zornoza, & Garcés, 2013).

The cement and concrete materials with CNFs can achieve elastic strain sensing of compression and tension region of beam and measure strains on the surface of a structural element (Han, Yu, & Ou, 2011; Howser, Dhonde, & Mo, 2011; Mo & Roberts, 2013; Saafi, 2009).

Nano-Scale Graphite

NGPs are carbon-based conductive nanoparticles. Structurally, NGPs are, in general, layered and planar. The numbers of layers are from a few to several and the sum thickness of layers is in nanoscale. In addition, graphene oxide (GO) is single layer of NGPs.

1. **Mechanical Properties:** Huang et al. observed that the addition of NGPs increases the flexural strength of cementitious composites by 82% compared with the plain ones (Huang, 2012). Lv et al. found that GO nanosheets can remarkably increase the tensile/flexural strength of the corresponding cementitious composites. Especially, when the content of GO is 0.03%, the tensile, flexural and compressive strengths of cementitious composites increase by 78.6%, 60.7% and 38.9%, respectively (Lv et al., 2013).
2. **Durability:** Peyvandi et al. added the modified NGPs into four kinds of dry-cast concrete pipes in aggressive sanitary sewer environment to improve their durability characteristics including the moisture sorptivity and acid resistance. They observed that NGPs significantly improve the moisture transport performance and acid resistance of the composites at a low dosage (0.05 vol. % of cement-based composites) (Peyvandi, Soroushian, Balachandra, & Sobolev, 2013).
3. **Other Properties:** Electrical, electromagnetic interference and other properties of NGPs enhanced cement and concrete materials have also been investigated. For example, Du et al. studied the electrical resistance behavior of cementitious

composites with different contents of NGPs (0-4.8 wt. %), and observed that the composites exhibit a decrease in electrical resistivity and present a good sensing property when NGP content are in the range from 2.4 vol. % to 3.6 vol. % (Du, Quek, & Dai Pang, 2013).

Singh et al. studied the electromagnetic intereference shielding properties of GO-ferrofluid-cement-based composites. They observed the variation of shielding effectivness due to absorption (SE_A) and shielding effectivness due to reflection (SE_R) with frequency in the 8.2-12.4 GHz. The SE_T (total SE) of the GO-ferrofluid-cement-based composites is 46 dB, which is much higher than that of the pristine cement-based composites (4dB) (Singh, Mishra, Chandra, & Dhawan, 2011).

Nano-Scale Carbon Black

NCB, a form of amorphous carbon material, raised the interest of some concrete researchers because of their light weight, remarkable electrical conductivity, chemical and thermal properties, and low cost. The research mostly focuses on the electrical, thermal and sensing properties.

1. **Workability:** NCB, as a nanomaterial, has huge surface-area-to-volume ratio which leads to reduce the flowability of fresh cement and concrete materials. There are two methods to increase the workability of mixture. The first method is addition water-reducing. For example, Li et al. (Li, Xiao, & Ou, 2006,2008) used water-reducing agent in amount of 1.5% by weight of cement to increase the workability of cement paste. In addition, Han et al. (2007), Dai et al. (2010) and Xiao et al. (2011) also used water-reducing agent to increase the flowability of fresh mixture. The second method is using more water when the weight of binder is fixed, namely enlarge the water/binder ratio. For example, Dai et al. (2010) enhanced the water/cement ratio with increasing content of NCB to ensure the workability of mixture.
2. **Mechanical Properties:** The mechanical properties of cement and concrete materials with CNTs, CNFs and NGP have been widely researched and showed remarkable enhancement. However, the mechanical properties of cement and concrete materials with NCB are seldom studied and the results are inconsistent with each other.

Dai et al (2010) fabricated cement mortars containing NCB with average particle size of 33 nm. The water/cement ratio was 0.27-0.38 which increased with increasing content of NCB. They found that the compressive strengths decreased with increasing of NCB content, especially the compressive strength decreased substantially when the content was more than 3 wt. %.

Wang et al. (2008) prepared cement mortars containing 0.25%, 0.50%, 0.75%, 1.00%, 1.25%, 1.50%, 1.75% and 2.00% NCB with particle size of 33 nm. The fixed water/cement ratio was 0.5. They found that the compressive strengths of composites with 0.25%, 0.50% and 0.75% NCB at age of 28 d increased by 11.01%, 13.50% and 15.14%, respectively. In addition, the flexural strength enhanced by 17.76%, 34.94% and 44.32%, respectively, at age of 28 d.

The different results may be caused by different fabrication methods, different water/cement ratios and so on.

3. **Other Properties:**
   a. **Electrical Properties:** The relationship curves of resistivity of cement and concrete materials containing NCB versus the concentration of NCB have typical features of percolation phenomena (Dai, Sun, Liu, & Li, 2010; Li, Xiao, & Ou, 2006). Li et al. found the resistivity decreased obviously with increasing the amount of NCB from 7.22 to 11.39 vol. % which was called as percolation threshold zone (Li, Xiao, & Ou, 2006). Dai et al. reported that the resistivity decreased dramatically with the increasing NCB content from 0.36 to 1.34 vol. % (i.e. from 0.7 to 2.5 wt. %) (Dai, Sun, Liu, & Li, 2010).
   b. **Electromagnetic Properties:** Dai et al. (2010) studied the absorbing electromagnetic waves of NCB cement-based composites (CBCC). They found the CBCC exhibited good performance of absorbing electromagnetic waves when the frequency ranged from 8 to 26.5 GHz. When 2.5 wt. % of NCB was added into cement mortars, the minimum reflectivity reached -20.30 dB in the frequency of 20.6 GHZ.
   c. **Sensing Properties:** Li et al. (2006) firstly fabricated cement paste with NCB with the particle size of 120 nm. The amount of NCB were 5%, 10%, 12%, 15%, 20% and 25% by weight of cement, namely 3.11%, 6.04%, 7.22%, 8.79%, 11.39% and 13.85% by volume of composite. They observed a linear relationship between the fractional change in resistivity and compressive strain in cement paste with 15 wt. %, 20 wt. % and 25 wt. % of NCB. In addition, Li et al. also studied the facts affecting the resistance of NCB filled cement. They found that the resistance of NCB filled cement composite increased with water content and measure time which were attributed to a polarization effect (Li, Xiao, & Ou, 2008).

Han et al. (2007) explored the piezoresistivity of cement paste with CFs and NCB under single compressive loading and repeated compressive loads at different loading amplitudes, and found that the sensing property was reversible and stable within the elastic regime.

Xiao et al. (2011) fabricated NCB filled cement materials as cement-based strain sensor and embedded into concrete columns. They found that NCB filled cement materials had a good strain-sensing ability.

Other Nano-Scale Materials

Nano-CaCO$_3$

In recent years, nano-CaCO$_3$ (NC) has also been introduced in cement and concrete materials. The research mostly focused on using NC to improve the mechanical properties at early age and durability of cement and concrete materials with high volume of fly ash.

1. **Workability:** The results about the workability of NC filled cement and concrete materials had disagreement with each other. For example, Camiletti et al. (2013) found that the flowability of ultra high performance concrete increased with the addition of NC. However, Liu et al. (2012) found that addition of 1 wt. %, 2 wt. % and 3 wt. % NC to cement paste decreased the flowability. Stupit et al. (2014) found that addition of NC (1%, 2%, 3% and 4% by weight of binder) could slightly reduce the workability of ordinary and high volume fly ash mortar/concrete.
2. **Hydration:**
    a. **Hydration Heat:** Kawashima et al. (2013) measured the hydration heat of cement past and cement past with 5 wt. % NC, a water/binder fixed at 0.43. They found that 5 wt. % NC accelerated hydration rate, i.e. the peak was higher and occurred earlier compared with plain cement paste.
    b. **Setting Time:** Liu et al (2012) prepared the cement paste (water/cement=0.45) with NC at levels of 0%, 1%, 2% and 3% by weight of cement. They reported that the initial and final setting time of NC filled cement paste was shorted. Camiletti et al (2013) found that addition of NC can significantly accelerate the setting time of composites cured at simulated cold and normal field conditions. Kawashima et al. (2013) studied the setting time of cement paste containing 50 wt. % fly ash and 5 wt. % NC. They found that the addition of 5% NC helped accelerate both initial and final setting time.
3. **Mechanical Properties:** Researchers found that the NC can improve the compressive strength of cement and concrete materials (Camiletti, Soliman, & Nehdi, 2013; Kawashima, Hou, Corr, & Shah, 2013; Shaikh & Supit, 2014; Supit, Shaikh, Singh, & Ahalawat, 2014). For example, Stupit et al. (2014) fabricated concrete and mortar containing high volume of fly ash and with

NC at levels of 0%, 1%, 2%, 3% and 4% by weight of cement. NC particle size ranged 15-40 nm. They found that the compressive strength increased with NC up to 1% replacement and then it decreased at both 7 d and 28 d. The compressive strength of cement mortars containing 60% fly ash and 1% NC enhanced by 100% and 111% compared with mortar containing 60% fly ash at age of 7 d and 28 d, respectively. Skaikh et al. found that the compressive strength reached its maximum when the content of NC was 1% (Shaikh & Supit, 2014). Camiletti et al. (2013) fabricated ultra high performance concrete with NC at levels of 0%, 2.5% and 5% replacing cement by volume. They found that adding 2.5-5% NC induced about 32-75% improvement in the 24h compressive strength relative to the plain concrete, respectively.

4. **Durability:** Skaikh et al. (2014) studied the chloride penetration of concrete with NC at levels of 0%, 1%, 2%, 3% and 4%. They found that the chloride ion permeability of Portland cement concrete with 1% NC decreased by approximately 20% and 50% at age of 28 and 90 d, respectively. Additionally, reductions in chloride ion penetration of high volume fly ash concretes containing 40% and 60% fly ash and 1% NC were about 19% and 12%, respectively after 28 d of curing. However, no significant improvement in above concretes was observed at age of 90 d. Qian et al. (2009) also found that addition of NC to fly ash concrete enhanced the impenetrability for chloride.

Nano-Clay

1. **Mechanical Properties:** Chang et al. (2007) fabricated cement paste with nano-clay, about 100 nm, at levels of 0%, 0.2%, 0.4%, 0.6% and 0.8% by weight of cement. The water to cement ratio for all the mixture was 0.55. By comparing the compressive strength of the specimens cured for 7, 14, 28 and 56 d, it could be observed that the compressive strengths increased with nano-clay particles up to 0.6% and then decreased. The compressive strength of cement paste with 0.6% nano-clay increased by 13.24% compared with plain cement paste at age of 56 d.

Morsy et al. (2007) partially replaced white Portland cement with nano-clay at levels of 0%, 2% and 4% by weight. The optimal level of nano-clay content was achieved at 2% replacement. The indirect tensile strength of composites with 2% nano-clay increased by 28% and 25% at age of 7 and 28 d, respectively. In addition, Morsy et al. (2011) also fabricated cement mortars with 6% nano-clay by weight of cement. The water to binder ratio used for all the mixtures was 0.50. They found that addition of 6% nano-clay to cement mortars enhanced the compressive strength by 18% compared with plain cement mortars at age of 28 d.

Yeganeh et al. (2008) prepared cement mortars containing 10% high impact poly styrene (HIPS) in powder form and 5% nano-clay, and found that addition of nano-clay dramatically improved adhesion strength and compression modulus.

The cement mortars were produced with 5%, 20%, 30% and 50% waste glass powders in weight, and 2.5% nano-clay, replacing cement (Aly, Hashmi, Olabi, Messeiry, & Hussain, 2011). The results showed that adding nano-clay to cement mortars could improve the compressive strength, flexural strength and fracture energy at age of 28 d.

2. **Durability:**
   a. **Permeability:** Chang et al. (2007) studied permeability property of cement past with different dosage of nano-clay. They observed that the optimum value of nano-clay was achieved at 0.4% for all age. The permeability coefdcient of cement paste with 0.4% nano-clay decreased about 49.95% relative to plain cement paste at age of 56 d.
   b. **Alkali-Silica Reaction:** Aly et al. (2011) studied the alkali-silica reaction of cement mortars containing waster-glass powders and nano-clay at level of 2.5%, and found addition nano-clay could reduce the expansion due to alkali-silica reaction.

MODIFICATION MECHANISMS

Modification Mechanisms of Workability

In most situations, addition of nanomaterials to cement and concrete materials could reduce the flowability and the possibility of bleeding and segregation. Due to having ultra high specific surface area, nanomaterials have stronger attraction to water than that of cement particles. As a result, the amount of lubricating water available is decreased in the fresh mixture.

Accelerate Hydration Mechanisms

Due to great surface energy of naomaterials, the hydration production will deposit on the nanomaterials that located in cement and concrete materials. The nanomaterials just like 'nucleus' which are surrounded by the hydration production. What's more, the nano-particles or nano-fibers have high activity which will further speed up the hydration process of cement (Li Li, Xiao, Yuan, & Ou, 2004). Makar et al. observed that the CNTs appear to act as nucleating sites for the C_3S hydration products (Makar

& Chan, 2009), thus accelerating the hydration reaction of the C_3S. Chen et al. (2012) reported that the rapid consumption of free water by nanomaterials accelerated the bridging process of gaps so that solididcation occurred earlier.

Enhancement/Modification Mechanisms of Mechanical Properties

The contributing factors for enhancement/modification mechanisms of mechanical properties may include: nucleus effect, filler effect, pozzolanic activity, crack bridging, pinning effect, fiber pull-out, crack deflection, fiber debonding, fiber breaking and so on. Nucleus effect and filler effect widely existed in nanomaterials filled cement and concrete materials. Pozzolanic activity only shows in nanomaterials which can react with CH or CaO. In addition, there are some enhancement/modification mechanisms only for nano-fibers.

Nucleus Effect

Nucleus effect enhances the mechanical properties mainly in following three aspects.

1. **Accelerating the Hydration:** Accelerating the hydration can increase the content of CH at early age. Jalal et al. (2013) found that addition of NT to concrete containing fly ash could accelerate C-S-H gel formation due to increasing CH amount at the early age. As a result, split tensile strength of concrete specimens increases.
2. **Reducing the Size of Crystal:** The nanomaterials among the hydration productions can prevent the crystal from forming big size, such as CH and AFt (Li, Xiao, Yuan, & Ou, 2004). The fine crystals are benefit to enhance the mechanical properties of cement and concrete materials.
3. **Changing Tendency of Crystal:** The orientation index has effect on the mechanical properties of cement and concrete materials, especially of the CH in the ITZ. Meng et al. (2012) observed that addition NT could change tendency of CH crystal. They reported, because of the nucleus effect, that the decrease and modification of orientation index of CH crystal may increase strength rather than increase the amount of hydration products.

Filler Effect

The nanomaterials, located in cement and concrete materials, will fill the pore and enhance the strength (Gaitero, Campillo, & Guerrero, 2008; Li, Xiao, Yuan, & Ou, 2004). Said et al. (2012) used TG to analysis the hydration products and found that increasing NS from 3% to 6% did not decrease the content of CH, which indicated that the improvement of performance associated with increase the amount of NS from 3% to 6% may be mainly attributed to the filler effect. The CNTs/CNFs could improve pore size distribution and decrease the porosity by filling the gaps (or pores) between the hydration products. Therefore, the cement and concrete materials get a much more compacted structure (Han, Yu, & Ou, 2011; Konsta-Gdoutos, Metaxa, & Shah, 2010; Li, Wang, & Zhao, 2005; Nochaiya & Chaipanich, 2011).

Pozzolanic Activity

The pozzolanic activity of NS can consume CH crystal, decrease the orientation of CH crystal and reduce the size of CH at interfacial transition zone (ITZ) (Ye et al., 2007). In addtion, the NS reacting with CH can not only generate C-S-H, but also reduce the particle of aggregating NS which will form a weak zone (Gaitero, Campillo, & Guerrero, 2008; Li, Xiao, Yuan, & Ou, 2004). All of above mentioned makes contribution to improving the mechanical properties of cement and concrete materials.

Fiber Reinforcement

Nanofibers own some enhancement mechanisms besides nucleus effect and filler effect, as follow: (1) CNTs/CNFs have excellent mechanical properties, as shown in Table 4; (2) The bonding between CNTs/CNFs and the matrix is strong, especially for the covalence-modified CNTs/CNFs. For example, Peyvandi et al. (2013) added hydrophilic groups (-COOH) covalence-modified CNFs to cement paste and found that the -COOH groups could form strong coordinate bonds with the Ca^{2+} in cement hydrates, thus enhancing the mechanical properties of composites; (3) Crack bridging, pinning effect, fiber pull-out, crack deflection, fiber debonding and fiber breaking improve the mechanical properties (Han, Yu, & Ou, 2011; Al-Rub, Ashour, & Tyson, 2012).

Enhancement/Modification Mechanisms of Durability

Some mechanisms are also right for mechanical properties and durability, such as nucleus effects, filler effect and pozzolanic activity which make the matrix more homogeneous and compact. As a result, durability and strength are improved. Gaiteor et al. (2008) found that the addition of NS can modify the structure of the C-S-H gel, thus increasing the length of the chain which can increase the calcium stabilization.

Modification Mechanisms of Functional Properties

Conductive and Sensing Mechanisms

The conductive and sensing mechanisms of NCMs and nano-scale metallic oxides filled cement and concrete materials mainly include the intrinsic electrical conductivity, contacting conduction, tunneling conduction and field emission conduction (Grujicic, Cao, & Gersten, 2003;Han et al., 2011;Li & Chou, 2008;Wang, Ding, & Wang, 2009). In fact, the conduction mechanisms are very complex in nature and interrelated with each other. For a specific nano cement and concrete materials, one or several factors among these aspects may occupy leading position. Moreover, the electrical resistivity of nano conductive cement and concrete would change under different loads. In this condition, the electrical mechanisms mainly include two aspects: electronic conduction and hole conduction through nanomaterials by the tunneling effect; electronic conduction and hole conduction through contacting conduction of nanomaterials (Huang, 2012). The two factors are both responsible for the electrical property of the NCMs and nano-scale metallic filled cementitious composites while different dosages of nanomaterials make dominant factor imparity.

There are mainly four factors that may contribute to the sensing property. Under different external environment, different changes happen in the mixtures which include the intrinsic resistance of nanomaterials; the bonding between nanomaterials and matrix; the contact between nanomaterials; and the tunneling distance between nanomaterials. It is worth noting that the four factors usually work together for the contribution to sensing property while only one or several of the factors are dominant at certain percolation zones, namely content levels of nanomaterials. In a general way, the content of nanomaterials above the percolation threshold is beneficial to the sensing sensitivity under tension, while that below the percolation threshold is beneficial for the sensing sensitivity under compression. Therefore, the percolation threshold is an important parameter for designing and optimizing the sensing property of nano cement and concrete materials (Han, Yu, & Ou, 2011).

Microwave Absorption Mechanisms

The presence of conductive nanomaterials, such as CNTs/CNFs, in the insulative cementitious matrix results in space charge accumulating at the interfaces, which contributes to the higher microwave absorption in the mixture (Singh et al., 2013). Increasing MWCNT contents leads to increase of dielectric constant, improving the capability of the composites to shield electromagnetic waves, a reduction of skin depth and an increase in AC conductivity along with an improvement of input impedance. This enhances both the amount of electromagnetic radiation penetrating inside the shield and the effective absorption capability. Therefore, addition conductive nanomaterials, such as CNTs/CNFs, to concrete matrix increases the interfacial polarization and the effective anisotropy energy of composites which lead to the high SE of nano cement and concrete materials (Guo, 2013; Singh et al., 2013).

Air Purifying Mechanisms

Anatase TiO_2 is popular as photocatalyst and used in cement and concrete materials. The basic decomposition mechanism includes two steps: in the first step, photocatalyst irradiation generates radicals which have strong oxidation; in the second step, pollutants reach the surface of NT, then react with radicals and produce innocuous compounds (Ballari, Hunger, Hüsken, & Brouwers, 2010). Anatase phase TiO_2 has higher band gap (3.2 eV) compared with the phase of rutile (3.0 eV). The NT will form high activity electrons (e-) and "holes" under irradiation of a specific wavelength of light. The H_2O and O_2 on the surface of NT react with the electrons and "holes" and produce $HO_2^{\cdot}$ and OH· (as described in Equations (1)-(5)). The photocatalytic conversion of NO_x, namely nitric oxide (NO) and nitrogendioxide (NO_2), is as follows (Equations (6)-(8)) (Chen & Poon, 2009; Hüsken, Hunger, & Brouwers, 2009). The last produce is NO_3^- which is harmless to the environment.

$$photocatalyst + h\mu \rightarrow photocatalyst + e^- + h^+ \tag{1}$$

$$H_2O \leftrightarrow H^+ + OH^- \tag{2}$$

$$h^+ + OH^- \rightarrow OH^{\cdot} \tag{3}$$

$$e^- + O_2 \rightarrow O_2^- \tag{4}$$

$$H^+ + O_2^- \rightarrow HO_2^{\cdot} \tag{5}$$

$$NO + HO_2^{\cdot} \rightarrow NO_2 + OH^{\cdot} \tag{6}$$

$$NO + OH^· \rightarrow NO_2 + H^+ \tag{7}$$

$$NO_2 + OH^· \rightarrow NO_3^- + H^+ \tag{8}$$

CONCLUSION

Nanotechonolgy and nanomaterials play significant roles in understanding the behavior of cement and concrete materials at nano-scale and dealing with the problems and challenges of cement and concrete materials. The advantages of addition nanomaterials to cement and concrete materials mainly include:

1. Fabricating of high performance concrete;
2. Achieving multifuntionlities;
3. Ultilizing more admixture with littler reducing even increasing in early age strength;
4. Achieving the same strength grade with fewer cement;
5. Reducing the curing time to speed up the construction progress.

Although a great deal of research has been conducted about the properties of cement and concrete materials by adding different kinds of nanomaterials over the past decade, there still exists many challenges needed to be addressed. Future work may include:

1. Investigating the behavior of hydration production at nano-scale and learning more about the relationship between the nano-scale behavior and marco-properties, namely finding the nano genomic code and the nano behavior blueprint of cement and concrete materials;
2. Paying more attention to multifunctional cement and concrete materials;
3. Studying about nanomaterials filled cement and concrete materials in a more systematical and long-term way;
4. Adding two or more types of nanomaterials to enhance the properties of cement and concrete materials more effectively;
5. Matching the type and dosage of naomaterials with the type and dosage of admixture to find a balance between environment and economy;
6. Dispersion, not mentioned in this chapter, is very beneficial to effectively use the nanomaterials.

ACKNOWLEDGMENT

The authors thank the funding supported fromthe National Science Foundation of China (grant Nos. 51578110, 51428801 and 51178148), the Program for New Century Excellent Talents in University of China (grant No. NCET-11-0798), and the Ministry of Science and Technology of China (grant No. 2011BAK02B01). The authors also thank Ruiqi Su, Tianjiao Wu and Tian Zhang of Dalian University of Technology for their assistance.

REFERENCES

Al-Rub, R. K. A., Ashour, A. I., & Tyson, B. M. (2012). On the aspect ratio effect of multi-walled carbon nanotube reinforcements on the mechanical properties of cementitious nanocomposites. *Construction & Building Materials, 35*, 647–655.

Al-Rub, R. K. A., Tyson, B. M., Yazdanbakhsh, A., & Grasley, Z. (2012). Mechanical Properties of Nanocomposite Cement Incorporating Surface-Treated and Untreated Carbon Nanotubes and Carbon Nanofibers. *Journal of Nanomechanics and Micromechanics, 2*(1), 1–6.

Aly, M., Hashmi, M. S. J., Olabi, A. G., Messeiry, M., & Hussain, A. I. (2011). Effect of nano clay particles on mechanical, thermal and physical behaviours of waste-glass cement mortars. *Materials Science and Engineering A, 528*(27), 7991–7998.

Arefi, M. R., Javeri, M. R., & Mollaahmadi, E. (2011). To study the effect of adding Al_2O_3 nanoparticles on the mechanical properties and microstructure of cement mortar. *Life Science Journal, 8*(4), 613–617.

Arulraj, G. P., & Carmichael, M. J. (2011). Effect of nano-fly ash on strength of concrete. *International Journal of Civil & Structural Engineering, 2*(2), 475–482.

Baeza, F. J., Galao, O., Zornoza, E., & Garcés, P. (2013). Multifunctional Cement Composites Strain and Damage Sensors Applied on Reinforced Concrete (RC) Structural Elements. *Materials (Basel), 6*(3), 841–855.

Ballari, M. M., Hunger, M., Hüsken, G., & Brouwers, H. (2010). Modelling and experimental study of the NOx photocatalytic degradation employing concrete pavement with titanium dioxide. *Catalysis Today, 151*(1), 71–76.

Barbhuiya, S., Mukherjee, S., & Nikraz, H. (2014). Effects of nano-Al_2O_3 on early-age microstructural properties of cement paste. *Construction & Building Materials, 52*, 189–193.

Bartos, P. J. (2009). Nanotechnology in construction: a roadmap for development. In Z. Bittnar, P. J. M. Bartos, J. Nemecek, V. Smilauer, & J. Zeman (Eds.), *Nanotechnology in Construction 3* (pp. 15–26). Berlin, Germany: Springer.

Behfarnia, K., & Salemi, N. (2013). The effects of nano-silica and nano-alumina on frost resistance of normal concrete. *Construction & Building Materials, 48*, 580–584.

Berra, M., Carassiti, F., Mangialardi, T., Paolini, A. E., & Sebastiani, M. (2012). Effects of nanosilica addition on workability and compressive strength of Portland cement pastes. *Construction & Building Materials, 35*, 666–675.

Brown, L., Sanchez, F., Kosson, D., & Arnold, J. (2013). Performance of carbon nanofiber-cement composites subjected to accelerated decalcification. In *Proceedings of EPJ Web of Conferences*. London, UK: EDP Sciences.

Camiletti, J., Soliman, A. M., & Nehdi, M. L. (2013). Effects of nano- and micro-limestone addition on early-age properties of ultra-high-performance concrete. *Materials and Structures, 46*(6), 881–898.

Campillo, I., Guerrero, A., Dolado, J. S., Porro, A., Ibáñez, J. A., & Goñi, S. (2007). Improvement of initial mechanical strength by nanoalumina in belite cements. *Materials Letters, 61*(8), 1889–1892.

Cárdenas, C., Tobón, J. I., García, C., & Vila, J. (2012). Functionalized building materials: Photocatalytic abatement of NOx by cement pastes blended with TiO_2 nanoparticles. *Construction & Building Materials, 36*, 820–825.

Chang, T., Shih, J., Yang, K., & Hsiao, T. (2007). Material properties of Portland cement paste with nano-montmorillonite. *Journal of Materials Science, 42*(17), 7478–7487.

Chen, J., Kou, S., & Poon, C. (2012). Hydration and properties of nano-TiO_2 blended cement composites. *Cement and Concrete Composites, 34*(5), 642–649.

Chen, J., & Poon, C. (2009). Photocatalytic construction and building materials: From fundamentals to applications. *Building and Environment, 44*(9), 1899–1906.

Collepardi, M., Olagot, J. O., Skarp, U., & Troli, R. (2002). Influence of amorphous colloidal silica on the properties of self-compacting concretes. In *Proceedings of the International Conference in Concrete Constructions - innovations and developments in concrete materials and constructions*. Dundee, UK: ICE.

Constantinides, G., & Ulm, F. (2004). The effect of two types of C-S-H on the elasticity of cement-based materials: Results from nanoindentation and micromechanical modeling. *Cement and Concrete Research, 34*(1), 67–80.

Constantinides, G., & Ulm, F. (2007). The nanogranular nature of C-S-H. *Journal of the Mechanics and Physics of Solids, 55*(1), 64–90.

Dai, Y., Sun, M., Liu, C., & Li, Z. (2010). Electromagnetic wave absorbing characteristics of carbon black cement-based composites. *Cement and Concrete Composites, 32*(7), 508–513.

DeJong, M. J., & Ulm, F. (2007). The nanogranular behavior of C-S-H at elevated temperatures (up to 700°C). *Cement and Concrete Research, 37*(1), 1–12.

Du, H., Quek, S. T., & Dai Pang, S. (2013, April). Smart multifunctional cement mortar containing graphite nanoplatelet. In *Proceedings of SPIE Smart Structures and Materials + Nondestructive Evaluation and Health Monitoring*. Orlando, FL: International Society for Optics and Photonics.

Fan, J. J., Tang, J. Y., Cong, L. Q., & Mcolm, I. J. (2004). Influence of synthetic nano-ZrO2 powder on the strength property of portland cement. *Jianzhu Cailiao Xuebao, 7*(4), 462–467.

Gaitero, J. J., Campillo, I., & Guerrero, A. (2008). Reduction of the calcium leaching rate of cement paste by addition of silica nanoparticles. *Cement and Concrete Research, 38*(8-9), 1112–1118.

Galao, O., Baeza, F. J., Zornoza, E., & Garcés, P. (2013). Strain and damage sensing properties on multifunctional cement composites with CNF admixture. *Cement and Concrete Composites, 46*, 90–98.

Gao, D., Sturm, M., & Mo, Y. L. (2009). Electrical resistance of carbon-nanofiber concrete. *Smart Materials and Structures, 18*(9), 095039.

Gay, C., & Sanchez, F. (2010). Performance of Carbon Nanofiber-Cement Composites with a High-Range Water Reducer. *Transportation Research Record: Journal of the Transportation Research Board, 2142*(1), 109-113.

Ghafari, E., Costa, H., Julio, E., Portugal, A., & Duraes, L. (2014). The effect of nanosilica addition on flowability, strength and transport properties of ultra high performance concrete. *Materials & Design, 59*, 1–9.

Gong, H., Zhang, Y., Quan, J., & Che, S. (2011). Preparation and properties of cement based piezoelectric composites modified by CNTs. *Current Applied Physics, 11*(3), 653–656.

Groves, G. W., Sueur, P. J., & Sinclair, W. (1986). Transmission Electron Microscopy and Microanalytical Studies of Ion-Beam-Thinned Sections of Tricalcium Silicate Paste. *Journal of the American Ceramic Society, 69*(4), 353–356.

Grujicic, M., Cao, G., & Gersten, B. (2003). Enhancement of field emission in carbon nanotubes through adsorption of polar molecules. *Applied Surface Science*, *206*(1), 167–177.

Guo, Z. (2013). *Study on the Electromagnetic Wave Absorbing Properties of Multi-walled Carbon Nanotube/Cement Composites.* (Master's Dissertation). Dalian University of Technology, Dalian, China.

Habermehl-Cwirzen, K., Penttala, V., & Cwirzen, A. (2008). Surface decoration of carbon nanotubes and mechanical properties of cement/carbon nanotube composites. *Advances in Cement Research*, *20*(2), 65–73.

Han, B., & Ou, J. (2007). Embedded piezoresistive cement-based stress/strain sensor. *Sensors and Actuators. A, Physical*, *138*(2), 294–298.

Han, B., Sun, S., Ding, S., Zhang, L., Yu, X., & Ou, J. (2015). Review of nanocarbon-engineered multifunctional cementitious composites. *Composites. Part A, Applied Science and Manufacturing*, *70*, 69–81.

Han, B., Yang, Z., Shi, X., & Yu, X. (2013). Transport Properties of Carbon-Nanotube/Cement Composites. *Journal of Materials Engineering and Performance*, *22*(1), 184–189.

Han, B., Yu, X., & Kwon, E. (2009). A self-sensing carbon nanotube/cement composite for traffic monitoring. *Nanotechnology*, *20*(44), 445501.

Han, B., Yu, X., & Ou, J. (2011). Multifunctional and smart carbon nanotube reinforced cement-based materials. In K. Gopalakrishnan, B. Birgisson, P. Taylor, & N. O. Attoh-Okine (Eds.), *Nanotechnology in civil infrastructure* (pp. 1–47). Berlin, Germany: Springer.

Han, B., Yu, X., & Ou, J. (2014). Self-sensing concrete in smart structures. Waltham, MA: Butterworth Heinemann (Elsevier).

Han, B., Zhang, K., Burnham, T., Kwon, E., & Yu, X. (2013). Integration and road tests of a self-sensing CNT concrete pavement system for traffic detection. *Smart Materials and Structures*, *22*(1), 15020.

Han, B., Zhang, K., Yu, X., Kwon, E., & Ou, J. (2012). Electrical characteristics and pressure-sensitive response measurements of carboxyl MWNT/cement composites. *Cement and Concrete Composites*, *34*(6), 794–800.

Hannesson, G., Kuder, K., Shogren, R., & Lehman, D. (2012). The influence of high volume of fly ash and slag on the compressive strength of self-consolidating concrete. *Construction & Building Materials*, *30*, 161–168.

Haruehansapong, S., Pulngern, T., & Chucheepsakul, S. (2014). Effect of the particle size of nanosilica on the compressive strength and the optimum replacement content of cement mortar containing nano-SiO$_2$. *Construction & Building Materials, 50*, 471–477.

He, X., & Shi, X. (2008). Chloride Permeability and Microstructure of Portland Cement Mortars Incorporating Nanomaterials. *Transportation Research Record: Journal of the Transportation Research Board, 2070*(1), 13-21.

Hosseini, P., Booshehrian, A., & Madari, A. (2011). Developing Concrete Recycling Strategies by Utilization of Nano-SiO$_2$ Particles. *Waster and Biomass Valorization, 2*(3), 347–355.

Hou, D., Zhu, Y., Lu, Y., & Li, Z. (2014). Mechanical properties of calcium silicate hydrate (C–S–H) at nano-scale: A molecular dynamics study. *Materials Chemistry and Physics, 146*(3), 503–511.

Hou, P., Kawashima, S., Wang, K., Corr, D. J., Qian, J., & Shah, S. P. (2013). Effects of colloidal nanosilica on rheological and mechanical properties of fly ash-cement mortar. *Cement and Concrete Composites, 35*(1), 12–22.

Howser, R. N., Dhonde, H. B., & Mo, Y. L. (2011). Self-sensing of carbon nanofiber concrete columns subjected to reversed cyclic loading. *Smart Materials and Structures, 20*(8), 085031.

Huang, S. (2012). *Multifunctional Graphite Nanoplatelets (GNP) Reinforced Cementitious Composites*. (Master's Dissertation). National University of Singapore.

Hunashyal, A. M., Sundeep, G. V., Quadri, S. S., & Banapurmath, N. R. (2011). Experimental investigations to study the effect of carbon nanotubes reinforced in cement-based matrix composite beams. *Proceedings of the Institution of Mechanical Engineers. Part N, Journal of Nanoengineering and Nanosystems, 225*(1), 17–22.

Hüsken, G., Hunger, M., & Brouwers, H. (2009). Experimental study of photocatalytic concrete products for air purification. *Building and Environment, 44*(12), 2463–2474.

Ibrahim, R. K., Hamid, R., & Taha, M. R. (2012). Fire resistance of high-volume fly ash mortars with nanosilica addition. *Construction & Building Materials, 36*, 779–786.

Jalal, M., Ramezanianpour, A. A., & Pool, M. K. (2013). Split tensile strength of binary blended self compacting concrete containing low volume fly ash and TiO$_2$ nanoparticles. *Composites. Part B, Engineering, 55*, 324–337.

Jennings, H. M. (2003). Colloid model of C-S-H and implications to the problem of creep and shrinkage. *Materials and Structures, 37*(265), 59–70.

Jo, B., Chakraborty, S., Kim, K. H., & Lee, Y. S. (2014). Effectiveness of the Top-Down Nanotechnology in the Production of Ultrafine Cement. *Journal of Nanomaterials, 2014*, 1–9.

Jo, B., Kim, C., Tae, G., & Park, J. (2007). Characteristics of cement mortar with nano-SiO_2 particles. *Construction & Building Materials, 21*(6), 1351–1355.

Jo, B. W., Chakraborty, S., & Kim, K. H. (2014). Investigation on the effectiveness of chemically synthesized nano cement in controlling the physical and mechanical performances of concrete. *Construction & Building Materials, 70*, 1–8.

Kawashima, S., Hou, P., Corr, D. J., & Shah, S. P. (2013). Modification of cement-based materials with nanoparticles. *Cement and Concrete Composites, 36*, 8–15.

Kerienė, J., Kligys, M., Laukaitis, A., Yakovlev, G., Špokauskas, A., & Aleknevičius, M. (2013). The influence of multi-walled carbon nanotubes additive on properties of non-autoclaved and autoclaved aerated concretes. *Construction & Building Materials, 49*, 527–535.

Khoshakhlagh, A., Nazari, A., & Khalaj, G. (2012). Effects of Fe_2O_3 Nanoparticles on Water Permeability and Strength Assessments of High Strength Self-Compacting Concrete. *Journal of Materials Science and Technology, 28*(1), 73–82.

Kjellsen, K. O., Jennings, H. M., & Lagerblad, B. (1996). Evidence of hollow shells in the microstructure of cement paste. *Cement and Concrete Research, 26*(4), 593–599.

Kong, D., Du, X., Wei, S., Zhang, H., Yang, Y., & Shah, S. P. (2012). Influence of nano-silica agglomeration on microstructure and properties of the hardened cement-based materials. *Construction & Building Materials, 37*, 707–715.

Konsta-Gdoutos, M. S., Metaxa, Z. S., & Shah, S. P. (2010). Multi-scale mechanical and fracture characteristics and early-age strain capacity of high performance carbon nanotube/cement nanocomposites. *Cement and Concrete Composites, 32*(2), 110–115.

Li, C., & Chou, T. (2008). Modeling of damage sensing in fiber composites using carbon nanotube networks. *Composites Science and Technology, 68*(15), 3373–3379.

Li, G. (2004). Properties of high-volume fly ash concrete incorporating nano-SiO_2. *Cement and Concrete Research, 34*(6), 1043–1049.

Li, G., Wang, P., & Zhao, X. (2005). Mechanical behavior and microstructure of cement composites incorporating surface-treated multi-walled carbon nanotubes. *Carbon, 43*(6), 1239–1245.

Li, H., Xiao, H., Guan, X., Wang, Z., & Yu, L. (2014). Chloride diffusion in concrete containing nano-TiO_2 under coupled effect of scouring. *Composites. Part B, Engineering, 56*, 698–704.

Li, H., Xiao, H., & Ou, J. (2006). Effect of compressive strain on electrical resistivity of carbon black-filled cement-based composites. *Cement and Concrete Composites, 28*(9), 824–828.

Li, H., Xiao, H., & Ou, J. (2008). Electrical property of cement-based composites filled with carbon black under long-term wet and loading condition. *Composites Science and Technology, 68*(9), 2114–2119.

Li, H., Xiao, H., Yuan, J., & Ou, J. (2004). Microstructure of cement mortar with nano-particles. *Composites. Part B, Engineering, 35*(2), 185–189.

Li, H., Xiao, H. G., & Ou, J. P. (2004). A study on mechanical and pressure-sensitive properties of cement mortar with nanophase materials. *Cement and Concrete Research, 34*(3), 435–438.

Li, H., Xiao, H. G., Yuan, J., & Ou, J. P. (2004). Microstructure of cement mortar with nano-particles. *Composites. Part B, Engineering, 35*(2), 185–189.

Li, H., Zhang, M., & Ou, J. (2006). Abrasion resistance of concrete containing nano-particles for pavement. *Wear, 260*(11-12), 1262–1266.

Li, Z., Wang, H., He, S., Lu, Y., & Wang, M. (2006). Investigations on the preparation and mechanical properties of the nano-alumina reinforced cement composite. *Materials Letters, 60*(3), 356–359.

Lim, S., & Mondal, P. (2014). Micro- and nano-scale characterization to study the thermal degradation of cement-based materials. *Materials Characterization, 92*, 15–25.

Lin, D. F., Lin, K. L., Chang, W. C., Luo, H. L., & Cai, M. Q. (2008). Improvements of nano-SiO_2 on sludge/fly ash mortar. *Waste Management (New York, N.Y.), 28*(6), 1081–1087.

Liu, X., Chen, L., Liu, A., & Wang, X. (2012). Effect of Nano-$CaCO_3$ on Properties of Cement Paste. *Energy Procedia, 16*, 991–996.

Lucas, S. S., Ferreira, V. M., & de Aguiar, J. L. B. (2013). Incorporation of titanium dioxide nanoparticles in mortars -Influence of microstructure in the hardened state properties and photocatalytic activity. *Cement and Concrete Research, 43*, 112–120.

Ludvig, P., Calixto, J. M., Ladeira, L. O., & Gaspar, I. C. P. (2011). Using Converter Dust to Produce Low Cost Cementitious Composites by in situ Carbon Nanotube and Nanofiber Synthesis. *Materials (Basel), 4*(12), 575–584.

Luo, J., Duan, Z., & Li, H. (2009). The influence of surfactants on the processing of multi-walled carbon nanotubes in reinforced cement matrix composites. [a]. *Physica Status Solidi, 206*(12), 2783–2790.

Lv, S., Ma, Y., Qiu, C., Sun, T., Liu, J., & Zhou, Q. (2013). Effect of graphene oxide nanosheets of microstructure and mechanical properties of cement composites. *Construction & Building Materials, 49*, 121–127.

Makar, J. M., & Chan, G. W. (2009). Growth of Cement Hydration Products on Single-Walled Carbon Nanotubes. *Journal of the American Ceramic Society, 92*(6), 1303–1310.

Makar, J. M., Margeson, J. C., & Luh, J. (2005). *Carbon nanotube/cement composites-early results and potential applications.* Paper presented at 3rd International Conference on Construction Materials: Performance, Innovations and Structural Implications, Vancouver, Canada.

Materazzi, A. L., Ubertini, F., & D'Alessandro, A. (2013). Carbon nanotube cement-based transducers for dynamic sensing of strain. *Cement and Concrete Composites, 37*, 2–11.

McCarthy, M. J. &. D. (1999). Towards maximising the use of fly ash as a binder. *Fuel, 78*(2), 121–132.

Meng, T., Yu, Y., Qian, X., Zhan, S., & Qian, K. (2012). Effect of nano-TiO_2 on the mechanical properties of cement mortar. *Construction & Building Materials, 29*, 241–245.

Mo, Y. L., & Roberts, R. H. (2013). Carbon Nanofiber Concrete for Damage Detection of Infrastructure. In R. Maguire (Ed.), *Advances in Nanofibers* (pp. 125–143). Rijeka, Croatia: InTech.

Morsy, M. S., & Aglan, H. A. (2007). Development and characterization of nanostructured-perlite-cementitious surface compounds. *Journal of Materials Science, 42*(24), 10188–10195.

Morsy, M. S., Alsayed, S. H., & Aqel, M. (2011). Hybrid effect of carbon nanotube and nano-clay on physico-mechanical properties of cement mortar. *Construction & Building Materials, 25*(1), 145–149.

Musso, S., Tulliani, J., Ferro, G., & Tagliaferro, A. (2009). Influence of carbon nanotubes structure on the mechanical behavior of cement composites. *Composites Science and Technology, 69*(11-12), 1985–1990.

Nasibulin, A. G., Tapper, U., Tian, Y., Penttala, V., Karppinen, M. J., Kauppinen, E. I. S., & Malm, J. E. M. (2009). A novel cement-based hybrid material. *New Journal of Physics, 11*(2), 23013.

Nasibulina, L. I., Anoshkin, I. V., Semencha, A. V., Tolochko, O. V., Malm, J. E., Karppinen, M. J., & Kauppinen, E. I. (2012). Carbon nanofiber/clinker hybrid material as a highly efficient modificator of mortar mechanical properties. *Materials Physics and Mechanics, 13*, 77–84.

Nazari, A., Rafieipour, M. H., & Riahi, S. (2011). The effects of CuO nanoparticles on properties of self compacting concrete with GGBFS as binder. *Materials Research-Ibero-American Journal of Materials, 14*(3), 307–316.

Nazari, A., & Riahi, S. (2010). The effect of TiO_2 nanoparticles on water permeability and thermal and mechanical properties of high strength self-compacting concrete. *Materials Science and Engineering A, 528*(2), 756–763.

Nazari, A., & Riahi, S. (2011). The Effects of TiO_2 Nanoparticles on Flexural Damage of Self-compacting Concrete. *International Journal of Damage Mechanics, 20*(7), 1049–1072.

Nazari, A., & Riahi, S. (2011a). The effects of SiO_2 nanoparticles on physical and mechanical properties of high strength compacting concrete. *Composites. Part B, Engineering, 42*(3), 570–578.

Nazari, A., & Riahi, S. (2011b). Abrasion resistance of concrete containing SiO_2 and Al_2O_3 nanoparticles in different curing media. *Energy and Building, 43*(10), 2939–2946.

Nazari, A., & Riahi, S. (2011c). Compressive strength and abrasion resistance of concrete containing SiO_2 and Cr_2O_3 nanoparticles in different curing media. *Magazine of Concrete Research, 64*(2), 177–188.

Nazari, A., & Riahi, S. (2011d). Al_2O_3 nanoparticles in concrete and different curing media. *Energy and Building, 43*(6), 1480–1488.

Nazari, A., & Riahi, S. (2011e). Improvement compressive strength of concrete in different curing media by Al_2O_3 nanoparticles. *Materials Science and Engineering A, 528*(3), 1183–1191.

Nazari, A., & Riahi, S. (2011f). Abrasion resistance of concrete containing SiO_2 and Al_2O_3 nanoparticles in different curing media. *Energy and Building, 43*(10), 2939–2946.

Nazari, A., & Riahi, S. (2011g). TiO_2 nanoparticles effects on physical, thermal and mechanical properties of self compacting concrete with ground granulated blast furnace slag as binder. *Energy and Building, 43*(4), 995–1002.

Nazari, A., & Riahi, S. (2011h). Computer-aided design of the effects of Fe_2O_3 nanoparticles on split tensile strength and water permeability of high strength concrete. *Materials & Design, 32*(7), 3966–3979.

Nazari, A., & Riahi, S. (2011i). The effects of zinc dioxide nanoparticles on flexural strength of self-compacting concrete. *Composites. Part B, Engineering, 42*(2), 167–175.

Nazari, A., & Riahi, S. (2011j). Effects of CuO nanoparticles on compressive strength of self-compacting concrete. *Sadhana, 36*(3), 371–391.

Nazari, A., Riahi, S., Riahi, S., Shamekhi, S. F., & Khademno, A. (2010). Mechanical properties of cement mortar with Al_2O_3 nanoparticles. *Journal of American Science, 6*(4), 94–97.

Nazari, A., Sh, R., Sh, R., Shamekhi, S. F., & Khademno, A. (2010a). Influence of Al_2O_3 nanoparticles on the compressive strength and workability of blended concrete. *Journal of American Science, 6*(5), 6–9.

Nazari, A., Sh, R., Sh, R., Shamekhi, S. F., & Khademno, A. (2010b). Benefits of Fe_2O_3 nanoparticles in concrete mixing matrix. *Journal of American Science, 6*(4), 102–106.

Nazari, A., Sh, R., Sh, R., Shamekhi, S. F., & Khademno, A. (2010c). The effects of incorporation Fe_2O_3 nanoparticles on tensile and flexural strength of concrete. *Journal of American Science, 6*(4), 90–93.

Nochaiya, T., & Chaipanich, A. (2011). Behavior of multi-walled carbon nanotubes on the porosity and microstructure of cement-based materials. *Applied Surface Science, 257*(6), 1941–1945.

Noorvand, H., Abang Ali, A. A., Demirboga, R., Farzadnia, N., & Noorvand, H. (2013). Incorporation of nano TiO2 in black rice husk ash mortars. *Construction & Building Materials, 47*, 1350–1361.

Oltulu, M., & Şahin, R. (2011). Single and combined effects of nano-SiO_2, nano-Al_2O_3 and nano-Fe_2O_3 powders on compressive strength and capillary permeability of cement mortar containing silica fume. *Materials Science and Engineering A, 528*(22-23), 7012–7019.

Oltulu, M., & Şahin, R. (2013). Effect of nano-SiO_2, nano-Al_2O_3 and nano-Fe_2O_3 powders on compressive strengths and capillary water absorption of cement mortar containing fly ash: A comparative study. *Energy and Building, 58*, 292–301.

Oltulu, M., & Şahin, R. (2014). Pore structure analysis of hardened cement mortars containing silica fume and different nano-powders. *Construction & Building Materials, 53*, 658–664.

Peyvandi, A., Sbia, L. A., Soroushian, P., & Sobolev, K. (2013). Effect of the cementitious paste density on the performance efficiency of carbon nanofiber in concrete nanocomposite. *Construction & Building Materials, 48*, 265–269.

Peyvandi, A., Soroushian, P., Abdol, N., & Balachandra, A. M. (2013). Surface-modified graphite nanomaterials for improved reinforcement efficiency in cementitious paste. *Carbon, 63*, 175–186.

Peyvandi, A., Soroushian, P., Balachandra, A. M., & Sobolev, K. (2013). Enhancement of the durability characteristics of concrete nanocomposite pipes with modified graphite nanoplatelets. *Construction & Building Materials, 47*, 111–117.

Qian, K. L., Meng, T., Qian, X. Q., & Zhan, S. L. (2009). Research on some properties of fly ash concrete with nano-$CaCO_3$ middle slurry. *Key Engineering Materials, 405*, 186–190.

Quercia, G., Spiesz, P., Husken, G., & Brouwers, J. (2012). *Effects of amorphous nano-silica additions on mechanical and durability performance of SCC mixtures*. Paper presented at the International congress on durability of concrete, Trondheim, Norway.

Research, C. A. C. (2011). The Effects of ZnO_2 Nanoparticles on Strength Assessments and Water Permeability of Concrete in Different Curing Media. *Materials Research-Ibero-American Jouranl of Materials, 14*(2), 178–188.

Riahi, S., & Nazari, A. (2011a). Compressive strength and abrasion resistance of concrete containing SiO$_2$ and CuO nanoparticles in different curing media. *Science China-Technological Sciences, 54*(9), 2349–2357.

Riahi, S., & Nazari, A. (2011b). Physical, mechanical and thermal properties of concrete in different curing media containing ZnO$_2$ nanoparticles. *Energy and Building, 43*(8), 1977–1984.

Richardson, I. G. (1999). The nature of C-S-H in hardened cements. *Cement and Concrete Research, 29*(8), 1131–1147.

Richardson, I. G., & Groves, G. W. (1992). Microstructure and microanalysis of hardened cement pastes involving ground granulated blast-furnace slag. *Journal of Materials Science, 27*(22), 6204–6212.

Saafi, M. (2009). Wireless and embedded carbon nanotube networks for damage detection in concrete structures. *Nanotechnology, 20*(39), 395502.

Sáez De Ibarra, Y., Gaitero, J. J., Erkizia, E., & Campillo, I. (2006). Atomic force microscopy and nanoindentation of cement pastes with nanotube dispersions. *Physica Status Solidi (a), 203*(6), 1076-1081.

Said, A. M., Zeidan, M. S., Bassuoni, M. T., & Tian, Y. (2012). Properties of concrete incorporating nano-silica. *Construction & Building Materials, 36*, 838–844.

Scrivener, K. L., & Kirkpatrick, R. J. (2008). Innovation in use and research on cementitious material. *Cement and Concrete Research, 38*(2), 128–136.

Senff, L., Labrincha, J. A., Ferreira, V. M., Hotza, D., & Repette, W. L. (2009). Effect of nano-silica on rheology and fresh properties of cement pastes and mortars. *Construction & Building Materials, 23*(7), 2487–2491.

Shah, A. H., Sharma, U. K., Roy, D. A., & Bhargava, P. (2013). Spalling behaviour of nano SiO2 high strength concrete at elevated temperatures. In *Proceedings of MATEC Web of Conferences*. London, UK: EDP Sciences.

Shah, S. P., Konsta-Gdoutos, M. S., & Metaxa, Z. S. (2010). Exploration of fracture characteristics, nanoscale properties and nanostructure of cementitious matrices with carbon nanotubes and carbon nanofibers. In *Proceedings of the 7th international conference on fracture mechanics of concrete and concrete structures*. Seoul, Korea: Korea Concrete Institute.

Shaikh, F. U. A., & Supit, S. W. M. (2014). Mechanical and durability properties of high volume fly ash (HVFA) concrete containing calcium carbonate (CaCO$_3$) nanoparticles. *Construction & Building Materials, 70*, 309–321.

Shekari, A. H., & Razzaghi, M. S. (2011). Influence of Nano Particles on Durability and Mechanical Properties of High Performance Concrete. *Procedia Engineering, 14*, 3036–3041.

Shih, J. Y., Chang, T. P., & Hsiao, T. C. (2006). Effect of nanosilica on characterization of Portland cement composite. *Materials Science and Engineering A-structural Materials Properties Microstructure and Processing, 424*(1-2), 266–274.

Shon, H., Phuntsho, S., Okour, Y., Cho, D. L., Kim, K. S., Li, H. J., & Kim, J. H. et al. (2008). Visible light responsive titanium dioxide (TiO_2). *Journal of the Korean Industrial and Engineering Chemistry, 19*(1), 1–16.

Shukla, P., Bhatia, V., Gaur, V., Basniwal, R. K., Singh, B. K., & Jain, V. K. (2012). Multiwalled carbon nanotubes reinforced portland cement composites for smoke detection. *Solid State Phenomena, 185*, 21–24.

Sindu, B. S., Sasmal, S., & Gopinath, S. (2014). A multi-scale approach for evaluating the mechanical characteristics of carbon nanotube incorporated cementitious composites. *Construction & Building Materials, 50*, 317–327.

Singh, A. P., Gupta, B. K., Mishra, M., Chandra, A., Mathur, R. B., & Dhawan, S. K. (2013). Multiwalled carbon nanotube/cement composites with exceptional electromagnetic interference shielding properties. *Carbon, 56*(0), 86–96.

Singh, A. P., Mishra, M., Chandra, A., & Dhawan, S. K. (2011). Graphene oxide/ferrofluid/cement composites for electromagnetic interference shielding application. *Nanotechnology, 22*(46), 465701.

Sumer, M. (2012). Compressive strength and sulfate resistance properties of concretes containing Class F and Class C fly ashes. *Construction & Building Materials, 34*, 531–536.

Supit, S. W., Shaikh, F. U., Singh, L. P., & Ahalawat, S. K. B. S. (2014). Effect of nano-$CaCO_3$ on compressive strength development of high volume fly ash mortars and concretes. *Journal of Advanced Concrete Technology, 12*, 178–186.

Swamy, R. N., Ali, S. A., & Theodorakopoulos, D. D. (1983, September). Early strength fly ash concrete for structural applications. *ACI Journal Proceedings, 80*(5), 414–423.

Taplin, J. H. (1959). A method for following the hydration reaction in portland cement paste. *Australian Journal of Applied Science, 10*(3), 329–345.

Thomas, J. J., & Jennings, H. M. (2006). A colloidal interpretation of chemical aging of the C-S-H gel and its effects on the properties of cement paste. *Cement and Concrete Research, 36*(1), 30–38.

Thomas, M. D., & Bamforth, P. B. (1999). Modelling chloride diffusion in concrete: Effect of fly ash and slag. *Cement and Concrete Research, 29*(4), 487–495.

Wang, B., Wang, L., & Lai, F. C. (2008). Freezing resistance of HPC with nano-SiO_2. *Journal of Wuhan University of Technology-Mater, 23*(1), 85–88.

Wang, L., Ding, T., & Wang, P. (2009). Influence of carbon black concentration on piezoresistivity for carbon-black-filled silicone rubber composite. *Carbon, 47*(14), 3151–3157.

Wang, Y., Zhao, X., Du, J., & Lan, S. (2008). Study on improving mechanical property and pressure sensibility of cement-based composites with nano-sized carbon black. *New Building Materials, 35*(12), 6–9.

Wansom, S., Kidner, N. J., Woo, L. Y., & Mason, T. O. (2006). AC-impedance response of multi-walled carbon nanotube/cement composites. *Cement and Concrete Composites, 28*(6), 509–519.

Williamson, R. B. (1972). Solidification of Portland cement. *Progress in Materials Science, 15*(3), 189–286.

Xiao, H., Li, H., & Ou, J. (2011). Self-monitoring Properties of Concrete Columns with Embedded Cement-based Strain Sensors. *Journal of Intelligent Material Systems and Structures, 22*(2), 191–200.

Yakovlev, G., Kerienė, J., Gailius, A., & Girnienė, I. (2006). Cement based foam concrete reinforced by carbon nanotubes. *Materials Science, 12*(2), 147–151.

Yao, W., Zuo, J., & Wu, K. (2013). Microstructure and thermoelectric properties of carbon nanotube-carbon fiber/cement composites. *Journal of Functional Materials, 13*.

Ye, Q., Zhang, Z., Kong, D., & Chen, R. (2007). Influence of nano-SiO_2 addition on properties of hardened cement paste as compared with silica fume. *Construction & Building Materials, 21*(3SI), 539–545.

Yeganeh, J. K., Sadeghi, M., & Kourki, H. (2008). Recycled HIPS and nanoclay in improvement of cement mortar properties. *Malaysian Polym J, 3*(2), 32–38.

Yousefi, A., Allahverdi, A., & Hejazi, P. (2013). Effective dispersion of nano-TiO_2 powder for enhancement of photocatalytic properties in cement mixes. *Construction & Building Materials, 41*, 224–230.

Zhang, M., & Li, H. (2011). Pore structure and chloride permeability of concrete containing nano-particles for pavement. *Construction & Building Materials, 25*(2), 608–616.

Zhang, M. H., & Islam, J. (2012). Use of nano-silica to reduce setting time and increase early strength of concretes with high volumes of fly ash or slag. *Construction & Building Materials, 29*, 573–580.

Zhang, M. H., Islam, J., & Peethamparan, S. (2012). Use of nano-silica to increase early strength and reduce setting time of concretes with high volumes of slag. *Cement and Concrete Composites, 34*(5), 650–662.

Chapter 3
Trends in Nanoscopy in Materials Research:
Nano-Scale Microscopic Characterization of Cements

Ahmed Sharif
Bangladesh University of Engineering and Technology (BUET), Bangladesh

ABSTRACT

Nanotechnology has become one of the most emerging research areas to the researchers of the present world due to the wide application of nanomaterials including structures and buildings. With the rapid advancement of nanotechnology, manipulation and characterization of materials in nano scale have become an obvious part of construction related technology. This chapter will focus on some of the nano characterization techniques that are most frequently used in current research of nano materials. In particular scanning electron microscopy, transmission electron microscopy, atomic force microscopy, scanning tunneling microscopy, tomography, scanning transmission X-ray microscopy and laser scanning confocal microscopy are addressed. The basic principle of these characterization techniques and their limitations were briefly discussed in this chapter. In addition, a number of case studies related to microscopic characterization of nano materials utilizing the aforementioned techniques from the published literature were discussed.

DOI: 10.4018/978-1-5225-0344-6.ch003

INTRODUCTION

Nanotechnology, a multidisciplinary research area of the current science and technology which includes material science, biological science, organic chemistry, molecular biology, surface science, micro fabrication and so on (R. Saini, S. Saini, & Sharma, 2010). It is become essential to adopt new tactics of making things through understanding and control over the fundamental building blocks (i.e. atoms, molecules and nanostructures) of all physical things. Over the few decades many researchers have shown their research interests on nano materials and are investing their time, skill and knowledge with a view to providing a new dimension in modern science and technology for the betterment of the world. Nano particles are typically smaller than large biological molecules such as enzymes, receptors, and antibodies. Being hundred to ten thousand times smaller than human cell, nanoparticles can offer novel interactions with biomolecules both on the surface of and inside the cells which may revolutionize cancer diagnosis and treatment (Cai, Gao, Hong, & Sun, 2008). The well-studied nanoparticles include quantum dots (Cai, Hsu, Li, & Chen, 2007; Cai et al., 2006), carbon nanotubes (Liu et al., 2006), paramagnetic nanoparticles (Thorek, Chen, Czupryna, & Tsourkas, 2006), liposomes (Park, Benz, & Martin, 2004), gold nanoparticles (Huang, Jain, I. H. El-Sayed, & M. A. El-Sayed, 2007), and many others (Grodzinski, Silver, & Molnar, 2006).

Nanotechnology based application in construction related technology is still very insignificant. Following after evolving nanotechnology applications in biomedical and electronic industries, the construction industry recently started seeking out a way to advance conventional construction materials using a variety of nanomaterials. Several nanomaterials can enhance fundamental characteristics of construction materials such as strength, durability, and thermal properties and perform as key sensing components to monitor structural safety and health. Despite the current relatively high cost of nanostructured products, their use in construction materials is likely to increase because of imparting highly valuable properties at relatively low additive ratios in nanostructured materials.

A range of nanomaterials can be incorporated for favorable applications in construction that encompass superior structural properties, functional paints and coatings, and high-resolution sensing/actuating devices. Carbon nanotubes (CNTs) can extra ordinarily enhance mechanical properties by bonding concrete mixtures (i.e. cementitious agents and concrete aggregates), reduce their fragility, prevent crack propagation and as well as improve their thermal properties (Lee, Mahendra, & Alvarez, 2010). Nanoparticles (e.g. SiO_2, TiO_2 and Fe_2O_3) can penetrate nano- or micropores that developed during cement hydration of concrete (Becher, 1991). Addition of magnetic nickel nanoparticles during concrete formation increases the compressive strength by over 15% as the magnetic interaction enhances the mechanical properties of cement mortars. Copper nanoparticles (NPs) mitigate the surface roughness of steel to promote the weldability and render the steel surface corrosion-resistant (Becher, 1991).

However, a little commercial activity has started to emerge and some nano-based materials are now adopted by the construction industry. There seems to be a lack of consciousness of nanotechnology among construction professionals (Zhu, Bartos, & Porro, 2004). To fix this skepticism, integrated actions are required for focused R&D for appropriate knowledge transfer in construction industry. Even minor developments in materials and practices could bring large accumulated benefits. The maximum impact is expected to come from enhancement in performance of materials with much improved energy efficiency, sustainability and adaptability to changing environment with nanotechnological development (Campillo, Dolado, & Porro, 2004).

Nano particles show outstanding properties due to their large surface area compared to volume (as shown in Figure 1). The surface area imparts a serious change of surface energy and surface morphology. All these factors alter the basic properties and the chemical reactivity of the nanomaterials. The change in properties causes improved catalytic ability, tunable wavelength-sensing ability and better-designed pigments and paints with self-cleaning and self-healing features. In order to manipulate materials' property in nano scale the morphology, particle size, dispersion, chemical and physical characteristics of the constituent must be known through various characterization techniques (Sobolev & Gutiérrez, 2005). Characterization of nano materials poses enormous challenges to researchers. Thus nano characterization of materials has become a significant field of research now-a-days. Though research in synthesis of nano particles, manufacturing of nano materials viz. nano composites, nano medicine, nano bio implant and their characterization is going on, very few articles on nano characterization of materials are found.

In this book chapter I have chosen some frequently used characterization techniques of nano materials, briefly mentioned their basic principles and current research trends with these methods. Here I focused on various case studies published in the literature where the characterization techniques of our present interest were instrumented. In addition the findings of the researchers in these studies are discussed in this review.

Scanning Electron Microscopy (SEM)

For research concerning nano-materials, it is necessary to observe the morphology and composition of samples. At present, scanning electron microscopy (SEM) is heavily used in the pursuit of the further understanding of nano-materials, the combination of higher magnification, larger depth of field, greater resolution and

Figure 1. Particle-size and specific-surface-area scale related to concrete materials
© 2005 The American Ceramic Society. Used with permission.

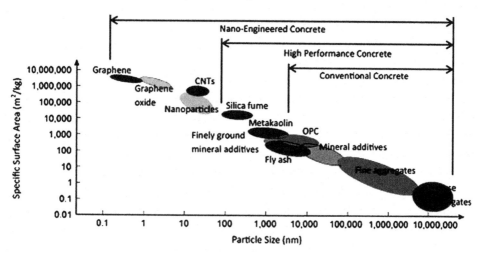

compositional and crystallographic information makes the SEM one of the most heavily used instruments in research areas and industries (Liu et al., 2013; Stevens et al., 2009). Main Applications of SEM

- **Topography:** The surface features of an object and its texture.
- **Morphology:** The shape and size of the particles making up the object.
- **Composition:** The elements and compounds that the object is composed of and the relative amounts of them.
- **Crystallographic Information:** How the grains are arranged in the object.

In SEM, electron beam is formed by an electron gun and speeded up by anode voltage downward to the sample target. The cathode and wehnelt electrode are drifting on a negative potential from kV to several tens kV, while anode is grounded. The grounded anode accelerates emitted electrons from cathode (thermoionicly or field emission). The wehnelt modulates the intensity of electron beam. Since the electron beam is a small geometrical volume in which are confined elementary charges of the same signs, the coulomb forces cause the divergence of electron beam. Therefore such a divergent electron beam has to be focused using electron optical lenses. In Figure 2, a magnetic focusing electron lens is located nearby the anode to focus the electron (Australian Microscopy and Microanalysis Research Facility [AMMRF], 2015). The condenser lens is used to adjust the spot size of the electron beam. In addition to these two condensers there is a magnetic lens that finally focuses the e-beam on specimen. This last magnetic lens is called objective.

The scanning generator powers x and y coils for both the electron beam probe and electron beam of cathode ray tube (CRT). These two scans are synchronized. The secondary electron signal collected by a detector is processed, amplified and supplied to the wehnelt electrode of CRT to modulate the intensity of electron beam for the image formation.

In cement-based building materials, nano-SiO_2 can behave as a nucleus to tightly bond with cement hydrates. Earlier the stable gel structures were formed and the mechanical properties of hardened cement paste were found to be improved when a smaller amount of nano-SiO_2 was added (Jo, Kim, Tae, & Park, 2007). It was found that nano-SiO_2 could improve the pressure-sensitive properties of cement mortar (Li, Xiao, & Ou, 2004a; Li, Xiao, Yuan, & Ou, 2004b). Fly ash concrete with nano-SiO_2 showed better density and strength (Ji, Huang, & Zheng, 2003). Another research indicated that high-strength concrete with nano-SiO_2 had higher flexural strength (Ye, Zhang, Kong, Chen, & Ma, 2003). It was also proved that

Figure 2. Schematic of an SEM
Courtesy: Australian Microscopy and Microanalysis Research Facility (AMMRF).

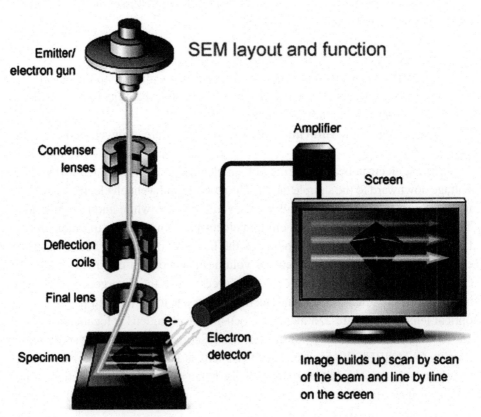

nano-SiO_2 improved the microstructure and water permeability resistant capacity of concrete. SEM imaging of the nano particles added concrete revealed that the texture of hydrate products (C–S–H and others) of nano-SiO_2 concrete was very dense and compact, and big crystals such as $Ca(OH)_2$ and Aft (i.e. alumina ferric oxide tri-sulfate) disappeared (Figure 3) (Ji, 2005).

Carbon nanotubes (CNTs) are one of the strongest materials that have ever been found. The superior mechanical, thermal and electrical properties of CNTs have been studied, and CNTs have been extensively utilized in the production of a wide range of composite materials (Ji, 2005; Makar & Chan, 2009; Chen et al., 2011; Koo et al., 2014). CNTs show promising potential as ordinary Portland cement (OPC) reinforcements since CNTs are possibly the strongest and smallest reinforcing fibers that can be incorporated in OPC. Dispersion of CNTs is difficult within most aqueous media because of the highly attractive van der Waals forces between the CNT particles that tend to cause agglomeration (Makar & Chan, 2009). A better dispersion of the CNTs may result in the higher interfacial contact area between the CNTs and the matrix as well as more evenly distributed stresses in the composites. Therefore, it is essential to integrate dispersion techniques into the fabrication of the CNT–cement composites for more effective use of CNTs as reinforcements. Chen et al. (2011) characterized the surface morphologies of concrete/CNTs by SEM. In a recent response to the issue of workability and dispersant compatibility, the dispersion of CNT and workability of CNT–OPC composites using several types of cement-compatible dispersants was investigated. Subsequently, the dispersion effect was characterized by SEM as shown in Figure 4 (Chen et al., 2011). The polycarboxylate provided reasonably uniform dispersion of CNTs within the paste (Figure 4a). White arrows indicate the locations of CNTs dispersed within the sample in Figure 4a.

Figure 3. Microstructure of nano-SiO_2 concrete at curing age of (a) 28 days and (b) 180 days
© 2005 Elsevier Ltd. Used with permission.

Figure 4. SEM images showing the effects of admixtures on dispersion of CNTs; (i)–(ii) represent no admixture and polycarboxylate, respectively.
© 2011 Taylor & Francis. Used with permission.

The use of multipoles in SEM has significantly improved lens performance through the reduction of third-order spherical aberrations and such devices are now commercially available as aberration correctors (Rose, 1990; Haider, Braunshausen, & Schwan, 1995; Krivanek, Dellby, Lupini, 1999). In addition, higher-order aberration correctors are developing as a topic of important research (Sawada et al., 2009; Krivanek, Nellist, Dellby, Murfitt, & Szilagyi, 2003). The development of monochromators for the electron-source was already reported for the purpose of narrowing the energy spread of source electrons, a major contributing factor to chromatic aberration, which itself is a limiting factor for SEM resolution and low landing energies. In the last decade there has been a significant dive forward in the utility of SEM, which is now labeled as high-resolution SEM (HRSEM). Improvement of objective lens; precise control of the landing energy; dramatic reduction in charging (the main contributor to loss of information in observation of insulating materials); retaining a suitably small probe size; and obtaining selective information by tuning proper electron energy ranges make promising possibilities of the resolving-power of low voltage HRSEM in the near future (Mackie et al., 1992).

Transmission Electron Microscopy (TEM)

Transmission agrees microscopy (TEM) permits a qualitative evaluation of the internal structure and spatial distribution of the various phases with fine details. In scientific community TEM is now being widely used by the researchers for its outstanding performance in high resolution imaging capability. This enables the instrument's user to direct visualization—even as small as a single column of atoms, which is thousands of times smaller than the smallest resolvable object in a light microscope. With the development of TEM, the associated technique of scanning transmission electron microscopy (STEM) was re-investigated with development of the field emission gun and adding a high quality objective lens. Using this design, Crewe et al. (1970) demonstrated the ability to image atoms using annular dark-field imaging. Crewe et al. (1970) at the University of Chicago built a STEM able to visualize single heavy atoms on thin carbon substrates. Later, Meyer et al. (2008) described the direct visualization of light atoms such as carbon and even hydrogen using TEM and a clean single-layer graphene substrate.

TEM consists of an assembly of electromagnetic lenses arranged in a vertical column with an electron source and fluorescent screen at opposing end of the column. In most cases, the microscope has electron source (gun) at top of the column as depicted in Figure 5. Electrons produced at the gun are accelerated down the column at a vacuum level of 10^{-6} Torr towards the screen by a high voltage, typically between 100 and 300 kV. These electrons are accumulated and converged by electromagnetic lenses located immediately below the electron source. Varying the lenses current and the aperture size, the electron beam diameter can be altered from about 1 mm to 2 nm or less at the specimen surface. The specimen to be considered is fixed in a special holder and is inserted into the evacuated column below the condenser lenses (see Figure 5) (Department of Physics, The Chinese University of Hong Kong, 2015). The electrons emerging from the lower side of the specimen are focused by the objective lens, which surrounds the specimen. In the lower focal plane of this lens, a diffraction pattern of the specimen is formed. At a position below the lower focal plane, an inverted image of the specimen is created. The sole purpose of the subsequent post-specimen lenses is to magnify one of these two plane images. By varying the currents in these later lenses, either a magnified diffraction pattern or a magnified image of the specimen can be obtained on the fluorescent screen. A camera, normally located under moveable fluorescent screen, is used for making a permanent record of the diffraction pattern or image. The final image magnification can be varied by changing the currents of the post - specimen lenses from typical 10^2 to 10^6 times.

TEM has now become a part and parcel in nano scale characterization of materials in the field of cancer research, virology, materials science as well as pollution, nanotechnology, semiconductor and construction materials research. A few studies carried out by some researchers and their findings will be discussed.

Nanomaterials are being considered for various uses in the construction and related infrastructure industries. To achieve environmentally responsible nanotechnology in construction, it is important to consider the lifecycle impacts of nanomaterials on the health of construction workers and dwellers, as well as unintended environmental effects at all stages of manufacturing, construction, use, demolition, and disposal. The size distributions of the nanomaterials can be visualized with TEM. Commercially available colloidal nano-SiO_2 (CNS) with a mean particle size of 20 nm and a

Figure 5. The schematic outline of a TEM
Courtesy: Department of Physics, The Chinese University of Hong Kong.

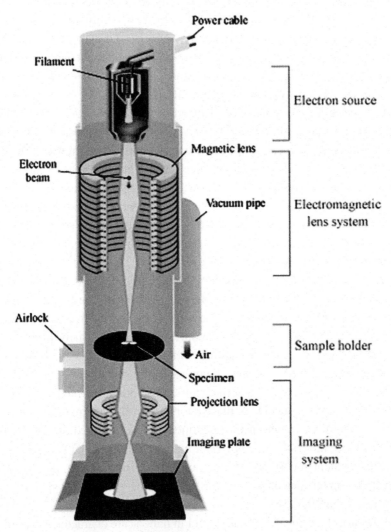

solids content of 30% was used in a research (Hou et al., 2015). Morphologic graphs shown in Figure 6 indicate that CNS particles are generally round in shape. Although most of the particles are well-dispersed, agglomerations are found. The dark spot in the TEM image indicated the presence of seriously agglomerated clusters in CNS.

Endurance of concrete depends strongly on its pore structure, especially porosity and pore size distribution. These features determine the transport properties of the harmful agents (water, ions and gases) that may cause concrete deterioration by physical and/or chemical attack (Scarfato, Di Maio, Fariello, Russo, & Incarnato,

Figure 6. TEM image of (a) colloidal nano-SiO$_2$ (CNS) and (b) higher magnification view
© 2014 Elsevier Ltd. Used with permission.

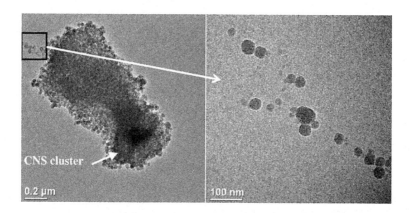

2012). Regarding this issue, a new class of polymeric nanocomposites based surface treatments based on was recently investigated by Scarfato et al. (2012). The study was carried out on hybrid organic–inorganic systems prepared by solvent intercalation of a modified montmorillonite into commercial resins. The results demonstrated that the nanoclay addition had significantly upgraded the protection quality of the resins. TEM assessment was carried out on the hybrid films to have a better demonstration of the hierarchical structure of the hybrid nanocomposites. As in Figure 7, TEM image of the nanocomposite system shows the clay particles homogenously distributed into the resin matrix and really dispersed on a nano-scale; a mixed exfoliated/intercalated morphology was obtained, as demonstrated by the presence of both completely delaminated and individually dispersed silicate sheets and intercalated structures with small stack dimensions (<15–20 nm).

Transmission electron microscopy (TEM) was a technique that was used to identify crystalline phases and morphological and structural characteristics (Chaunsali & Peethamparan, 2013). The TEM nano graphs in bright field mode and selected area of electron diffraction patterns (SAEDP) for cement paste at age of 90 days revealed specific phases (i.e. hexagonal portlandite, $Ca(OH)_2$ crystalline cubes and aragonite, $CaCO_3$).

The goal of forming smaller electron probes is not achievable solely by improving the aberration-corrected optics. The brightness of the electron source and the geometrical source size are crucial parameters that ultimately limit the size of the electron probe. The current high-resolution electron microscopy (HREM) and Z-contrast techniques are very powerful for resolving the arrangement of columns of atoms in end-on images parallel to the beam. The brightness of the electron source

Figure 7. TEM image of the nanocomposite (i.e. montmorillonite incorporated resin film): A: delaminated silicate sheets; B: intercalated silicate structures
© 2012 Elsevier Ltd. Used with permission.

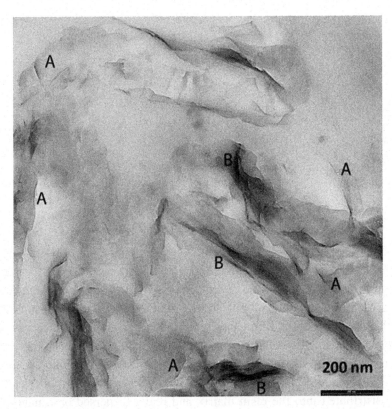

and the geometrical source size are crucial parameters that ultimately limit the size of the electron probe. The techniques are therefore well suited to study the interfaces with nanomaterials and defects, because the images can distinguish between different atomic species.

Atomic Force Microscopy (AFM)

The atomic force microscope (AFM) is a scanning probe imaging instrument that provides spatial resolution at a picometer (depending on specimen type and surface roughness) with a sharpened probe of less than 10 nm in diameter typically (Meyer, 1992). The AFM is one of the leading tools for imaging, measuring, and manipulating matter at the nanoscale. The data is accumulated by "sensing" the surface with a mechanical probe. Tiny piezoelectric elements facilitate the precise and accurate movements on (electronic) command enable the very specific scanning. According to the nature of the tip motion AFM process is usually described as one of three modes named contact mode/static mode, tapping mode/intermittent contact/vibrating mode and non-contact mode.

In contact –AFM mode, also known as repulsive mode, an AFM tip makes soft "physical contact" with the sample. The tip is attached to the end of a cantilever with a low spring constant, lower than the effective spring constant holding the atoms of sample together. As the scanner gently traces the tip across the sample (or the sample under tip), the contact force causes the cantilever to bend to accommodate changes in topography.

Non-contact AFM (NC-AFM) is one of several vibrating cantilever techniques in which an AFM cantilever is vibrated near the surf ace of a sample. The spacing between the tip and the sample for NCAFM is on the order of tens to hundreds of angstroms. NC-AFM is desirable because it provides a means for measuring sample topography with little or no contact between the tip and the sample. Like contact AFM, non- contact AFM can be used to measure the topography of insulators and semiconductors as well as electrical conductors. The total force between the tip and sample in non -contact regime is very low, generally about 10^{-12} N. This force is advantageous for studying soft or elastic samples. A further advantage is that samples like silicon wafers are not contaminated through contact with the tip.

Intermittent –contact AFM (IC-AFM) is similar to NC-AFM, except that for IC-AFM the vibrating cantilever tip is brought closer to the sample so that at the bottom of its travel is just barely hits, or "taps", the sample. The IC-AFM operating region is indicated on the Van der Waals curve in Figure 8 (Department of Physics, University of California at Santa Barbara, 2015). As for NC-AFM, for IC-AFM the cantilever's oscillation amplitude changes in response to sample spacing. An image representing surface topography is obtained by monitoring these changes.

To investigate the use of hybrid recycled powder as a supplementary cementing material, the pozzolanic property of hybrid powder was comprehensively studied by utilizing advanced tools including SEM, AFM, LPS and XRD (Liu, Tong, Liu, Yang, & Yu, 2014). Recycled powder of different concrete–clay brick ratios was first studied with a focus on its fineness, loss on ignition, strength activity index and water requirement. Then, the microstructure characteristics and chemical composition of the hybrid recycled powder, obtained directly from a dust collection system, were qualitatively and quantitatively probed. To develop a deeper understanding of the activity mechanism of hybrid powder, nanoscale characterization was employed to scan and analyze the microstructure of cement paste supplemented with hybrid recycled powder. It was found that the activity mechanism of hybrid powder was strongly correlated with its unique microstructure morphology and chemical composition. In order to study the effect of hybrid recycled powder on cement paste by AFM, a representative region containing the hybrid recycled

Figure 8. The schematic outline of an atomic force microscopy (AFM) showing its deferent operation mode
Courtesy: Department of Physics, University of California at Santa Barbara, USA.

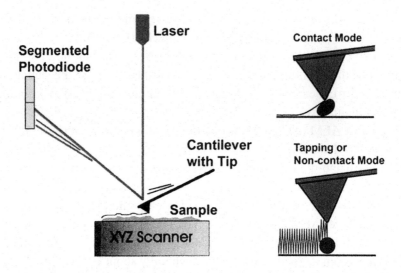

particles must be identified for the nano-scale probe. This was accomplished by searching for the clay brick particles in the hybrid powder, which displayed a color very different from other constituents in the mortar. Considering that finer concrete powder prefered to be attached on coarser clay brick particles, it was more appropriate to treat them as hybrid particles here than just as pure clay brick particles. In this study, a representative region containing 1 hybrid particle and 1 fine aggregate (sand), which was denoted as Region I, was selected (Figure 9).

AFM has become a vital tool for studying the micro- and nano worlds. AFM defies most other surface instrumentation in ease of use, sensitivity and versatility. An encouraging way is to combine the AFM with two or three complementary techniques in one apparatus. In this way morphological, chemical, mechanical and kinetic information can be generated simultaneously in one experiment.

Scanning Tunneling Microscopy (STM)

A scanning tunneling microscope (STM) is an instrument for imaging surfaces at the atomic level. For an STM, good resolution is considered to be 100 pm lateral resolution and 10 pm depth resolution through which individual atoms within materials can be routinely imaged and manipulated (Bai, 2000). The process of scanning tunneling microscopy is based on the concept of quantum tunneling. In this technique a fine probe tip is scanned over the conducting surface of the material to

Figure 9. AFM image of Region I
© 2014 Elsevier Ltd. Used with permission.

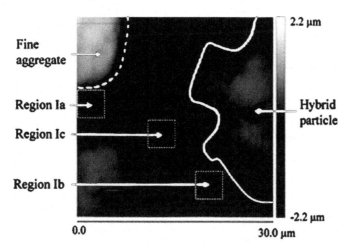

be examined with the help of a piezoelectric crystal at a distance of 0.5-1 nm and the resulting current or the position of the tip required to maintain a constant tunneling current is monitored as shown in Figure 10 (Institute of Experimental and Applied Physics, University of Kiel, 2015). When a conducting tip is brought very near to the surface to be examined, a bias (voltage difference) applied between the two can allow electrons to tunnel through the vacuum between them. The resulting tunneling current is a function of tip position, applied voltage, and the local density of states (LDOS) of the sample (Chen, 1993). Information acquired by monitoring the current as the tip's position scans across the surface is usually displayed in image form. The use of STM in nano characterization is very much advantageous as it can be used not only in ultra-high vacuum but also in air, water, and various other liquid or gas ambient, and at temperatures ranging from near zero Kelvin to a few hundred degrees Celsius (Chen, 1993).

Recently, Alkhateb et al. established a framework for evaluating the performance of cementitious materials reinforced with functionalized graphene nanoplatelets (2013). In their research graphene cementitious nanocomposites showed promising results in improving the overall properties of cementitious nanocomposites. Scanning tunneling microscopy (STM) provided a probe for the morphology and electrical properties at atomic resolution in all three dimensions. Atomically resolved STM topographs of graphene was reported (Ishigami, Chen, Cullen, Fuhrer, & Williams, 2007; Zhang et al., 2008; Deshpande, Bao, Miao, Lau, & LeRoy, 2009) from which the height of graphene ripples was determined to be 3-5 Å. Xu, et al. reported on the scanning tunneling microscopy study of a new class of corrugations in exfoli-

Figure 10. Principle of scanning tunneling microscopy
Courtesy: Institute of Experimental and Applied Physics, University of Kiel.

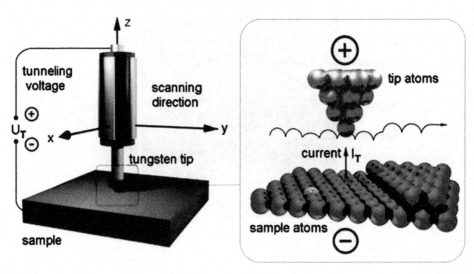

ated monolayer graphene sheets with wrinkles ~10 nm in width and ~3 nm in height (2009). Wrinkle-like structures was seen before in TEM images of suspended graphene sheets (Ferrari et al., 2006) and in STM and high resolution SEM images of graphene grown on conducting substrates (Biedermann, Bolen, Capano, Zemlyanov, & Reifenberger, 2009; Li et al., 2009), however their structure and properties was carefully characterized. Through scanning tunneling spectroscopy (STS), they further demonstrated that the wrinkles had lower electrical conductance when compared to other regions of grapheme and were characterized by the presence of mid-gap states, which was in agreement with theoretical predictions.

It was surprising that atomically resolved topographs (Figure 11a and b) revealed very different structures for the wrinkles in comparison with other parts of graphene. A triangular pattern was observed over the entire graphene wrinkle (Figure 11a and b) and the distance between adjacent bright spots was ~2.5 Å. In comparison, the topograph taken on the same graphene sheet adjacent to the wrinkle (Figure 11c) revealed the honeycomb structure that was consistently observed on the "flat" (by "flat" meant only the ~4 Å ripples were present) parts of the monolayer graphene sheets investigated in their study.

Although the theoretical methods used to simulate STM can vary widely, the most important simplifications include the inability to handle truly the real structure of the tip. Even though hypothetical analysis has proved to be very useful in determining the tip sample separation in STM operation, it still remains challenging. Tip

Figure 11. STM topographs of graphene wrinkles: (a) atomically resolved topograph obtained on the top of the wrinkle; (b) a close-up of the observed "three-for-six" triangular pattern on the wrinkle, corresponding to the green square in (a)
© 2009 American Chemical Society. Used with permission.

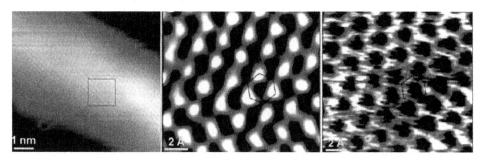

and surface atomic relaxation have proved to be so crucial to imaging that they are increasingly included for the distance dependent contrast observed in STM.

Focused Ion Beam (FIB)/SEM Tomography

Full understanding of nano materials' behavior, control in manufacturing process and monitoring while in service need a characterization technique providing a three dimensional vision. A Characterization technique would be highly advantageous when it reveals the richness of microstructure in three dimensions rather than just providing two dimensional projections. Various techniques e.g. X ray tomography and mechanical sectioning for three dimensional reconstruction of microstructures have been carried out in the field of materials science for decades. Tomography in a wider meaning, which includes all techniques that use sectional views (not only projections) as an intermediate step from which the three-dimensional object is re-assembled (Bushby et al., 2011).

Focused Ion Beam Scanning electron microscopy (FIB/SEM) is a tomography technique where a focused gallium ion beam used to sequentially mill away the sample surface to expose newly generated fresh surface, and a backscattered electron (BSE) detector, used to image the milled surfaces, generates a large series of images that can be reconstructed into a 3D rendered image of samples under investigation Bushby et al., 2011). Nanometer scale milling with simultaneous imaging is continued alternately with multiple repetitions to acquire an entire three-dimensional stack of fresh surface images. Finally 3D exclamation and organization of the fresh surface images recorded previously is obtained with distinct software to achieve detail information of a selected volume of material (Uchic, Holzer, Inkson, Principe, & Munroe, 2007; Sakamoto, Cheng, Takahashi, Owari, & Nihei, 1998; Tomiyasu,

Fukuju, Komatsubara, Owari, & Nihei, 1998; Dunn & Hull, 1999; Inkson, Steer, Möbus, & Wagner, 2001). Its resolution depends on the accuracy of individual FIB incisions and as well as on the SEM resolution. Back scatter electrons (BSE) are employed to improve the identification of phases and their 3D-reconstruction (Schaffer et al., 2007). Volumes up to $(10–30)^3$ μm^3 can be three dimensionally reconstructed with FIB/SEM with voxel dimensions approaching tens of nanometers (Uchic et al., 2007). After preparation of the sample FIB tomography was realized by a series of moves in iteration as shown in Figure 12. At first, the uppermost surface was protected by a sputter-resistant metal coating (e.g. Pt) to attain well-defined edges. And then a trench was incised to a definite depth, which liberated up a adequate cross-sectional area to view. The first image was then obtained and the 'internal microstructure' of the material was revealed. Subsequently, cutting and portraying of a fixed volume was interchanged to gain an entire three-dimensional pile of fresh topographic images. Specialist software was employed to achieve a continuous 3D volume of material with three-dimensional interpolation and collation (Inkson et al., 2001).

An example is in visualizing nano- and microstructures of particle granularity in cement (Figure 13) (Holzer, Muench, Wegmann, Gasser, & Flatt, 2006; Münch, Gasser, Holzer, & Flatt, 2006). After compaction, the particle morphology and statistical distribution was evaluated from the three-dimensional FIB stack. Grains from an ordinary Portland cement (OPC) sample appeared dark and were impregnated with a metal alloy (bright) after compacting the dry powder in order to improve image contrast and assist segmentation. Contrast enhancement by filling pores with high-Z metal was employed successfully in this particular case.

More recently, FIB tomography has been combined with Energy Dispersive Spectroscopy analyzes (EDS), and three dimensional information of the sample chemistry was obtained (SEM-EDS/FIB) (F. Lasagni, A. Lasagni, Holzapfel, Mücklich, & Degischer, 2006). Another well-established technique for the study of the crystalline structure of materials includes Electron backscatter diffraction (EBSD). The quantitative description of the grains(i.e. grain size, phases, texture, subgrains, pores) in three dimensions marks SEM-EBSD/FIB a very powerful technique for 3D-characterization of materials (Zaafarani, Raabe, Singh, Roters, & Zaefferer, 2006).

Scanning Transmission X ray Microscopy (STXM)

Scanning transmission X ray Microscope (STXM) is a very versatile microscope which in its simplest form can provide chemically specific information on nm-size areas of a thin sample. The principle of STXM is very simple one. The fundamental principle utilizes a coherent, monochromatic beam of X ray focused on the sample

Figure 12. Principle of FIB tomography: (a) specimen setup with Pt protection layer on the region of interest; (b) iterative three-dimensional sectioning and SEM imaging in dual-beam instruments; and (c) optional step for scanning ion microscopy imaging after specimen tilt, θ
© 2001 John Wiley & Sons, Inc. Used with permission.

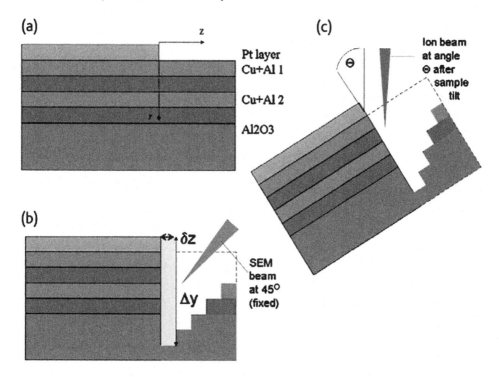

Figure 13. FIB tomography applied to a cement particulate system: (a) raw data after alignment and preprocessing; and (b) semitransparent visualization, where each particle is labeled with a specific color value; Cube size: 23.5 x 19.2 x 9.5 μm
© 2006 John Wiley & Sons, Inc. Used with permission.

by a Fresnel zone plate (ZP). The sample is scanned by the focused beam in the focal plane of the zone plate, then the X ray signal transmitted through the sample is monitored and thus an image of the sample is obtained as shown in Figure 14 (Tyliszczak & Chou, 2015). Only selected elements will absorb X ray. The image difference is formed by the absorbance of an X ray photon containing a specific energy by the respective elements. In general, STXM refers to the microscopes working with beams in soft X rays range. Although there exist some other methods in focusing X ray beams, zone plate has been proved to be the most useful and to give smallest focal spot. STXM being a microscope finds its primary application in X-ray absorption spectroscopy (XAS) which is a powerful method in characterization of chemical state, especially of light elements such as C, N, O etc. (Monteiro et al., 2013). Currently, zone plates with 20 nm outer zone width for STXM and 15 nm for full field TXM were state of art (Tyliszczak & Chou, 2015).

In a study of synchrotron STXM by scanning the sample of hydrated cement pastes (made with a blend of Portland cement and Calera carbonates) in the x–y direction or x direction of selected sample areas at energy increments of 0.1–0.25 eV (Figure 15), various sorts of images was captured systematically (Monteiro et al., 2013). The stack images provided two-dimensional resolution with a NEXAFS spectrum for a specific element on each pixel of the image. In this case, the best attainable spatial resolution was 25 nm (Monteiro et al., 2013).

Radiation damage of the sample is one of the most important aspects of X ray microscopy (Beetz & Jacobsen, 2003), where the scanning x ray microscope causes

Figure 14. Schematic of the optical elements used in a STXM equipped with an annular quadrant detector (AQD)
Courtesy: Lawrence Berkeley National Laboratory.

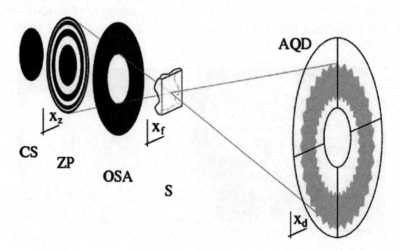

Figure 15. STXM image of selected areas taken at Ca LII, III-edge for NEXAFS analysis
© 2013 Elsevier Ltd. Used with permission.

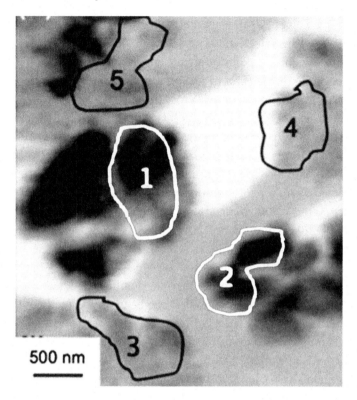

minimum damage among all X ray microscopy techniques (Beetz & Jacobsen, 2003). However, there are some measures to be taken to minimize radiation damage such as using a fast photon shutter which expose the samples only during data collection, using a highly efficient photon detector (Coffey, Urquhart and Ade, 2002). To utilize phase or other type of contrast in STXM imaging a multi-segment detector (Tyliszczak et al., 2004) or high speed, low resolution CCD camera with a scintillator (Feser et al., 2006) are developed recently.

Laser Scanning Confocal Microscopy (LSCM)

The technique of laser scanning with rotating disk confocal fluorescence microscopy is an essential means in materials science due to its special contrast approach for offering features that are not readily available using other contrast modes (Minsky,

1988; Yoo et al., 2006; Brakenhoff, Voort, Spronsen, & Nanninga, 1989; Carlsson et al., 1985; Claxton, Fellers, & Davidson, 2006). The confocal principle in laser scanning microscope is schematically presented in Figure 16 (Bio-imaging Unit, Newcastle Biomedicine, 2015). Coherent light emitted by the laser system (excitation source) passes through a pinhole aperture that is situated in a conjugate plane (confocal) with a scanning point on the specimen and a second pinhole aperture positioned in front of the detector (a photomultiplier tube). As the laser is reflected by a dichromatic mirror and scanned across the specimen in a defined focal plane, secondary fluorescence emitted from points on the specimen (in the same focal plane) pass back through the dichromatic mirror and are focused as a confocal point at the detector pinhole aperture. In this form of confocal microscopy, excitation light (laser) is focused onto a single point of the specimen and light returning from that point is collected using a photo multiplier tube (PMT). The point of illumination is scanned across the specimen in a raster fashion (rectangular pattern of scanning line by line) to build up the final image.

Coherent light emitted by the laser system scans across the specimen and secondary fluorescence emitted from points on the specimen pass back through the dichromatic mirror and is focused at the photomultiplier detector. The resulting image of secondary fluorescence emission can be viewed directly in the eyepieces or projected onto the surface of an electronic array detector or traditional film plane.

Concrete Durability Group of Imperial College London used LSCM for imaging pores and voids in hardened mortar and concrete as shown in Figure 17 (Head, Wong, & Buenfeld, 2006). In their study hollow shell hydration grains was imaged to observe the different forms of capillary pores between hydration grains and hardened cement paste (HCP).Several connecting pore channels (labeled 'cc' in Figure 17b and c) can be observed between different grains and capillary pores facilitating the transportation of fluids and ions. It was also reported that hollow shell cement grains, not visible on the surface, could be observed below specimen surfaces due to the semi-transparency of C–S–H and other reaction products. The laser probe with sufficient power could penetrate several micrometers of overlying material. Several grains were identified that were not observed to lie in the vicinity of micro-cracks and/or air voids (Head et al., 2006).

Recent advances in fluorophore design have led to improved synthetic and naturally occurring molecular probes which exhibit a high level of photostability and target specificity. Continued advances in fluorophore design, dual laser scanning and multispectral imaging will also be important in the coming years and will dramatically improve the collection and analysis of data obtained from complex experiments in construction related materials imaging.

Figure 16. Schematic diagram of the optical pathway and principal components in a laser scanning confocal microscope
Courtesy: Bio-imaging Unit, Newcastle Biomedicine, Newcastle University.

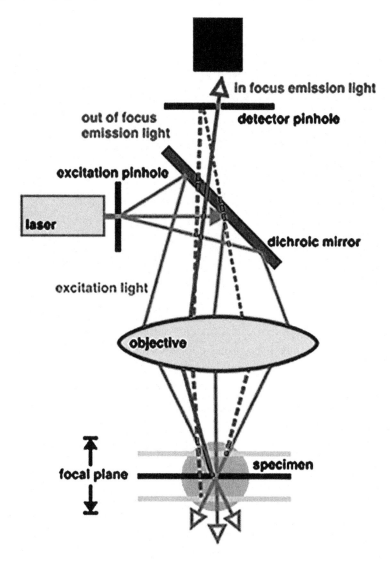

CONCLUSION

The characterization methods must provide structural information at all appropriate scales to describe where every atom starts, where each ends up and how and why they get there in a transformation processes. The further shrinkage of the dimensions

Figure 17. Series of images showing very fine connecting channels (cc) between capillary pores 'p' and hollow shell cement grains 'h' below specimen surface (0.6 w / c, 28 days): (a) top image in stack; (b) area 1 in (a) (3.08 μm below surface); (c) area 2 in (a) (1.76 μm below surface))
© 2006 Elsevier Ltd. Used with permission.

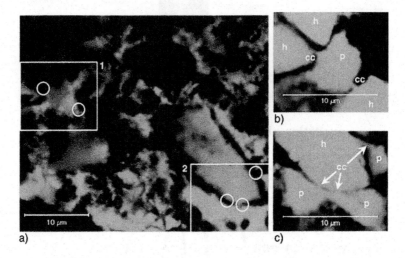

and the introduction of new materials concepts demand for ever better resolution and stability. Nanoscopic analysis is essential for the characterization of nanomaterials added structures. This chapter discusses selected case studies in nanoscale microscopic observation of cement based materials with the help of various advanced characterization techniques. Recent progress in the performance of SEM helped to observe directly fine surface structures even those that are electrical insulators, especially in low energy imaging. Except the difficulties in sample preparation TEM provides greater advantage in materials characterization allowing nanometer level resolution. A huge scope to work with both the AFM and STM to characterize the nanostructures including detailed morphology characteristics including the shape and size of features at the nanoscale of harden cement paste, mortar and concrete is available. Furthermore, three-dimensional imaging becomes a crucial property of the nanoscopic methods and needs further development in the future. However, the instrumentation and above all understanding and interpreting the results require sufficient skill and expertise of the respective users. The potential properties of nanomaterials have not been fully brought out in the construction industries. This is mainly due to the insufficient control of interfaces and edges of the nanomaterials with the conventional materials. Considering the recent rapid advances in technology, the control of interfaces and edges is expected to be possible in the near future.

ACKNOWLEDGMENT

The authors would like to thank the Bangladesh University of Engineering and Technology (BUET) in funding the present research.

REFERENCES

Alkhateb, H., Al-Ostaz, A., Cheng, A. H. D., & Li, X. (2013). Materials Genome for Graphene-Cement Nanocomposites. *Journal of Nanomechanics and Micromechanics*, *3*(3), 67–77. doi:10.1061/(ASCE)NM.2153-5477.0000055

Australian Microscopy and Microanalysis Research Facility (AMMRF). (n.d.). Retrieved from http://www.ammrf.org.au/myscope/sem/practice/principles/layout.php

Bai, C. (2000). *Scanning tunneling microscopy and its application* (Vol. 32). New York: Springer-Verlag Berlin Heidelberg.

Becher, P. F. (1991). Microstructural design of toughened ceramics. *Journal of the American Ceramic Society*, *74*(2), 255–269. doi:10.1111/j.1151-2916.1991.tb06872.x

Beetz, T., & Jacobsen, C. (2003). Soft X-ray radiation-damage studies in PMMA using a cryo-STXM. *Synchrotron Radiation*, *10*(3), 280–283. doi:10.1107/S0909049503003261 PMID:12714762

Biedermann, L. B., Bolen, M. L., Capano, M. A., Zemlyanov, D., & Reifenberger, R. G. (2009). Insights into few-layer epitaxial graphene growth on 4 H-SiC (000 1⁻) substrates from STM studies. *Physical Review B: Condensed Matter and Materials Physics*, *79*(12), 125411. doi:10.1103/PhysRevB.79.125411

Bio-imaging Unit. (n.d.). Newcastle Biomedicine, Newcastle University. Retrieved from http://www.ncl.ac.uk/bioimaging/techniques/confocal/

Brakenhoff, G. J., Voort, H. T. M., Spronsen, E. A., & Nanninga, N. (1989). Three-dimensional imaging in fluorescence by confocal scanning microscopy. *Journal of Microscopy*, *153*(2), 151–159. doi:10.1111/j.1365-2818.1989.tb00555.x PMID:2651673

Bushby, A. J., P'ng, K. M., Young, R. D., Pinali, C., Knupp, C., & Quantock, A. J. (2011). Imaging three-dimensional tissue architectures by focused ion beam scanning electron microscopy. *Nature Protocols*, *6*(6), 845–858. doi:10.1038/nprot.2011.332 PMID:21637203

Cai, W., Gao, T., Hong, H., & Sun, J. (2008). Applications of gold nanoparticles in cancer nanotechnology. *Nanotechnology, Science and Applications, 1*, 17. PMID:24198458

Cai, W., Hsu, A. R., Li, Z. B., & Chen, X. (2007). Are quantum dots ready for in vivo imaging in human subjects? *Nanoscale Research Letters, 2*(6), 265–281. doi:10.1007/s11671-007-9061-9 PMID:21394238

Cai, W., Shin, D. W., Chen, K., Gheysens, O., Cao, Q., Wang, S. X., & Chen, X. et al. (2006). Peptide-labeled near-infrared quantum dots for imaging tumor vasculature in living subjects. *Nano Letters, 6*(4), 669–676. doi:10.1021/nl052405t PMID:16608262

Campillo, I., Dolado, J. S., & Porro, A. (2004). High-performance nanostructured materials for construction. In Nanotechnology in Construction. Royal Society of Chemistry.

Carlsson, K., Danielsson, P. E., Liljeborg, A., Majlöf, L., Lenz, R., & Åslund, N. (1985). Three-dimensional microscopy using a confocal laser scanning microscope. *Optics Letters, 10*(2), 53–55. doi:10.1364/OL.10.000053 PMID:19724343

Chaunsali, P., & Peethamparan, S. (2013). Influence of the composition of cement kiln dust on its interaction with fly ash and slag. *Cement and Concrete Research, 54*, 106–113. doi:10.1016/j.cemconres.2013.09.001

Chen, C. J. (1993). *Introduction to scanning tunneling microscopy* (Vol. 227). New York: Oxford University Press.

Chen, S. J., Collins, F. G., Macleod, A. J. N., Pan, Z., Duan, W. H., & Wang, C. M. (2011). Carbon nanotube–cement composites: A retrospect. *The IES Journal Part A: Civil & Structural Engineering, 4*(4), 254–265.

Claxton, N. S., Fellers, T. J., & Davidson, M. W. (2006). Laser scanning confocal microscopy. Tallahassee, FL: Department of Optical Microscopy and Digital Imaging, Florida State University. Retrieved from http://www.olympusconfocal.com/theory/LSCMIntro.pdf

Coffey, T., Urquhart, S. G., & Ade, H. (2002). Characterization of the effects of soft X-ray irradiation on polymers. *Journal of Electron Spectroscopy and Related Phenomena, 122*(1), 65–78. doi:10.1016/S0368-2048(01)00342-5

Coleman, J. N., Khan, U., Blau, W. J., & Gun'ko, Y. K. (2006). Small but strong: A review of the mechanical properties of carbon nanotube–polymer composites. *Carbon, 44*(9), 1624–1652. doi:10.1016/j.carbon.2006.02.038

Crewe, A. V., Wall, J., & Langmore, J. (1970). Visibility of single atoms. *Science, 168*(3937), 1338–1340. doi:10.1126/science.168.3937.1338 PMID:17731040

Department of Physics, The Chinese University of Hong Kong. (n.d.). Retrieved from http://www.hk-phy.org/atomic_world/tem/tem02_e.html

Department of Physics, University of California at Santa Barbara. (n.d.). Retrieved from http://web.physics.ucsb.edu/~hhansma/biomolecules.htm

Deshpande, A., Bao, W., Miao, F., Lau, C. N., & LeRoy, B. J. (2009). Spatially resolved spectroscopy of monolayer graphene on SiO_2. *Physical Review B: Condensed Matter and Materials Physics, 79*(20), 205411. doi:10.1103/PhysRevB.79.205411

Dunn, D. N., & Hull, R. (1999). Reconstruction of three-dimensional chemistry and geometry using focused ion beam microscopy. *Applied Physics Letters, 75*(21), 3414–3416. doi:10.1063/1.125311

Ferrari, A. C., Meyer, J. C., Scardaci, V., Casiraghi, C., Lazzeri, M., Mauri, F., & Geim, A. K. et al. (2006). Raman spectrum of graphene and graphene layers. *Physical Review Letters, 97*(18), 187401. doi:10.1103/PhysRevLett.97.187401 PMID:17155573

Feser, M., Hornberger, B., Jacobsen, C., De Geronimo, G., Rehak, P., Holl, P., & Strüder, L. (2006). Integrating Silicon detector with segmentation for scanning transmission X-ray microscopy. *Nuclear Instruments & Methods in Physics Research. Section A, Accelerators, Spectrometers, Detectors and Associated Equipment, 565*(2), 841–854. doi:10.1016/j.nima.2006.05.086

Grodzinski, P., Silver, M., & Molnar, L. K. (2006). *Nanotechnology for cancer diagnostics: promises and challenges*. Academic Press.

Haider, M., Braunshausen, G., & Schwan, E. (1995). Correction of the spherical aberration of a 200 kV TEM by means of a hexapole-corrector. *Optik (Stuttgart), 99*(4), 167–179.

Head, M. K., Wong, H. S., & Buenfeld, N. R. (2006). Characterisation of 'Hadley' grains by confocal microscopy. *Cement and Concrete Research, 36*(8), 1483–1489. doi:10.1016/j.cemconres.2005.12.020

Holzer, L., Muench, B., Wegmann, M., Gasser, P., & Flatt, R. J. (2006). FIB-Nanotomography of Particulate Systems—Part I: Particle Shape and Topology of Interfaces. *Journal of the American Ceramic Society, 89*(8), 2577–2585. doi:10.1111/j.1551-2916.2006.00974.x

Hou, P., Cheng, X., Qian, J., Zhang, R., Cao, W., & Shah, S. P. (2015). Characteristics of surface-treatment of nano-SiO2 on the transport properties of hardened cement pastes with different water-to-cement ratios. *Cement and Concrete Composites, 55,* 26–33. doi:10.1016/j.cemconcomp.2014.07.022

Huang, X., Jain, P. K., El-Sayed, I. H., & El-Sayed, M. A. (2007). Gold nanoparticles: Interesting optical properties and recent applications in cancer diagnostics and therapy. *Nanomedicine (London), 2*(5), 681–693. doi:10.2217/17435889.2.5.681 PMID:17976030

Inkson, B. J., Steer, T., Möbus, G., & Wagner, T. (2001). Subsurface nanoindentation deformation of Cu–Al multilayers mapped in 3D by focused ion beam microscopy. *Journal of Microscopy, 201*(2), 256–269. doi:10.1046/j.1365-2818.2001.00767.x PMID:11207928

Institute of Experimental and Applied Physics, University of Kiel. (n.d.). Retrieved from http://www.ieap.uni-kiel.de/surface/ag-kipp/stm/stm.htm

Ishigami, M., Chen, J. H., Cullen, W. G., Fuhrer, M. S., & Williams, E. D. (2007). Atomic structure of graphene on SiO_2. *Nano Letters, 7*(6), 1643–1648. doi:10.1021/nl070613a PMID:17497819

Ji, T. (2005). Preliminary study on the water permeability and microstructure of concrete incorporating nano-SiO_2. *Cement and Concrete Research, 35*(10), 1943–1947. doi:10.1016/j.cemconres.2005.07.004

Ji, T., Huang, Y. Z., & Zheng, Z. Q. (2003). Primary investigation of physics and mechanics properties of nano-concrete. *Concrete (London), 3*(48), 13–14.

Jo, B. W., Kim, C. H., Tae, G. H., & Park, J. B. (2007). Characteristics of cement mortar with nano-SiO2 particles. *Construction & Building Materials, 21*(6), 1351–1355. doi:10.1016/j.conbuildmat.2005.12.020

Koo, Y., Littlejohn, G., Collins, B., Yun, Y., Shanov, V. N., Schulz, M., & Sankar, J. et al. (2014). Synthesis and characterization of Ag–TiO 2–CNT nanoparticle composites with high photocatalytic activity under artificial light. *Composites. Part B, Engineering, 57,* 105–111. doi:10.1016/j.compositesb.2013.09.004

Krivanek, O. L., Dellby, N., & Lupini, A. R. (1999). Towards sub-electron beams. *Ultramicroscopy, 78*(1-4), 1–11. doi:10.1016/S0304-3991(99)00013-3

Krivanek, O. L., Nellist, P. D., Dellby, N., Murfitt, M. F., & Szilagyi, Z. (2003). Towards sub-0.5 Å electron beams. *Ultramicroscopy, 96*(3), 229–237. doi:10.1016/S0304-3991(03)00090-1 PMID:12871791

Lasagni, F., Lasagni, A., Holzapfel, C., Mücklich, F., & Degischer, H. P. (2006). Three Dimensional Characterization of Unmodified and Sr-Modified Al-Si Eutectics by FIB and FIB EDX Tomography. *Advanced Engineering Materials, 8*(8), 719–723. doi:10.1002/adem.200500276

Lee, J., Mahendra, S., & Alvarez, P. J. (2010). Nanomaterials in the construction industry: A review of their applications and environmental health and safety considerations. *ACS Nano, 4*(7), 3580–3590. doi:10.1021/nn100866w PMID:20695513

Li, H., Xiao, H. G., & Ou, J. P. (2004a). A study on mechanical and pressure-sensitive properties of cement mortar with nanophase materials. *Cement and Concrete Research, 34*(3), 435–438. doi:10.1016/j.cemconres.2003.08.025

Li, H., Xiao, H. G., Yuan, J., & Ou, J. (2004b). Microstructure of cement mortar with nano-particles. *Composites. Part B, Engineering, 35*(2), 185–189. doi:10.1016/S1359-8368(03)00052-0

Li, X., Cai, W., An, J., Kim, S., Nah, J., Yang, D., & Ruoff, R. S. et al. (2009). Large-area synthesis of high-quality and uniform graphene films on copper foils. *Science, 324*(5932), 1312–1314. doi:10.1126/science.1171245 PMID:19423775

Liu, Q., Tong, T., Liu, S., Yang, D., & Yu, Q. (2014). Investigation of using hybrid recycled powder from demolished concrete solids and clay bricks as a pozzolanic supplement for cement. *Construction & Building Materials, 73*, 754–763. doi:10.1016/j.conbuildmat.2014.09.066

Liu, Z., Cai, W., He, L., Nakayama, N., Chen, K., Sun, X., & Dai, H. et al. (2006). In vivo biodistribution and highly efficient tumour targeting of carbon nanotubes in mice. *Nature Nanotechnology, 2*(1), 47–52. doi:10.1038/nnano.2006.170 PMID:18654207

Liu, Z., Fujita, N., Miyasaka, K., Han, L., Stevens, S. M., Suga, M., & Terasaki, O. (2013). A review of fine structures of nanoporous materials as evidenced by microscopic methods. *Microscopy, 62*(1), 109–146. doi:10.1093/jmicro/dfs098 PMID:23349242

Mackie, W. A., Morrissey, J. L., Hinrichs, C. H., & Davis, P. R. (1992). Field emission from hafnium carbide. *Journal of Vacuum Science & Technology. A, Vacuum, Surfaces, and Films, 10*(4), 2852–2856. doi:10.1116/1.577719

Makar, J. M., & Chan, G. W. (2009). Growth of Cement Hydration Products on Single-Walled Carbon Nanotubes. *Journal of the American Ceramic Society, 92*(6), 1303–1310. doi:10.1111/j.1551-2916.2009.03055.x

Meyer, E. (1992). Atomic force microscopy. *Progress in Surface Science, 41*(1), 3–49. doi:10.1016/0079-6816(92)90009-7

Meyer, J. C., Girit, C. O., Crommie, M. F., & Zettl, A. (2008). Imaging and dynamics of light atoms and molecules on graphene. *Nature, 454*(7202), 319–322. doi:10.1038/nature07094 PMID:18633414

Minsky, M. (1988). Memoir on inventing the confocal scanning microscope. *Scanning, 10*(4), 128–138. doi:10.1002/sca.4950100403

Monteiro, P. J., Clodic, L., Battocchio, F., Kanitpanyacharoen, W., Chae, S. R., Ha, J., & Wenk, H. R. (2013). Incorporating carbon sequestration materials in civil infrastructure: A micro and nano-structural analysis. *Cement and Concrete Composites, 40*, 14–20. doi:10.1016/j.cemconcomp.2013.03.013

Münch, B., Gasser, P., Holzer, L., & Flatt, R. (2006). FIB-Nanotomography of Particulate Systems—Part II: Particle Recognition and Effect of Boundary Truncation. *Journal of the American Ceramic Society, 89*(8), 2586–2595. doi:10.1111/j.1551-2916.2006.01121.x

Park, J. W., Benz, C. C., & Martin, F. J. (2004). Future directions of liposome-and immunoliposome-based cancer therapeutics. [). WB Saunders.]. *Seminars in Oncology, 31*, 196–205. doi:10.1053/j.seminoncol.2004.08.009 PMID:15717745

Rose, H. (1990). Outline of a spherically corrected semiaplanatic medium-voltage transmission electron-microscope. *Optik (Stuttgart), 85*(1), 19–24.

Saini, R., Saini, S., & Sharma, S. (2010). Nanotechnology: The future medicine. *Journal of Cutaneous and Aesthetic Surgery, 3*(1), 32.

Sakamoto, T., Cheng, Z., Takahashi, M., Owari, M., & Nihei, Y. (1998). Development of an ion and electron dual focused beam apparatus for three-dimensional microanalysis. *Japanese Journal of Applied Physics, 37*(4R), 2051–2056. doi:10.1143/JJAP.37.2051

Sawada, H., Sasaki, T., Hosokawa, F., Yuasa, S., Terao, M., Kawazoe, M., Nakamichi, T.,Kaneyama, T., Kondo, Y., Kimoto,K. & Suenaga, K. (2009). Correction of higher order geometrical aberration by triple 3-fold astigmatism field. *Journal of Electron Microscopy*, 1-7.

Scarfato, P., Di Maio, L., Fariello, M. L., Russo, P., & Incarnato, L. (2012). Preparation and evaluation of polymer/clay nanocomposite surface treatments for concrete durability enhancement. *Cement and Concrete Composites, 34*(3), 297–305. doi:10.1016/j.cemconcomp.2011.11.006

Schaffer, M., Wagner, J., Schaffer, B., Schmied, M., & Mulders, H. (2007). Automated three-dimensional X-ray analysis using a dual-beam FIB. *Ultramicroscopy*, *107*(8), 587–597. doi:10.1016/j.ultramic.2006.11.007 PMID:17267131

Sobolev, K., & Gutiérrez, M. F. (2005). How nanotechnology can change the concrete world. *American Ceramic Society Bulletin*, *84*(10), 14.

Stevens, S. M., Jansson, K., Xiao, C., Asahina, S., Klingstedt, M., Grüner, D., & Europe, J. E. O. L. (2009). An appraisal of high resolution scanning electron microscopy applied to porous materials. *JEOL News*, *44*(1), 17.

Thorek, D. L., Chen, A. K., Czupryna, J., & Tsourkas, A. (2006). Super paramagnetic iron oxide nanoparticle probes for molecular imaging. *Annals of Biomedical Engineering*, *34*(1), 23–38. doi:10.1007/s10439-005-9002-7 PMID:16496086

Tomiyasu, B., Fukuju, I., Komatsubara, H., Owari, M., & Nihei, Y. (1998). High spatial resolution 3D analysis of materials using gallium focused ion beam secondary ion mass spectrometry (FIB SIMS). *Nuclear Instruments & Methods in Physics Research. Section B, Beam Interactions with Materials and Atoms*, *136*, 1028–1033. doi:10.1016/S0168-583X(97)00790-8

Tyliszczak, T., & Chou, K. W. (n.d.). *STXM in nanoscience ver2*. Lawrence Berkeley National Laboratory. Retrieved from http://pdf.internationalx.net/STXM-in-nano-science-ver2---Lawrence-Berkeley-National-Laboratory-download-w18866.html#

Tyliszczak, T., Warwick, T., Kilcoyne, A. L. D., Fakra, S., Shuh, D. K., Yoon, T. H., . . . Acremann, Y. (2004). Soft X-ray scanning transmission microscope working in an extended energy range at the advanced light source. In Synchrotron Radiation Instrumentation (Vol. 705, pp. 1356-1359).

Uchic, M. D., Holzer, L., Inkson, B. J., Principe, E. L., & Munroe, P. (2007). Three-dimensional microstructural characterization using focused ion beam tomography. *MRS Bulletin*, *32*(05), 408–416. doi:10.1557/mrs2007.64

Xu, K., Cao, P., & Heath, J. R. (2009). Scanning tunneling microscopy characterization of the electrical properties of wrinkles in exfoliated graphene monolayers. *Nano Letters*, *9*(12), 4446–4451. doi:10.1021/nl902729p PMID:19852488

Ye, Q., Zhang, Z. N., Kong, D. Y., Chen, R. S., & Ma, C. C. (2003). Comparison of properties of high-strength concrete with nano-SiO2 and silica fume added. *Journal of Building Materials*, *6*(4), 281–285.

Yoo, H., Lee, S., Kang, D., Kim, T., Gweon, D., Lee, S., & Kim, K. (2006). Confocal Scanning Microscopy. *International Journal of Precision Engineering and Manufacturing, 7*(4), 3–7.

Zaafarani, N., Raabe, D., Singh, R. N., Roters, F., & Zaefferer, S. (2006). Three-dimensional investigation of the texture and microstructure below a nanoindent in a Cu single crystal using 3D EBSD and crystal plasticity finite element simulations. *Acta Materialia, 54*(7), 1863–1876. doi:10.1016/j.actamat.2005.12.014

Zhang, Y., Brar, V. W., Wang, F., Girit, C., Yayon, Y., Panlasigui, M., & Crommie, M. F. et al. (2008). Giant phonon-induced conductance in scanning tunnelling spectroscopy of gate-tunable graphene. *Nature Physics, 4*(8), 627–630. doi:10.1038/nphys1022

Zhu, W., Bartos, P. J. M., & Porro, A. (2004). Application of nanotechnology in construction. *Materials and Structures, 37*(9), 649–658. doi:10.1007/BF02483294

Chapter 4
Nanotechnology Applications in the Construction Industry

Iman Mansouri
Birjand University of Technology, Iran

Elaheh Esmaeili
Birjand University of Technology, Iran

ABSTRACT

Nanotechnology refers to the understanding and manipulation of materials on the nanoscale (<100 nm). This can lead to marked changes in material properties and can result in improved performance and new functionality. Nanomaterials with properties such as corrosion resistance, and strength and durability are of particular interests to construction professionals, because, these properties directly affect the selection of construction materials, erection methods, and on-site handling techniques. Applying nanotechnology to construction, in some cases, may result in visionary and paradigm-breaking advances. The incorporation of nanomaterials can improve structural efficiency, durability and strength of cementitious materials and can thereby assist in improving the quality and longevity of structures. This chapter tries to analyze nanotechnology in the context of construction and explores the current scenario of nanotechnology in the construction industry. In order to identify the potential benefits and existing barriers, an extensive literature review is conducted.

DOI: 10.4018/978-1-5225-0344-6.ch004

1. INTRODUCTION

Nanotechnology is the scientific examining, monitoring, and modifying of the behavior and practice of materials at nano-scale. It is also the process in which a material or device is created with constituents at the atomic and molecular scale. Therefore, it is an area of research and technology development aimed at understanding and controlling matter at its molecular level and thereby affecting the bulk properties of the material (Grove, Vanikar, & Crawford, 2010).

Nanotechnology encompasses a wide range of areas and many disciplines that motivate collaboration among engineers, scientists, innovators, and researchers in sustainable development. Research centers, consortiums, committees, and task forces that deal with it are vastly emerging globally, reflecting the interest of individuals, organizations, and governments (Khan, 2011). Recent developments led to a bang in nanotechnology-oriented materials in areas such as polymers, plastics, electronics, car manufacturing, and medicine (Birgisson, Taylor, Armaghani, & Shah, 2010).

Having many unrivaled characteristics of nanotechnology-derived products, newly developed nano-based products can meaningfully decrease existing civil engineering problems. Fundamentally, construction deals with high-tech materials and processes that have been used in construction. Hence, there is vast scope to apply nanotechnology in construction materials, which can exhibit, perhaps one of the most prominent, societal effects.

The rise of nanotechnology has greatly contributed to the development of materials by enhancing their properties. Cement-based materials are one class of these materials which benefited from this new technology. New studies have proved that there is a general tendency of improvement in the properties of cementitious materials using mineral admixtures like nanosilicon. A small amount of nanosilica, when dispersed in to a mortar mix properly, is able to increase the strength of the mortar over 15% (Flores et al., 2010; Hosseini, Booshehrian, & Farshchi, 2010).

Nanotechnology can also be used for design and construction processes in many areas since nanotechnology-generated products have lots of matchless features. These features can, again, significantly resolve current construction problems, and might change the requirement and organization of construction process.

2. NANOTECHNOLOGY

Nanotechnology has been evidently identified as one of the key, cross-disciplinary areas of study for the next 20 years. Noteworthy investments are being made in nanotechnology research all around the world. Recent studies have known the

Figure 1. Scanning electron microscope (SEM) and transmission electron microscope (TEM) images of 0D nanostructures: A) quantum dots; B) nanoparticles arrays; C) core-shell nanoparticles
Y. T. Kim, Han, Hong, & Kwon, 2010; Singh et al., 2012.

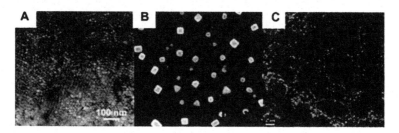

construction industry as one of the chief possible consumers of nano structured materials (Bartos, 2006).

Nanotechnology is the manipulation or study of matter on an atomic and molecular scale. Normally, nanotechnology works with materials, devices, and other structures with at least one dimension sized from 1 to 100 nanometers (P. N. Balaguru, 2005; Feynman, 1960; Sobolev, Flores, Hermosillo, & Torres-Martínez, 2008). In an original sense, nanotechnology denotes the ability to study or apply physical, chemical and biological systems at a scale ranging from few nanometers to submicron dimensions (Sobolev et al., 2008).

In general, the nanostructured materials are mainly classified to four groups; i.e. zero-dimentional (0D), one-dimentional (1D), two-dimentional (2D) and three-dimentional (3D) nanostructures, based on their dimensionality. Figure 1 exhibits some 0D nanostructured materials.

0D nanomaterials have been recently presented by many researchers (Janssen & Stouwdam, 2008; Lee et al., 2009; Mokerov, Fedorov, Velikovski, & Scherbakova, 2001), including uniform particles arrays (quantum dots), heterogeneous particles arrays, core–shell quantum dots, onions, hollow spheres and nanolenses which have many applications such as light emitting diodes (LEDs) (Janssen & Stouwdam, 2008), solar cells (Lee et al., 2009), single-electron transistors (Mokerov et al., 2001), etc.

The major important 1D nanostructures has been found by (Iijima, 1991) to be carbon nanotubes, owing considerable applications especially in manufacturing various nanocomposite materials. The other types of 1D nanomaterials contains nanowires, nanorods, nanotubes, nanobelts, nanoribbons, and hierarchical nanostructures, as some of them shown in Figure 2.

As known, 2D nanostructured materials composed of two dimensions with larger values than the nanometric size range. 2D nanostructures reveal unique char-

Figure 2. SEM images of 1D nanostructured materials: A) nanowires; B) nanorods; C) nanotubes
Singh et al., 2012; Okada, Kawashima, Nakata, & Ning, 2005; Xia, Feng, Wang, Lai, & Lu, 2010.

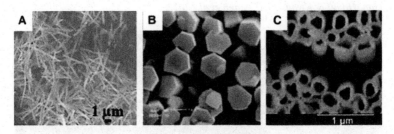

acteristics because of their distinct shapes and geometries that make them suitable for different applications such as building blocks, sensors, photocatalysts, nanocontainers, nanoreactors, and templates for manufacturing other 2D nanomaterials (Pradhan & Leung, 2008). In Figure 3, we show some examples of 2D nanostructures, such as junctions (continuous islands), branched structures and nanoprisms.

The last kind of nanostructured materials is attributed to 3D nanostructures which have recently attracted significant research interest. 3D nanostructures have interesting potential in catalysis, batteries and magnetic applications. Nanoballs, nanocoils, nanocones, nanopillers and nanoflowers are typical examples of 3D nanostructured materials. Figure 4 implies SEM and TEM images of some 3D nanostructured samples.

3. APPLICATIONS OF NANOTECHNOLOGY

Nanotechnology is a growing field of enquiry and its applications are seen in many areas of science and engineering. Many fields which nanotechnology could be applied are shown in Figure 5.

Figure 3. SEM and TEM images of 2D nanostructures: A) junctions (continuous islands); B) branched structures; C) nanoprisms
Singh et al., 2012; Nayak, Behera, & Mishra, 2010; Mann & Skrabalak, 2011.

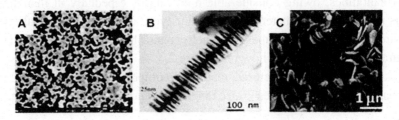

Figure 4. SEM and TEM images of 3D nanostructures: A) nanoballs; B) nanocoils; and C) nanocones
J. N. Wang, Su, & Wu, 2008; L. Wang & Yamauchi, 2009; Singh et al., 2012.

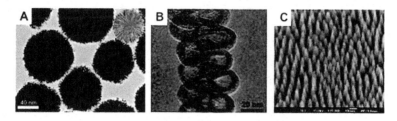

Construction industry is also one of the major fields adopting nanotechnology to lessen the shortcomings in construction materials and thereby improve the power and permanency of structures. Nanotechnology could be applied to different areas by either of the two approaches– Bottom-Up approach or Top-Down approach (Sanchez & Sobolev, 2010; Sobolev et al., 2008). Schematic representation of these approaches is shown in Figure 6.

These two approaches and their consequent applications are examined in detail below.

Figure 5. Fields of applications of nanotechnology
Rakesh, Renu, & Arun, 2011.

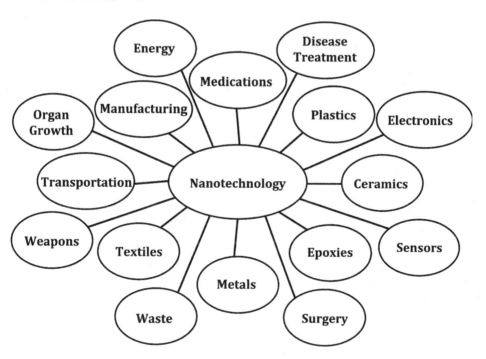

Figure 6. The top-down and bottom-up approaches in nanotechnology (1)
Sanchez & Sobolev, 2010.

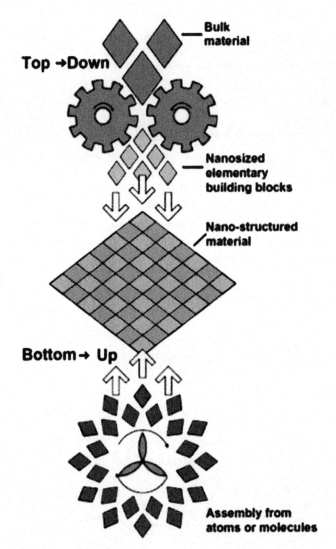

3.1 Nanotechnology and Concrete

Concrete is the most broadly used material in the world. It is predictable that its annual production is about 20 billion tons greater than any other material on the earth. It is considered to be the backbone of civil engineering.

Material-composed concrete are generally classified to normal-strength concrete (with 40–120 MPa compressive strength), high-strength concrete (with 640 MPa compressive strength), and ultra-high performance concrete (with >120 MPa compressive strength) (Habel, Viviani, Denarié, & Brühwiler, 2006; Wille, Naaman, El-Tawil, & Parra-Montesinos, 2012). Concrete materials of higher strength exhibit a growing tendency towards brittle modes of failure.

The distinct reinforcement systems used in concrete are generally fibers, having micrometer-scale diameters or larger (Barnett, Lataste, Parry, Millard, & Soutsos, 2010). At practical volume fractions (_1.0%), the spacing between those fibers is quite large. Microcracks can be formed and progress in the voids presented between conventional fibers (Gay & Sanchez, 2010; Hammel et al., 2004). Therefore, nano-scale reinforcement can fill the space between micro-scale fibers in order to prevent the formation and early growth of cracks in concrete (Abu Al-Rub, Tyson, Yazdanbakhsh, & Grasley, 2012; Sanchez & Sobolev, 2010). Examples of such nano-scale reinforcement include carbon nanotube (CNT) (Abu Al-Rub et al., 2012), carbon nanofiber (CNF) (Metaxa, Konsta-Gdoutos, & Shah, 2010; Yazdanbakhsh, Grasley, Tyson, & Abu Al-Rub, 2010), and graphite nanoplatelet (B. Li & Zhong, 2011; Peyvandi, Soroushian, Abdol, & Balachandra, 2013). Uniform dispersion and adequate interfacial interaction are adequate requirements for successful use of nano-scale reinforcement in cementitious materials (Konsta-Gdoutos, Metaxa, & Shah, 2010; Musso, Tulliani, Ferro, & Tagliaferro, 2009).

The application of nanotechnology in concrete is still in its early stages, but, it has been recognized that nano-particles can be used to produce nano-concrete with improved performance (Ashby, Ferreira, & Schodek, 2009). According to the researches in recent years, nano-concrete has been defined as a concrete made by Portland cement and cementitious particles with sizes smaller than 500nm (P. Balaguru & Chong, 2006; Sanchez & Sobolev, 2010). The reasons why nanotechnology is applied on concrete are to control material behavior, achieve superior mechanical and durability performance, and to provide new features in concrete such as low electrical resistivity, self-sensing capability, self-cleaning and self-healing abilities, high ductility and self-control of cracks (Sanchez & Sobolev, 2010). Due to the high surface area to volume ratio, nano-particles can act as nuclei for cement phases, while providing an excellent chemical reactivity to promote cement hydration. Figure 7 illustrates the range of specific surface area for different materials used in concrete. This figure also provides the definition of nano-concrete (nano-engineered concrete) (Ashby et al., 2009).

Concrete undertakes different phases during hydration where nano-materials can play an important role. The different hydration characteristics of concrete are listed in Table 1.

Calcium silicate hydrate (C–S–H) is the major product from the hydration process and presents a poorly crystalline or amorphous structure (Bensted & Barnes, 2002).

Figure 7. Specific surface area of different constituent materials used in conventional concrete, high-strength/high-performance concrete and nano-concrete
Ashby et al., 2009.

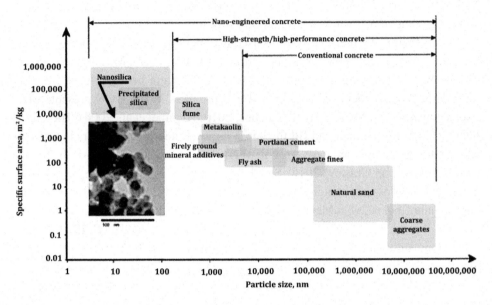

It is well known that C–S–H gel controls the strength and durability of concrete (Skinner, Chae, Benmore, Wenk, & Monteiro, 2010). Currently, there is no consensus concerning the atomic structure of C–S–H and how C–S–H is able to develop strength. Tylor (Taylor, 1997) described a model for characterizing the atomic structure of C–S–H based on X-ray diffraction and scanning electron microscope (SEM) testing. His research reported that C–S–H has an amorphous composition and its nano-structure accords with the 1.4nm tobermorite layer structure. In Also,

Table 1. Hydration characteristics of Portland cement concrete (w/c = 0.5)

Physical/Chemical Component	Approximate Volume (%)	Comments
C–S–H	50	Includes gel pores; poorly crystalline/amorphous structure
Calcium hydroxide (CH)	12	Crystalline structure
Ettringite and monosulphate phases (AFm and AFt phases)	13	Crystalline structure
Unreacted cement particles	5	Based on the hydration process
Capillary pores	20	Dependent of w/c

Bensted and Barnes 2002.

according to (Skinner et al., 2010), C–S–H can be characterised by a nanocrystalline structure, represented closely by 1.1 nm tobermorite (Safiuddin, Gonzalez, Cao, & Tighe, 2014).

Nanomaterials could create more effective interactions with the porous structure of cementitious composite than the microstructures because of their larger specific surface areas. Whereas, the pretty high capillary porosity (typically more than 10 vol.%) of cementitious paste and the presence of some formed micro-scale crystals such as calcium hydroxide, calcium aluminate hydrate, calcium monosulfate aluminate hydrate are not able to successfully interact with nanomaterials. Therefore, this prevents effective interactions of graphite nanomaterials with cement paste which in turn, limits advantageous effects in conventional cementitious matrices. Only the calcium silicate hydrate (C–S–H) forms enormous bonds with graphite nanomaterials (Peyvandi, Soroushian, & Jahangirnejad, 2013).

In general, nanotechnology lets us improve the understanding of the C–S–H atomic structure. According to (Skinner et al., 2010), this new knowledge allows scientists to manipulate the C–S–H structure and therefore, design concrete with enhanced properties. Based on nanotechnology (Selvam, Subramani, Murray, & Hall, 2009) clarified that electrostatic forces and bond forces in the silicate chains were mainly responsible for strength development in the C–S–H atomic structure. Therefore, nanotechnology can play a significant role in explaining the mechanisms of cement hydration in concrete. Moreover, nanotechnology may contribute to sustainable development. It has been reported that producing 1 ton of Portland cement produces about 0.8-1.0 ton of CO_2 that leads to adverse effect on the environment. Enhancement of concrete properties and durability would compensate this adverse effect and thus, the use of nano-concrete could be a strong motivation to environmental sustainability.

Nanotechnology is also used in studying its properties like hydration reaction, alkali silicate reaction (ASR) and fly ash reactivity. Alkali silicate reaction is caused due to alkali content of cement and silica present in reactive aggregates. The use of pozzolona in the concrete mix as a partial cement replacement can reduce the likelihood of ASR occurring as they reduce the alkalinity of a pore fluid.

Fly ash not only improves concrete durability and strength, but also sustainable factors through reducing the requirement for cement; however, the curing process of such concrete is slowed down due to the addition of fly ash, and early-stage strength is also low in comparison to normal concrete. Addition of Nano-silica leads to densifying the micro and nanostructure, resulting in improved mechanical properties (Jalal, Pouladkhan, Harandi, & Jafari, 2015). With the addition of nano-silica, part of the cement is substituted, but, the density and strength of the fly-ash concrete improves particularly in the early stages. For concrete containing a large volume of fly ash, at early age, it can improve pore size distribution by filling the pores between large fly ash and cement particles at Nano scale. The dispersal or slurry of amorphous nano-SiO_2 is used to improve segregation resistance for self-compacting concrete (Shaikh & Supit, 2015).

Finally, fiber wrapping of concrete is quite common nowadays for increasing the strength of pre-existing concrete structural elements. An advancement in the procedure involves the use of a fiber sheet (i.e. matrix) consisting of nano-silica particles and hardeners. These nanoparticles penetrate and close small cracks on the concrete surface and, in strengthening applications, the matrices form a strong tie between the surface of the concrete and the fiber reinforcement.

3.1.1 Carbon Nanotubes and Concrete

Carbon nanotubes and concretes (CNTs) composed of single or multiple graphene sheets rolled up in the form of tubes (J. Makar, Margeson, & Luh, 2005). The unique graphene structure presents CNTs with extraordinary characteristics. However, CNTs own lower density than steel or glass fiber, but, it has been found that CNTs show an elastic modulus in the terapascal range, yield strength of approximately 20–60 GPa, and yield strain of up to 10% (Walters et al., 1999).

Carbon nanotubes (CNTs) proved to be one of the most advantageous structure for the reinforcement at the nanoscale. The unique characteristics of CNTs in shape, chemical structure and mechanical properties, make them a good candidate for reinforcement of composite materials. The Young's modulus of an individual nanotube is reported to be around 1 TPa and its density is about 1.33 g/cm^3 (Salvetat et al., 1999). The CNTs' fracture strains were found to be around 10-15% for corresponding tensile stresses in the range of 65-93 GPa, respectively, as a result of molecular mechanic simulations (Belytschko, Xiao, Schatz, & Ruoff, 2002).

Many attempts have been made to add CNTs in cementitious matrices at an amount ranging from 0.5 to 2.0 wt.% (by weight of cement) (G. Y. Li, Wang, & Zhao, 2005, 2007; J. Makar et al., 2005). The results obtained from SEM and Vickers hardness measurements indicate that CNTs affect the early hydration progress by the enhancement of hydration rates.

According to literature (H. K. Kim, Nam, & Lee, 2012), CNTs are expected to provide mechanical reinforcement between hydration products of cement with nano-scale dimensions. Since, hydration products such as calcium–silicate–hydrates (C–S–H) and calcium hydroxide (CH) have similar or larger size than CNTs, hence, it seems that only a few CNTs are mainly attached to the hydration products (H. K. Kim et al., 2012).

Carbon nanotubes (CNTs) are composed of hollow tubular channels made of rolled graphene sheets, formed either by one wall (SWCNTs) or several walls (MWCNTs). Since, CNTs exhibit great mechanical properties along with extremely high aspect ratios (length-to-diameter ratio) ranging from 30 to more than many thousands, they are expected to produce significantly stronger and tougher cement composites than traditional reinforcing materials; e.g. glass or carbon fibers. In the better words, these unique characteristics make CNTs suitable for well- dispersing in the cement matrix than common fibers, which in turn, prevents crack propagation in the early stages within concrete. Furthermore, the amount of defects can also play an important role for cement reinforcement application. In fact, the results have been showed that defect-free CNTs are not able to present proper adhesion with the matrix (Sáez De Ibarra, Gaitero, Erkizia, & Campillo, 2006). On the other side, since, these defects can limit mechanical strength of the composite, the functionalization of CNTs may help to convert these reactive spots to strong chemical bonds between CNTs and cementitious matrix.

However, (Musso et al., 2009) observed from flexural experiments that the addition of carboxyl functionalized MWCNTs decreased mechanical strength by a factor of 2.5 with respect to the plain cement (Figure 8a), meanwhile, by using the pristine CNTs the enhancement of approximately 34% is gained. The average compression resistance of functionalized CNTs (f-CNTs) was even six times lower with respect to the concrete with no CNTs (Figure 8b); i.e. 15.53 ± 3.04 kN and 104 ± 20 kN, respectively. They justified this phenomenon to the results of mechanical characteristics, as reported in Figure 8. Poor mechanical properties observed after addition of f-CNTs were attributed to the lower values of Tobermorite gel as a consequence of a low cement paste hydration.

(Dweck, Ferreira da Silva, Büchler, & Cartledge, 2002) stated that tobermorite gel, which is formed during hydration of the calcium silicate phases such as C_3S ($3CaO.SiO_2$) and C_2S ($2CaO.SiO_2$), exerts a significant influence on cement strength. To realize this explanation, (Musso et al., 2009) applied thermogravimetric analysis

Figure 8. Modulus of rupture (a) and compression resistance (b) of concrete in the presence or absence of CNTs
Musso et al., 2009.

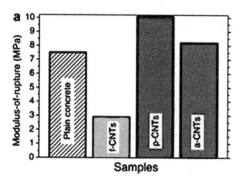
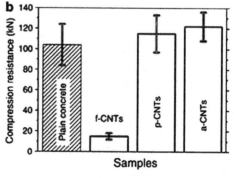

(TGA), as presented in Figure 9. According to Figure 9, the peak centroid at about 116 °C which is related to tobermorite gel ($3CaO.2SiO_2.xH_2O$) formed in presence of f-CNTs is significantly lower than that observed at around 102 °C for the plain concrete. They found that, due to the presence of hydrophilic groups on the f-CNT surface, the nanotubes absorb enormous amounts of water which in turn, hinders the cement paste hydration.

3.1.2 Nanoparticles and Concrete

Among all the nanomaterials, nanosilica is the most widely used material in the cement and concrete to improve the performance, because of both pozzolanic reactivity and the pore-filling characteristics. Nanosilica incorporated cement pastes are progressively studied to understand the hydration process and microstructure improvement. Basically, the major aim of the application of ultra-fine additives like nanosilica in cement-based composites is to improve the characteristics of the plastic and hardened material. Micro and nano-scaled silica particles play an important role as fillers, as they fill up the voids between the cement grains. By adjusting the right compositions, the higher packing density concludes lower demand of water consumption which in turn, it results in higher strength because of the reduced capillary porosity. Furthermore, nanosilica owns a higher pozzolanic reactivity compared to

Figure 9. DTG curves of anhydrous cement and concrete in the presence or absence of CNTs
Musso et al., 2009.

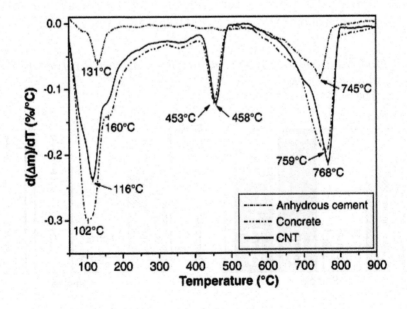

silica fume (Choolaei, Rashidi, Ardjmand, Yadegari, & Soltanian, 2012; P. Hou et al., 2013; Zapata, Portela, Suárez, & Carrasquillo, 2013).

The main reaction mechanism suggested for hydration of cement in the presence of nanosilica is the acceleration of hydration of cement. By the addition of nanosilica to the cement, $H_2SiO_4^{2-}$ forms which interacts with the available Ca^{2+} which results in an additional calcium–silicate–hydrate (C–S–H). C-S-H particles are distributed in the water between the cement as seeds for the formation of more values of C-S-H phase which in turn, accelerates early cement hydration. Figure 10 implies the effective performance of nano-particles/fibers addition to the concrete.

Figure 10. The effect of nano-particles/fibers addition to the concrete
Singh et al., 2012.

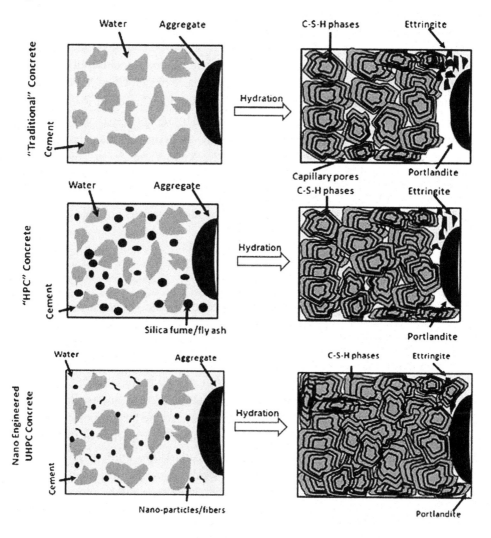

Cement mortar containing well-dispersed nanosilica has a dense microstructure even in the presence of small amounts of nanoparticles. However, not properly dispersing of the nanoparticles forms the voids and causes weak zones formation (H. Li, Xiao, & Ou, 2004). Accelerated hydration of cement paste and faster formation of calcium hydroxide at initial period was observed in the nanosilica added cement paste (P. Hou et al., 2013; Land & Stephan, 2012; Senff, Labrincha, Ferreira, Hotza, & Repette, 2009). That is why nanosilica particles act as nucleation sites to enhance the rate of hydration which depends upon the surface area of the applied nanosilica particles. However, in line with literature (Ltifi, Guefrech, Mounanga, & Khelidj, 2011; Qing, Zenan, Deyu, & Rongshen, 2007; Zapata et al., 2013), formation of C–S–H from calcium hydroxide by pozzolanic reaction was much higher and faster in cement paste containing nanosilica. It leads to the increase in compressive and flexural strength.

Besides nanosilica, the materials used in nanosize are nano-TiO_2, nano-Fe_2O_3, nano-Al_2O_3, carbon nanotubes/fibers. Nano-TiO_2 possess self-cleaning properties against environmental pollutants like NO_x, carbon monoxide, chlorophenols, etc. which performs based on its photocatalytic characteristics (J. Chen & Poon, 2009; Murata, Tawara, Obata, & Takeuchi, 1999). The Jubilee Church in Rome, Italy is an example for self-cleaning concrete structure. The recent researches have been showed that nano-Fe_2O_3 increases the flexural strength, electrical conductivity characteristics and self-sensing capability (H. Li, Xiao, Yuan, & Ou, 2004; Nazari, Riahi, Riahi, Shamekhi, & Khademno, 2010). According to (H. Li, Zhang, & Ou, 2006; Thiruchitrambalam, Palkar, & Gopinathan, 2004), nano-Al_2O_3 results in considerable improvement in modulus of elasticity, however, it is not able to enhance the compressive strength so much.

3.1.3 Nanofibers and Concrete

The improvement of mechanical properties of concrete using fibers has been widely investigated since 1960s (Brandt, 2008; Giner, Baeza, Ivorra, Zornoza, & Galao, 2012; Ivorra, Garcés, Catalá, Andión, & Zornoza, 2010). Fiber reinforced materials are mainly composed of two elements, the matrix and the fiber reinforcement. There are two important variables that conclude a critical role in the ultimate properties of the composite. The former is the fiber volumetric fraction, and the latter is the fiber's aspect ratio (fiber's length over its diameter). To the best of our knowledge, both parameters are directly proportional to mechanical properties of the fiber reinforced concrete (Fanella & Naaman, 1985; Köksal, Altun, Yiğit, & Şahin, 2008).

Since, cementitious materials are fairly weak in mechanical point of view, having low tensile strength and low strain capacity; fibers can be introduced into cement-based matrices to overcome these weaknesses. According to (Akkaya, Shah,

Nanotechnology Applications in the Construction Industry

& Ghandehari, 2003), pronounced reinforcement was observed following the use of microfibers in the mechanical properties of cementitious composites. However, microfibers delay the progression of the formed microcracks, they are not able to prevent their initiation.

New nanosized fibers with certain modifications have recently designed for the reinforcement within concrete (J. M. Makar & Beaudoin, 2003). The incorporation of these fibers at the nanoscale facilitates the control of the matrix cracks at the nanoscale level (Konsta-Gdoutos et al., 2010).

Fiber reinforced polymer (FRP) materials are currently produced in different configurations and are widely used for the strengthening and retrofitting of concrete structures and bridges. Recently, considerable research has been directed to characterize the use of FRP bars and strips as near surface mounted reinforcement, primarily for strengthening applications. Nevertheless, in-depth understanding of the bond mechanism is still a challenging issue. Investigations undertaken to evaluate bond characteristics of near surface mounted carbon FRP (CFRP) strips (Al-Mosawe, Al-Mahaidi, & Zhao, 2015; Haddad, Al-Rousan, Ghanma, & Nimri, 2015). Different embedment lengths were used to evaluate the development length needed for effective use of near surface mounted CFRP strips. A closed-form analytical solution is proposed to predict the interfacial shear stresses. The model is validated by comparing the predicted values with test results as well as nonlinear finite element modeling. A quantitative criterion governing the debonding failure of near surface mounted CFRP strips is established. The influence of various parameters including internal steel reinforcement ratio, concrete compressive strength, and groove width is discussed (Hassan & Rizkalla, 2003). Figure 11 shows three usual types of composite sheets.

Figure 11. FRPs: (a) aramid; (b) glass; (c) carbon

This review focuses on the recent progress of fabrication, properties, and structural applications of high-performance and multifunctional cementitious composites with nano carbon materials (NCMs) including carbon nanofibers, carbon nanotubes and nano graphite platelets. The improvement/modification mechanisms of these NCMs to composites are also discussed (Han et al., 2015). Figure 12 shows bonding between modified CNFs with cementitious material.

A number of studies have shown that the addition of polypropylene (PP) fibers can improve the abrasion resistance of concrete. Most of the test results indicated that the abrasion resistance of concrete containing PP fibers could increase by 20–60% according to different contents of PP fibers and different researchers (L. Chen, Mindess, & Morgan, 1995). The result of stimulant abrasion test carried out by the National Highway Laboratory of Norway showed that the abrasion resistance of concrete containing PP fibers increased by 52% (Bouhicha, Aouissi, & Kenai, 2005). However, the result tested by the U.S. Army Corps of Engineers using CRD-C52-54 method indicated that the abrasion resistance of concrete containing PP fibers could increase by 105% (B. Q. Zhong & Zhu, 2002). The abrasion resistance of concrete can be considerably improved with the addition of nano-particles or PP fibers. However, the indices of abrasion resistance of concrete containing nano-particles are much larger than that of concrete containing PP fibers. So, the nano-particles are more favorable to the abrasion resistance of concrete than PP fibers (Z. Li, Wang, He, Lu, & Wang, 2006).

Natural fibers today are a popular choice for applications in composite manufacturing. Based on the sustainability benefits, biofibers such as plant fibers are replacing synthetic fibers in composites. These fibers are used to manufacture several biocomposites. The chemical composition and properties of each of the fibers changes, which demands the detailed comparison of these fibers. The reinforcement potential of natural fibers and their properties have been described in numerous papers (Ramamoorthy, Skrifvars, & Persson, 2015). A detailed review of bamboo

Figure 12. Bonding between modified CNFs with cementitious material
Peyvandi, Soroushian, & Jahangirnejad, 2013.

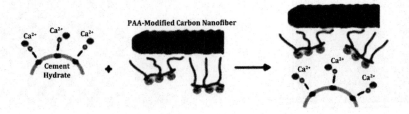

Nanotechnology Applications in the Construction Industry

fiber reinforced structural concrete elements was done by (Ghavami, 2005). Bagasse reinforced cement composites were prepared and analyzed by (Bilba, Arsene, & Ouensanga, 2003). Researchers also used barley straws reinforcement in biocomposites (Bouhicha et al., 2005). In recent years, both industrial and academic world are focusing their attention toward the development of sustainable composites, reinforced with natural fibers. In particular, among the natural fibers (i.e. animal, vegetable or mineral) that can be used as reinforcement, the basalt ones represent the most interesting for their properties. Basalt fibers (Figure 13) are mineral fibers, which offer several better features in comparison with glass fibers. Composites made of basalt fibers within polymer (both thermoplastic and thermoset), biodegradable, metallic and concrete matrices exhibit good properties (Fiore, Scalici, Di Bella, & Valenza, 2015).

The ISSC (also called self-monitoring, intrinsically smart, piezoresistive or pressure-sensitive concrete) is fabricated through incorporating some functional fillers such as carbon fiber (CF), carbon nanotubes (CNTs), and nickel powder (NP) into conventional concrete to increase its ability to sense the strain, stress, crack or damage in itself while maintaining or even improving its mechanical properties and durability. The functional fillers are dispersed in concrete matrix to form an extensive conductive network inside concrete composite. As this composite is deformed or stressed under external force or environmental action, the conductive network inside is changed, which affects its electrical behaviors. Strain (or deformation), stress (or external force), crack and damage under static and dynamic conditions

Figure 13. Structure of ISSC

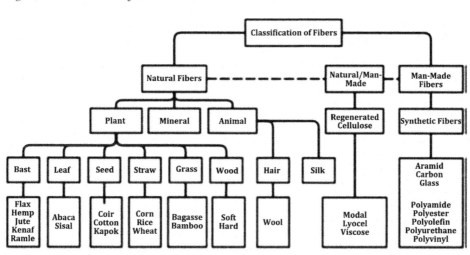

can therefore be detected through measuring the electrical properties of the composite. The ISSC has both structural and sensing functions, so, it replaces the need for additional sensors (Figure 13). It not only has potential in the field of structural health monitoring and condition evaluation for concrete structures, but also can be used for traffic detection, corrosion monitoring of rebar, military and border security, structural vibration control and so on (Chung, 1998; T. C. Hou & Lynch, 2005; Mao, Zhao, Sheng, Yang, & Shui, 1995; Ou & Han, 2009; Peyvandi, Soroushian, Abdol, et al., 2013).

Nanofibers are produced out of organic and inorganic polymers by electrospinning technology. Both polymer solutions and polymer melts can be electrospun. Fiber diameters of 50 to 500 nanometers are typical. An industrial–scale production method of nanofiber production has been developed. Small fiber diameters and great specific surface are the main specific features of nanofiber assemblies, i.e. nanofiber layers. A number of specific end-uses of nanofibers have been developed such as filters, sound absorbing materials, wound dressings, scaffolds for tissue engineering, etc. Machinery for nanofiber production is produced and offered in the Czech Republic. There is a great potential for utilizing of nanofibers in civil engineering.

Typical properties of nanofibersin comparison with conventional textile fibers and in special, extremely fine "melt blown" fibers are presented in Table 2.

Polymeric nanofibers are produced from a variety of organic and inorganic polymers such as polyvinyl alcohol, polyamides, polyurethanes, polyimides, polystyrene, p-HEMA, chitosan, various co-polymersand many others. According to the shape of collector electrode, nanofibers can be produced as planar layers, yarns, nanofiber-coated yarns, tubular bodies, 3D scaffolds for tissue engineering and others. Some of them are shown in Figure 14.

3.2 Nanotechnology and Steel

Since the second industrial revolution in late 19th and early 20th Century, steel has been extensively available and played a major role in the construction industry. The

Table 2. Typical dimensions of conventional fibers, melt-blown fibers and nanofibers

Fibres	Fibre Diameter (μ m)	Linear Density (dtex)	Specific Surface (m^2/g)
Conventional	10-40	1-30	ca. 0.2
Melt-blown	1-5	ca. 0.01	ca. 2
Nanofibres	0.05-0.5	ca. 0.0001	ca. 20

(Jirsak & Dao, 2009).

Figure 14. Planar nanofibre layer (A) and a yarn coated by nanofibres (B)
Jirsak & Dao, 2009.

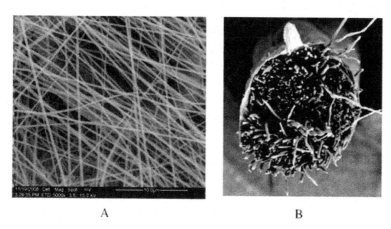

construction industry can benefit from the application of nanotechnology to steel; some of the promising areas under investigation or even available today would be explored in the following paragraphs (Zhang, Morsdorf, & Tasan, 2016; Y. Zhong, Liu, Wikman, Cui, & Shen, 2016).

Fatigue is a significant issue that could lead to the structural failure of steel subject to cyclic loading, such as in bridges or towers. It can occur at stresses meaningfully lower than the yield stress of the material and lead to a major shortening of useful life of the structure. The current design philosophy brings about one or more of three limiting measures: a design based on an intense decrease in the permissible stress, a shortened permissible service life or the need for a regular inspection regime. It would have a substantial influence on the life-cycle costs of structures and limit the effective use of resources; it is thus sustainability and a safety issue. Stress risers are in charge of initiating cracks which fatigue failure results from and research has shown that the addition of copper nanoparticles decreases the surface unevenness of steel which in turn, confines the number of stress risers and hence, fatigue cracking. Progressions in this technology would lead to better safety, less need for monitoring and more effective material use in construction prone to fatigue issues.

Two fairly new products available today are Sandvik Nanoflex and MMFX2 steel. Both of them are corrosion resistant, but they have different mechanical features and are the result of different applications of nanotechnology. Conventionally, the tradeoff between steel strength and ductility is an important issue for steel; the forces in modern construction require high strength, whereas safety (esp. in seismic areas) and stress redistribution require high ductility. This has led to the use of low strength ductile material in greater sizes than would otherwise be possible with high strength hard material and consequently, it is an issue of sustainability and efficient use of resources. Sandvik Nanoflex has both the desirable qualities of a high Young's Modulus and high strength, and it is also resistant to corrosion because of the presence of very stiff nanometer-sized particles in the steel matrix. It effectively matches high strength with special formability and currently it is being used in the production of some parts as diverse as medical instruments and bicycle components; however, its applications are increasing. The use of stainless steel reinforcement in concrete structures has usually been restricted to high risk environments as its use is cost prohibitive. However, MMFX2 steel, while having the mechanical properties of conventional steel, has a modified nano-structure that makes it corrosion resistant and it is an alternative to conventional stainless steel, but at a lower cost.

3.3. Nanotechnology and Green Building

Nanotechnology, as the manipulation of matter at the molecular scale, is bringing new materials and new possibilities to industries as varied as electronics, medicine, energy and aeronautics. Our ability to design new materials from the bottom up is influencing the building industry too. New materials and products based on nanotechnology can be found in building insulation, coatings, and solar technologies (Huang, Lin, & Hsu, 2015; Revel et al., 2013). Work now underway in nanotech labs will soon result in new products for lighting, structures, and energy. In the building industry, nanotechnology has by now brought to market self-cleaning windows, smog-eating concrete, and many other progresses. But, these developments and currently available products are slightly compared to those incubating in the world's nanotech labs these days. There, work is happening on illuminating walls that change color with the flip of a switch, nanocomposites as thin as glass yet capable of supporting entire buildings, and photosynthetic surfaces making any building frontage a source of free energy.

4. CONCLUSION

Study in nanotechnology related to construction is still in its initial stages; however, this chapter has revealed the main benefits and barriers that permit the effect of nanotechnology on construction to be defined.

However, the potential effects of nanotechnology on construction are mostly unidentified to the construction profession in general, although specific research is being done in universities and other institutes all over the world. These provide indicators to what would soon be available to industry. Many of these developments are in line for arriving within the next 5 years. In order to fully benefit from this new industrial revolution, a concerted effort is needed to overcome the vital barriers of lack of knowledge and conservatism in construction as regards nanotechnology. Nanotechnology is a multifaceted and deep subject and it is nearly impossible to grasp for those who are not actively involved; thus, awareness of research done can simply be increased by educating students and professionals through easily digestible information made available by universities, relevant institutions, journals and other sources.

Focused research into the timeous and directed research into nanotechnology for construction infrastructure should be pursued to ensure that the potential benefits of this technology can be harnessed to provide longer life and more economical infrastructure. This chapter concludes with a roadmap and strategic action plan on how nanotechnology can have its biggest impact on the field of civil engineering.

REFERENCES

Abu Al-Rub, R. K., Tyson, B. M., Yazdanbakhsh, A., & Grasley, Z. (2012). Mechanical Properties of Nanocomposite Cement Incorporating Surface-Treated and Untreated Carbon Nanotubes and Carbon Nanofibers. *Journal of Nanomechanics and Micromechanics*, *2*(1), 1–6. doi:10.1061/(ASCE)NM.2153-5477.0000041

Akkaya, Y., Shah, S. P., & Ghandehari, M. (2003). Influence of fiber dispersion on the performance of microfiber reinforced cement composites. *ACI Special Publications*, *216*, 1–18.

Al-Mosawe, A., Al-Mahaidi, R., & Zhao, X. L. (2015). Effect of CFRP properties, on the bond characteristics between steel and CFRP laminate under quasi-static loading. *Construction & Building Materials*, *98*, 489–501. doi:10.1016/j.conbuildmat.2015.08.130

Ashby, M. F., Ferreira, P. J., & Schodek, D. L. (2009). *Nanomaterials, Nanotechnologies and Design. An Introduction for Engineers and Architects*. Boston: Elsevier.

Balaguru, P., & Chong, K. (2006). Nanotechnology and concrete: Research opportunities. *Proceedings of ACI Session on Nanotechnology of Concrete: Recent Developments and Future Perspectives*.

Balaguru, P. N. (2005). *Nanotechnology and concrete: Background, opportunities and challenges.* Paper presented at the International Conference on Applications of Nanotechnology in Concrete Design. doi:10.1680/aonicd.34082.0012

Barnett, S. J., Lataste, J. F., Parry, T., Millard, S. G., & Soutsos, M. N. (2010). Assessment of fibre orientation in ultra high performance fibre reinforced concrete and its effect on flexural strength. *Materials and Structures/Materiaux et Constructions, 43*(7), 1009-1023.

Bartos, P. J. M. (2006). Nanotechnology in construction: A roadmap for development. *Proceedings of ACI Session on Nanotechnology of Concrete: Recent Developments and Future Perspectives.*

Belytschko, T., Xiao, S. P., Schatz, G. C., & Ruoff, R. S. (2002). Atomistic simulations of nanotube fracture. *Physical Review B: Condensed Matter and Materials Physics, 65*(23), 2354301–2354308. doi:10.1103/PhysRevB.65.235430

Bensted, J., & Barnes, P. (2002). *Structure and Performance of Cements.* Spon Press.

Bilba, K., Arsene, M. A., & Ouensanga, A. (2003). Sugar cane bagasse fibre reinforced cement composites. Part I. Influence of the botanical components of bagasse on the setting of bagasse/cement composite. *Cement and Concrete Composites, 25*(1), 91–96. doi:10.1016/S0958-9465(02)00003-3

Birgisson, B., Taylor, P., Armaghani, J., & Shah, S. P. (2010). American road map for research for nanotechnology-based concrete materials. *Transportation Research Record, 2142*, 130–137. doi:10.3141/2142-20

Bouhicha, M., Aouissi, F., & Kenai, S. (2005). Performance of composite soil reinforced with barley straw. *Cement and Concrete Composites, 27*(5), 617–621. doi:10.1016/j.cemconcomp.2004.09.013

Brandt, A. M. (2008). Fibre reinforced cement-based (FRC) composites after over 40 years of development in building and civil engineering. *Composite Structures, 86*(1-3), 3–9. doi:10.1016/j.compstruct.2008.03.006

Chen, J., & Poon, C. (2009). Photocatalytic construction and building materials: From fundamentals to applications. *Building and Environment, 44*(9), 1899–1906. doi:10.1016/j.buildenv.2009.01.002

Chen, L., Mindess, S., & Morgan, D. R. (1995). Comparative toughness testing of fiber reinforced concrete. *Testing of Fiber Reinforced Concrete, 47-75.*

Choolaei, M., Rashidi, A. M., Ardjmand, M., Yadegari, A., & Soltanian, H. (2012). The effect of nanosilica on the physical properties of oil well cement. *Materials Science and Engineering A, 538*, 288–294. doi:10.1016/j.msea.2012.01.045

Chung, D. D. L. (1998). Self-monitoring structural materials. *Materials Science and Engineering R Reports, 22*(2), 57–78. doi:10.1016/S0927-796X(97)00021-1

Dweck, J., Ferreira da Silva, P. F., Büchler, P. M., & Cartledge, F. K. (2002). Study by thermogravimetry of the evolution of ettringite phase during type II Portland cement hydration. *Journal of Thermal Analysis and Calorimetry, 69*(1), 179–186. doi:10.1023/A:1019950126184

Fanella, D. A., & Naaman, A. E. (1985). Stress–strain properties of fiber reinforced mortar in compression. *Journal of the American Concrete Institute, 82*(4), 475–483.

Feynman, R. P. (1960). There's plenty of room at the bottom. *Engineering and Science, 23*(5), 22–36.

Fiore, V., Scalici, T., Di Bella, G., & Valenza, A. (2015). A review on basalt fibre and its composites. *Composites. Part B, Engineering, 74*, 74–94. doi:10.1016/j.compositesb.2014.12.034

Flores, I., Sobolev, K., Torres-Martinez, L. M., Cuellar, E. L., Valdez, P. L., & Zarazua, E. (2010). Performance of cement systems with nano-SiO2 particles produced by using the sol-gel method. *Transportation Research Record, 2141*, 10–14. doi:10.3141/2141-03

Gay, C., & Sanchez, F. (2010). Performance of carbon nanofiber-cement composites with a high-range water reducer. *Transportation Research Record, 2142*, 109–113. doi:10.3141/2142-16

Ghavami, K. (2005). Bamboo as reinforcement in structural concrete elements. *Cement and Concrete Composites, 27*(6), 637–649. doi:10.1016/j.cemconcomp.2004.06.002

Giner, V. T., Baeza, F. J., Ivorra, S., Zornoza, E., & Galao, Ó. (2012). Effect of steel and carbon fiber additions on the dynamic properties of concrete containing silica fume. *Materials & Design, 34*, 332–339. doi:10.1016/j.matdes.2011.07.068

Grove, J., Vanikar, S., & Crawford, G. (2010). Nanotechnology new tools to address old problems. *Transportation Research Record, 2141*, 47–51. doi:10.3141/2141-09

Habel, K., Viviani, M., Denarié, E., & Brühwiler, E. (2006). Development of the mechanical properties of an Ultra-High Performance Fiber Reinforced Concrete (UHPFRC). *Cement and Concrete Research, 36*(7), 1362–1370. doi:10.1016/j.cemconres.2006.03.009

Haddad, R. H., Al-Rousan, R., Ghanma, L., & Nimri, Z. (2015). Modifying CFRP-concrete bond characteristics from pull-out testing. *Magazine of Concrete Research, 67*(13), 707–717. doi:10.1680/macr.14.00271

Hammel, E., Tang, X., Trampert, M., Schmitt, T., Mauthner, K., Eder, A., & Pötschke, P. (2004). Carbon nanofibers for composite applications. *Carbon, 42*(5-6), 1153–1158. doi:10.1016/j.carbon.2003.12.043

Han, B., Sun, S., Ding, S., Zhang, L., Yu, X., & Ou, J. (2015). Review of nanocarbon-engineered multifunctional cementitious composites. *Composites. Part A, Applied Science and Manufacturing, 70*, 69–81. doi:10.1016/j.compositesa.2014.12.002

Hassan, T., & Rizkalla, S. (2003). Investigation of bond in concrete structures strengthened with near surface mounted carbon fiber reinforced polymer strips. *Journal of Composites for Construction, 7*(3), 248–257. doi:10.1061/(ASCE)1090-0268(2003)7:3(248)

Hosseini, P., Booshehrian, A., & Farshchi, S. (2010). Influence of nano-SiO2 addition on microstructure and mechanical properties of cement mortars for ferrocement. *Transportation Research Record, 2141*, 15–20. doi:10.3141/2141-04

Hou, P., Kawashima, S., Kong, D., Corr, D. J., Qian, J., & Shah, S. P. (2013). Modification effects of colloidal nanoSiO2 on cement hydration and its gel property. *Composites. Part B, Engineering, 45*(1), 440–448. doi:10.1016/j.compositesb.2012.05.056

Hou, T. C., & Lynch, J. P. (2005). *Conductivity-based strain monitoring and damage characterization of fiber reinforced cementitious structural components.* Paper presented at the Proceedings of SPIE - The International Society for Optical Engineering. doi:10.1117/12.599955

Huang, H. L., Lin, C. C., & Hsu, K. (2015). Comparison of resistance improvement to fungal growth on green and conventional building materials by nano-metal impregnation. *Building and Environment, 93*(P2), 119–127. doi:10.1016/j.buildenv.2015.06.016

Iijima, S. (1991). Helical microtubules of graphitic carbon. *Nature, 354*(6348), 56–58. doi:10.1038/354056a0

Ivorra, S., Garcés, P., Catalá, G., Andión, L. G., & Zornoza, E. (2010). Effect of silica fume particle size on mechanical properties of short carbon fiber reinforced concrete. *Materials & Design, 31*(3), 1553–1558. doi:10.1016/j.matdes.2009.09.050

Jalal, M., Pouladkhan, A., Harandi, O. F., & Jafari, D. (2015). Comparative study on effects of Class F fly ash, nano silica and silica fume on properties of high performance self compacting concrete. *Construction & Building Materials, 94*, 90–104. doi:10.1016/j.conbuildmat.2015.07.001

Janssen, R. A. J., & Stouwdam, J. W. (2008). Red, green, and blue quantum dot LEDs with solution processable ZnO nanocrystal electron injection layers. *Journal of Materials Chemistry, 18*(16), 1889–1894. doi:10.1039/b800028j

Jirsak, O., & Dao, T. A. (2009). Production, properties and end-uses of nanofibres. *Proc. of the Nanotechnology in Construction, 3,* 95–100. doi:10.1007/978-3-642-00980-8_11

Khan, M. S. (2011). Nanotechnology in transportation: Evolution of a revolutionary technology. *TR News,* (277), 3-8.

Kim, H. K., Nam, I. W., & Lee, H. K. (2012). Microstructure and mechanical/EMI shielding characteristics of CNT/cement composites with various silica fume contents. *UKC 2012 on Science Technology, and Entrepreneurship,* 8-11.

Kim, Y. T., Han, J. H., Hong, B. H., & Kwon, Y. U. (2010). Electrochemical Synthesis of CdSe quantum-Dot arrays on a graphene basal plane using mesoporous silica thin-film templates. *Advanced Materials, 22*(4), 515–518. doi:10.1002/adma.200902736

Köksal, F., Altun, F., Yiğit, I., & Şahin, Y. (2008). Combined effect of silica fume and steel fiber on the mechanical properties of high strength concretes. *Construction & Building Materials, 22*(8), 1874–1880. doi:10.1016/j.conbuildmat.2007.04.017

Konsta-Gdoutos, M. S., Metaxa, Z. S., & Shah, S. P. (2010). Multi-scale mechanical and fracture characteristics and early-age strain capacity of high performance carbon nanotube/cement nanocomposites. *Cement and Concrete Composites, 32*(2), 110–115. doi:10.1016/j.cemconcomp.2009.10.007

Land, G., & Stephan, D. (2012). The influence of nano-silica on the hydration of ordinary Portland cement. *Journal of Materials Science, 47*(2), 1011–1017. doi:10.1007/s10853-011-5881-1

Lee, W., Kang, S. H., Kim, J. Y., Kolekar, G. B., Sung, Y. E., & Han, S. H. (2009). TiO2 nanotubes with a ZnO thin energy barrier for improved current efficiency of CdSe quantum-dot-sensitized solar cells. *Nanotechnology, 20*(33), 335706. doi:10.1088/0957-4484/20/33/335706

Li, B., & Zhong, W. H. (2011). Review on polymer/graphite nanoplatelet nanocomposites. *Journal of Materials Science, 46*(17), 5595–5614. doi:10.1007/s10853-011-5572-y

Li, G. Y., Wang, P. M., & Zhao, X. (2005). Mechanical behavior and microstructure of cement composites incorporating surface-treated multi-walled carbon nanotubes. *Carbon, 43*(6), 1239–1245. doi:10.1016/j.carbon.2004.12.017

Li, G. Y., Wang, P. M., & Zhao, X. (2007). Pressure-sensitive properties and microstructure of carbon nanotube reinforced cement composites. *Cement and Concrete Composites, 29*(5), 377–382. doi:10.1016/j.cemconcomp.2006.12.011

Li, H., Xiao, H. G., & Ou, J. P. (2004). A study on mechanical and pressure-sensitive properties of cement mortar with nanophase materials. *Cement and Concrete Research, 34*(3), 435–438. doi:10.1016/j.cemconres.2003.08.025

Li, H., Xiao, H. G., Yuan, J., & Ou, J. (2004). Microstructure of cement mortar with nano-particles. *Composites. Part B, Engineering, 35*(2), 185–189. doi:10.1016/S1359-8368(03)00052-0

Li, H., Zhang, M., & Ou, J. (2006). Abrasion resistance of concrete containing nano-particles for pavement. *Wear, 260*(11-12), 1262–1266. doi:10.1016/j.wear.2005.08.006

Li, Z., Wang, H., He, S., Lu, Y., & Wang, M. (2006). Investigations on the preparation and mechanical properties of the nano-alumina reinforced cement composite. *Materials Letters, 60*(3), 356–359. doi:10.1016/j.matlet.2005.08.061

Ltifi, M., Guefrech, A., Mounanga, P., & Khelidj, A. (2011). *Experimental study of the effect of addition of nano-silica on the behaviour of cement mortars.* Paper presented at the Procedia Engineering. doi:10.1016/j.proeng.2011.04.148

Makar, J., Margeson, J., & Luh, J. (2005). Carbon nanotube/cement composites - early results and potential applications. *NRCC-47643, 3rd International Conference on Construction Materials: Performance, Innovations and Structural Implications*, (pp. 1-10).

Makar, J. M., & Beaudoin, J. J. (2003). Carbon nanotubes and their application in the construction industry. *Proceedings, 1 Inter. Symp. on Nanotechnology in Constructionst*, (pp. 331-341).

Mann, A. K. P., & Skrabalak, S. E. (2011). Synthesis of single-crystalline nanoplates by spray pyrolysis: A metathesis route to Bi2WO6. *Chemistry of Materials, 23*(4), 1017–1022. doi:10.1021/cm103007v

Mao, Q., Zhao, B., Sheng, D., Yang, Y., & Shui, Z. (1995). Resistance Changement of Compression Sensible Cement Specimen Under Different Stresses. *Workshop on Smart Materials*, (pp. 35-39).

Metaxa, Z. S., Konsta-Gdoutos, M. S., & Shah, S. P. (2010). Carbon Nanofiber-Reinforced cement-based materials. *Transportation Research Record, 2142*, 114–118. doi:10.3141/2142-17

Mokerov, V. G., Fedorov, Y. V., Velikovski, L. E., & Scherbakova, M. Y. (2001). New quantum dot transistor. *Nanotechnology, 12*(4), 552–555. doi:10.1088/0957-4484/12/4/336

Murata, Y., Tawara, H., Obata, H., & Takeuchi, K. (1999). Air purifying pavement: Development of photocatalytic concrete blocks. *J Adv Oxidat Technol, 4*(2), 227–230.

Musso, S., Tulliani, J. M., Ferro, G., & Tagliaferro, A. (2009). Influence of carbon nanotubes structure on the mechanical behavior of cement composites. *Composites Science and Technology, 69*(11-12), 1985–1990. doi:10.1016/j.compscitech.2009.05.002

Nayak, B. B., Behera, D., & Mishra, B. K. (2010). Synthesis of silicon carbide dendrite by the arc plasma process and observation of nanorod bundles in the dendrite arm. *Journal of the American Ceramic Society, 93*(10), 3080–3083. doi:10.1111/j.1551-2916.2010.04060.x

Nazari, A., Riahi, S., Riahi, S., Shamekhi, S. F., & Khademno, A. (2010). Benefits of Fe_2O_3 nanoparticles in concrete mixing matrix. *J Am Sci, 6*(4), 102–106.

Okada, T., Kawashima, K., Nakata, Y., & Ning, X. (2005). Synthesis of ZnO nanorods by laser ablation of ZnO and Zn targets in He and O2 background gas. *Japanese Journal of Applied Physics, Part 1: Regular Papers and Short Notes and Review Papers, 44*(1 B), 688-691.

Ou, J., & Han, B. (2009). Piezoresistive cement-based strain sensors and self-sensing concrete components. *Journal of Intelligent Material Systems and Structures, 20*(3), 329–336.

Peyvandi, A., Soroushian, P., Abdol, N., & Balachandra, A. M. (2013). Surface-modified graphite nanomaterials for improved reinforcement efficiency in cementitious paste. *Carbon, 63,* 175–186. doi:10.1016/j.carbon.2013.06.069

Peyvandi, A., Soroushian, P., & Jahangirnejad, S. (2013). Enhancement of the structural efficiency and performance of concrete pipes through fiber reinforcement. *Construction & Building Materials, 45,* 36–44. doi:10.1016/j.conbuildmat.2013.03.084

Pradhan, D., & Leung, K. T. (2008). Vertical growth of two-dimensional zinc oxide nanostructures on ITO-coated glass: Effects of deposition temperature and deposition time. *The Journal of Physical Chemistry C, 112*(5), 1357–1364. doi:10.1021/jp076890n

Qing, Y., Zenan, Z., Deyu, K., & Rongshen, C. (2007). Influence of nano-SiO2 addition on properties of hardened cement paste as compared with silica fume. *Construction & Building Materials*, *21*(3), 539–545. doi:10.1016/j.conbuildmat.2005.09.001

Rakesh, K., Renu, M., & Arun, K. M. (2011). Opportunities & Challenges for Use of Nanotechnology in Cement-Based Materials. New Delhi: Rigid Pavements Division, Central Road Research Institute (CRRI).

Ramamoorthy, S. K., Skrifvars, M., & Persson, A. (2015). A review of natural fibers used in biocomposites: Plant, animal and regenerated cellulose fibers. *The Polish Review*, *55*(1), 107–162. doi:10.1080/15583724.2014.971124

Revel, G. M., Martarelli, M., Bengochea, M. Á., Gozalbo, A., Orts, M. J., Gaki, A., & Emiliani, M. et al. (2013). Nanobased coatings with improved NIR reflecting properties for building envelope materials: Development and natural aging effect measurement. *Cement and Concrete Composites*, *36*(1), 128–135. doi:10.1016/j.cemconcomp.2012.10.002

Sáez De Ibarra, Y., Gaitero, J. J., Erkizia, E., & Campillo, I. (2006). Atomic force microscopy and nanoindentation of cement pastes with nanotube dispersions. *Physica Status Solidi (A). Applications and Materials Science*, *203*(6), 1076–1081.

Safiuddin, M., Gonzalez, M., Cao, J., & Tighe, S. L. (2014). State-of-the-art report on use of nano-materials in concrete. *International Journal of Pavement Engineering*, *15*(10), 940–949. doi:10.1080/10298436.2014.893327

Salvetat, J. P., Bonard, J. M., Thomson, N. B., Kulik, A. J., Forró, L., Benoit, W., & Zuppiroli, L. (1999). Mechanical properties of carbon nanotubes. *Applied Physics. A, Materials Science & Processing*, *69*(3), 255–260. doi:10.1007/s003390050999

Sanchez, F., & Sobolev, K. (2010). Nanotechnology in concrete - A review. *Construction & Building Materials*, *24*(11), 2060–2071. doi:10.1016/j.conbuildmat.2010.03.014

Selvam, R. P., Subramani, V. J., Murray, S., & Hall, K. (2009). *Potential application of nanotechnology on cement based materials*. Potential Application of Nanotechnology on Cement Based Materials.

Senff, L., Labrincha, J. A., Ferreira, V. M., Hotza, D., & Repette, W. L. (2009). Effect of nano-silica on rheology and fresh properties of cement pastes and mortars. *Construction & Building Materials*, *23*(7), 2487–2491. doi:10.1016/j.conbuildmat.2009.02.005

Shaikh, F. U. A., & Supit, S. W. M. (2015). Chloride induced corrosion durability of high volume fly ash concretes containing nano particles. *Construction & Building Materials, 99,* 208–225. doi:10.1016/j.conbuildmat.2015.09.030

Singh, J. P., Dixit, G., Srivastava, R. C., Agrawal, H. M., Reddy, V. R., & Gupta, A. (2012). Observation of bulk like magnetic ordering below the blocking temperature in nanosized zinc ferrite. *Journal of Magnetism and Magnetic Materials, 324*(16), 2553–2559. doi:10.1016/j.jmmm.2012.03.045

Skinner, L. B., Chae, S. R., Benmore, C. J., Wenk, H. R., & Monteiro, P. J. M. (2010). Nanostructure of calcium silicate hydrates in cements. *Physical Review Letters, 104*(19), 195502. doi:10.1103/PhysRevLett.104.195502

Sobolev, K., Flores, I., Hermosillo, R., & Torres-Martínez, L. M. (2008). *Nanomaterials and nanotechnology for high-performance cement composites.* Paper presented at the American Concrete Institute, ACI Special Publication.

Taylor, H. F. W. (1997). Cement Chemistry (2nd ed.). London, UK: Thomas Telford. doi:10.1680/cc.25929

Thiruchitrambalam, M., Palkar, V. R., & Gopinathan, V. (2004). Hydrolysis of aluminium metal and sol-gel processing of nano alumina. *Materials Letters, 58*(24), 3063–3066. doi:10.1016/j.matlet.2004.05.043

Walters, D. A., Ericson, L. M., Casavant, M. J., Liu, J., Colbert, D. T., Smith, K. A., & Smalley, R. E. (1999). Elastic strain of freely suspended single-wall carbon nanotube ropes. *Applied Physics Letters, 74*(25), 3803–3805. doi:10.1063/1.124185

Wang, J. N., Su, L. F., & Wu, Z. P. (2008). Growth of highly compressed and regular coiled carbon nanotubes by a spray-pyrolysis method. *Crystal Growth & Design, 8*(5), 1741–1747. doi:10.1021/cg700671p

Wang, L., & Yamauchi, Y. (2009). Facile synthesis of three-dimensional dendritic platinum nanoelectrocatalyst. *Chemistry of Materials, 21*(15), 3562–3569. doi:10.1021/cm901161g

Wille, K., Naaman, A. E., El-Tawil, S., & Parra-Montesinos, G. J. (2012). Ultra-high performance concrete and fiber reinforced concrete: Achieving strength and ductility without heat curing. *Materials and Structures/Materiaux et Constructions, 45*(3), 309-324.

Xia, H., Feng, J., Wang, H., Lai, M. O., & Lu, L. (2010). MnO2 nanotube and nanowire arrays by electrochemical deposition for supercapacitors. *Journal of Power Sources, 195*(13), 4410–4413. doi:10.1016/j.jpowsour.2010.01.075

Yazdanbakhsh, A., Grasley, Z., Tyson, B., & Abu Al-Rub, R. K. (2010). Distribution of carbon nanofibers and nanotubes in cementitious composites. *Transportation Research Record, 2142*, 89–95. doi:10.3141/2142-13

Zapata, L. E., Portela, G., Suárez, O. M., & Carrasquillo, O. (2013). Rheological performance and compressive strength of superplasticized cementitious mixtures with micro/nano-SiO2 additions. *Construction & Building Materials, 41*, 708–716. doi:10.1016/j.conbuildmat.2012.12.025

Zhang, J., Morsdorf, L., & Tasan, C. C. (2016). Multi-probe microstructure tracking during heat treatment without an in-situ setup: Case studies on martensitic steel, dual phase steel and β-Ti alloy. *Materials Characterization, 111*, 137–146. doi:10.1016/j.matchar.2015.11.019

Zhong, B. Q., & Zhu, Q. (2002). Study on the application of polypropylene fibre concrete in water and hydropower projects. *Water Res. Plann. Design, 1*, 54–58.

Zhong, Y., Liu, L., Wikman, S., Cui, D., & Shen, Z. (2016). Intragranular cellular segregation network structure strengthening 316L stainless steel prepared by selective laser melting. *Journal of Nuclear Materials, 470*, 170–178. doi:10.1016/j.jnucmat.2015.12.034

Chapter 5

Nanotechnology Future and Present in Construction Industry:
Applications in Geotechnical Engineering

Umair Hasan
Curtin University of Technology, Australia

Amin Chegenizadeh
Curtin University, Australia

Hamid Nikraz
Curtin University, Australia

ABSTRACT

After the introduction of nanotechnology, it has been widely researched in geotechnical engineering field. This chapter aims to study these advancements with specific focus on geotechnical applications. In-situ probing of soil and rock masses through nanomaterials may help in providing better safeguards against natural hazards. The molecular dynamics and finite element methods may also be used for the modelling of the nanostructures to better understand the material behavior, causing a bottom-up approach from nano to macroscopic simulations. Nanoclays, nano-metallic oxides and fibers (carbon nanotubes) can enhance the mechanical characteristics of weak, reactive and soft soils. Nanomaterials may also be used for improving the performance of reinforced concrete pavements by enhancing the thermal, mechanical and electrical characteristics of the concrete mixes. The chapter presents a review of the current researches and practices in the nano-probing, nanoscale modelling and application of nanomaterials for soil, pavement concrete mortar and subgrade stabilization.

INTRODUCTION

The recent advancements in microscopic sciences have enabled us to understand the behavior of materials on a nano scale. These new discoveries on the nanomaterial have yielded the development of the revolutionizing field of nanotechnology. Nanomaterial is a crystalline or non-crystalline material that has a size in the nanometer scale ranging from 1 nm to 100 nm. This technology has been attracting attention as due to the nano size, the manipulation of material properties can be more profitable and can be used to produce templates or models useful for the large scale applications. According to Hull et al. (2014), these advancements and understanding on the nanoscale have redefined human achievements in the fields of mechanization, medicine and engineering.

In essence, by carefully understanding the quantum effects which dictate the material behavior at the nanoscale and engineering the nanoparticles materials; we can basically control the size, shape, and the morphology; as well as the arrangement of materials on the nanometer level which in turns determines the characteristics of the synthesized nanoparticles. The reason behind the controlling influence of quantum effects on the nanometer scale is that the material dimensions are analogous to quanta of energy that produce fundamental excitations in materials. These phenomena are responsible for the transformation of material properties (Picraux, 2014). The fact that the characteristics of the material can be changed after converting to nanomaterial is governed by two principles. Firstly, the nanomaterials have a surface area greater than the parent material. This can increase the chemical reactivity and increase the strength of the electronic properties. Secondly, the quantum effects of nanoscale materials are dominated mainly on the effect of optical and magnetic properties of the material.

The chapter is focused on applications of nanotechnology in geotechnical engineering. The following sections are designed to address the issue appropriately.

- Investigation of nano-sensors for soil characterisation.
- Consideration of empirical relations as part of nanotechnology impact to predict the behaviour of soil and rock.

Figure 1. Structure of the chapter

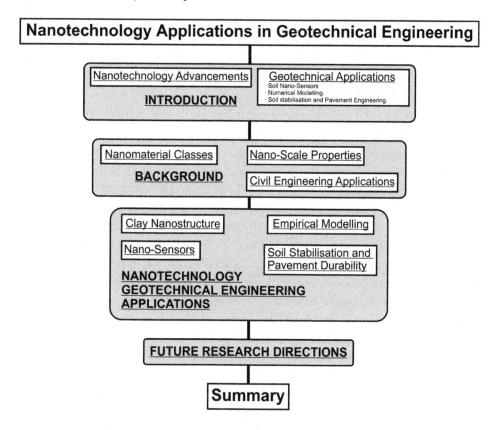

- Applications of nanotechnology for geotechnical applications such as soil stabilisation and pavement engineering.
- Future applications of nanotechnology in geotechnical engineering.

The structure of this chapter has also been illustrated in Figure 1.

BACKGROUND

The smaller size of the nanomaterials, i.e., having dimensions less than 100nm. The nanomaterials can be either naturally existing or may be the result of human activities. The naturally occurring nanomaterials include materials formed as a result of climatic or environmental factors like lunar and volcanic dusts, minerals, magnetotactic bacteria or MTB. The nanomaterials that are available through human interventions or actions are further categorized into two classes; incidental and engineered nanomaterials. The products of industrial activities like fumes from burning or consumption of fossil fuels and welding activities are called the incidental nanoparticles such as the anthropogenic materials. Due to the origin of the natural and the accidental nanomaterials, they have different types of shapes which may range from largely irregular to partially regular shapes. On the other hand the nanomaterials that are manufactured in the laboratories mostly have regular shapes like spherical, circular or ring-shape and tubes etc. (A. P. Kumar, Depan, Singh Tomer, & Singh, 2009).

Several researches in the fields of nanotechnology and nanoscience are being focused to produce newer and more efficient nanomaterials (Kuhlbusch, Asbach, Fissan, Gohler, & Stintz, 2011) (Yilbas, 2014). There are two different techniques that can be used for the production of nanomaterials; top-down and bottom-up (Hashmi, 2014). The top-down approach is basically the creation of small materials based on the larger material, thus moving from "top" to "down" size. While the bottom-up approach is to combine or merge atoms or molecules into larger particles; therefore, going from "bottom" to the "upper" level. The nanomaterial itself can be divided into four groups based upon how many dimensions of the nanomaterial are in the nanometric scale. The choice of the type or class of any nanoparticle for the research or commercial applications mostly depends upon the desired results. The shapes are further elaborated in the illustration provided Figure 2. The four classes are namely:

- **Zero-Dimensional:** These nanomaterials have all three dimensions at the nanometric scale and are therefore equiaxed. The examples are nanoparticles; for example metal oxides like silica, semiconductors and fullerenes.
- **One-Dimensional:** The formation of the elongated tube-like microstructure due to the presence of one larger dimension and two nanometric dimensions are labelled as one-dimensional. The examples are nanotubes like CNTs, nanorods and the nanowires.
- **Two-Dimensional:** These nanomaterials have two larger dimensions with only one dimension in the nanometric scale. The examples are thin filmed

Figure 2. Classes of nanomaterials
Sources: A. P. Kumar et al., 2009; Schodek, Ferreira, and Ashby (2009); and Sajanlal, Sreeprasad, Samal, and Pradeep (2011).

2 Dimensional Nanolayers

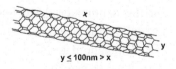
1 Dimensional Nanotubes

Eqiuaxed (0D) Nanoparticles

nano-layers or nano-sheets (multilayer, monolayer, self-assembled, mesoporous). Clay is one of the most naturally existing types of this class of nanomaterials due to the presence of various layered silicate minerals.
- **Three-Dimensional Nanocomposites:** Nanocomposite, nanograined, microporous, mesoporous, intercalation, organic-inorganic hybrids are the examples of three-dimensional nanomaterials. These materials exist in the form of multi-phased solid materials and have one of these phases containing the nanometric characteristics of the one-, two- or three-dimensional nanomaterials.

NANO-SCALE PROPERTIES

The surface area per mass of the nanomaterials is considerably higher than the surface area of the large scale particles producing far improved reactivity (Oberdürster, 2000) and the structure and composition of material are no longer the controlling parameters of material properties. Therefore, the nanomaterials exhibit novel phenomena dictated by quantum effects and the higher surface interactions.

Many studies have focused on identifying the nano-scale properties (Virji & Stefaniak, 2014) as the control of quantum effects can aid to tune the material properties to become more suitable for the desired function. The advantages of the properties of nano-sized materials are:

Electrical Nature

As stated earlier, the nanomaterial has a larger surface area; therefore, the nanomaterial can have a greater energy than the regular size material. The energy band gradually turns to molecular orbitals. The electric resistivity will reduce with the availability of more contact points.

Examples of applications: In cement mortars, for example in case of concrete pavements, study has shown the inclusion of carbon nanotube fiber in the mixture may result in the reduced mortar electrical resistivity as further points of contact are present (G. Y. Li, Wang, & Zhao, 2005).

Magnetic Nature

Magnetic strength is the measurement of the level of magnetism. The level of magnetism increases with grain size decrement and increased particles' surface area per unit volume. Thus nanomaterials have better characteristics for the improvement of magnetic properties.

Examples of applications: Application on the ship's engines, ultra-sensitive instruments and magnetic resonance imaging (MRI) diagnostic tool (Shi et al., 2003). CNTs can also provide electro-magnetic field shielding to cementitious composites (Sanchez & Sobolev, 2010).

Mechanical Properties

Nanomaterial has greater hardness, high strength and elasticity moduli, high aspect ratio and scratch resistance when compared with the regular-sized material. Nanomaterials like nano-silica and nano-ferric oxide can be inducted in cement matrix for crack-bridging (H. Li, Xiao, Yuan, & Ou, 2004).

Examples of applications: CNTs due to their higher aspect ratios, smaller sizes and high strengths have been used in cement mortars or as reinforcements in cementitious composites to strengthen the mortar matric, such as for reinforced concrete pavements (Makar, Margeson, & Luh, 2005; Salemi & Behfarnia, 2013). Another nano material, nano-montmorillonite was investigated in laboratory to cause increment in the Portland cement mixture compressive strength (Chang, Shih, Yang, & Hsiao, 2007).

Optical Properties

Nanocrystalline systems have interesting optical properties, which varies with the nature of conventional crystals.

Examples of applications: in optoelectronics, electro chromic for liquid crystal displays (LCDs).

Chemistry of Nanoparticles

The addition of surface area will increase the chemical activity of the material.

Examples of applications: Increased pozzolanic reactions and hydration of cementitious materials through adding nanoscale materials like nano-silica (Jo, Kim, Tae, & Park, 2007). The applications are for concrete pavements and soils stabilized with nanomaterial and cementitious compounds.

ROLE OF NANOTECHNOLOGY IN CIVIL ENGINEERING APPLICATIONS

Nano technology is among the most rapidly developing technologies in the world. The advancements and applications to practical engineering problems have come a long from the initial lecture about the "room at the bottom" by Richard Feynman (Feynman, 1961) on nanotechnology. The civil engineering world is also get involved in the applications of nanotechnologies for various purposes including development of nano construction materials and soil stabilization techniques (Marzieh Kadivar, Kazem Barkhordari, & Mehdi Kadivar, 2011). Nano technology is a technology that allows modelling and manipulation of material particles that have a very small size on the nanometer scale, which is comparable to 1 nanometer or 10^{-9} meters. Therefore, this technology is applied to construction materials, in order to achieve more solid material composition and produce a greater compressive strength.

Nanotechnology is often applied in the construction industry to develop newer and more improved materials that can be durable. For instance, to obtain a more dense material of mortar, fly ash as can be used as supplementary material in the mixing. Fly ash as by product of coal combustion waste material is also expected to be more valuable and reducing the adverse effects of the of coal waste. Nano technology is also applied to the fly ash material, and form a new material which is fly ash nano. The fly ash nano addition during mortar mixing can affect the compressive strength of the mortar. Another example is the use of carbon nanotubes (CNTs) for producing stronger and more durable concrete.

Carbon Nanotubes have a much higher strength when compared with steel, prepared from carbon manipulation and modification. Structurally they are composed of single or more graphene layers, which is a sheet shaped by carbon atoms bonded in hexagonal honeycomb shape, forming a cylinder structure (S. Kumar, Kolay, Malla, & Mishra, 2012). Due to the higher strength and Young's modulus of CNT, they have been found to cause enhancement of flexural as well as compression strengths of cement-based materials (Nochaiya & Chaipanich, 2011). However the applications of nanotechnology extends well beyond concrete or cement strengthening and hydration acceleration to the development of corrosion resistant steel (Abdel Hameed, Abu-Nawwasb, & Shehataa, 2013), self-healing and antipollution TiO_2 coatings (W. Li, Li, & Wang, 2010; Stankiewicz, Szczygieł, & Szczygieł, 2013) and fire-resistant steel and glass products (Z. Wang, Han, & Ke, 2006). Although, the uses of the nanotechnology in the fields of civil engineering range from material engineering to structural applications, geotechnical engineers have also been utilizing nanotechnology advancements for enhancing the behaviour of the geotechnical structures to achieve the desired satisfactory safety, durability, strength and performance results.

NANOTECHNOLOGY FOR GEOTECHNICAL ENGINEERING APPLICATIONS

The modern geotechnical engineering has arrived at the point of induction of state-of-the-art technologies to find solutions to common engineering problems including development of new materials to achieve more stable soil composition and produce a greater compressive strength as well as to increase the durability of the soil mass. The novel approach of nanotechnology can be applied to perform nanoscale monitoring and investigation of the soil and rock masses. It can be used to not only gain a more accurate insight of the mineralogy and morphological composition of the soil. Furthermore, the manipulation of the soil and rock fabric at the molecular or atomic level can be facilitated.

The applications also extend towards the improvement of the engineering characteristics of the underlying soil strata for pavement engineering. Many of the problems that arise in the field of pavement engineering are due to the issues related with the engineering properties of the soil. The soil-cement mixtures are often used as sub-base or base for rigid or flexible pavements respectively. In order to meet the safety and operational requirements, the cement-soil mixtures are required to pass the threshold of the expected strength, flexural and tensile; and operational, durability and life-expectancy requirements (Little, Males, Prusinski, & Stewart, 2000). Another application of cement in pavement construction is the paving of concrete pavements which can be paved in four different types (Portland Cement Association, n.d.):

- **Dowel-Plain Concrete Pavement:** Utilize dowels for the transference of loads and prevention of faulting.
- **Plain Concrete Pavement without Dowel:** The load transference and faulting preventions is achieved through the interlocking mechanism of the aggregates.
- **Reinforced Pavement with Contraction Joints:** They have steel reinforcements and dowels in the contraction joints.
- **Reinforced Pavement without Contraction Joints:** They have continuous longitudinal steel reinforcements.

The durability of the concrete pavements is largely influenced by factors such as the soil matrix, aggregates and the air voids bonds in the concrete (Pospíchal, Kucharczyková, Misák, & Vymazal, 2010). Therefore, the design life and durability of the concrete pavements can be increased if the mechanical properties such as the strength and the microstructure of the concrete composition can be modified to cause better inter-particle interactions through the grafting of particles on aggregates, admixtures and or cement particles. This can be achieved through the use of nanomaterials which may induce novel characteristics in the matrix or improve the existing properties to promote targeted mechanical or microstructural properties (Sanchez & Sobolev, 2010).

Even before the establishment of a proper nanotechnology discipline in geotechnical engineering, engineers and researchers have focused on clay particles with sizes less than 2 μm. In other words, clays tend to behave "nano-mechanically".

As it is with other materials, the soil particles change properties and behavior with the variations in the scale such as changing to clay and platy mica from primarily large feldspar or quartz. This variation in the properties and the behavior increasingly become difficult to monitor and manipulate as the sizes reduce from large to significantly smaller scales. This falls in line with the principal idea of controlling nanoparticles properties and behavior to achieve desired outcomes (Drexler, 1996). Therefore, a carefully controlled manipulation of properties like size, particle size distribution (PSD), texture, specific surface, inter-particle interactions and chemistry and composition on the nanoscale can serve as blocks for building improved materials on the large scale (National Research Council, 2006). Based on the desired outcomes, the new or modified materials may possess novel properties due to their chemical composition, assembly, sizes and morphology that may be engineered to behave in the designed fashion.

NANOSTRUCTURE OF CLAYS

The chemical and physiological interactions between the particles are significantly important at the nanoscale. This phenomenon is observable in the clay particles, which are structured as hexagonal sheets of mineral silicate, due to their smaller sizes. The behavior of the clay is dictated by its mineralogical and morphological composition. The clay minerals fall in the hydrous phyllosicicates group, as significant amount of water can be found between their sheets which can be further increased or decreased due to the changes in atmospheric conditions. Controlled further addition of water increases the plastic behavior of the clay particles and renders their remolding significantly easier. However, the addition or absorption of water by the clay particles usually results in the expansion of the clay as the water particles penetrate the voids between the silicate layers of the micro- to submicroscopic crystals. In nature, it is quite rare to find virgin clay particles as they often found mixed with other clays and minerals like feldspar, mica, carbonates and quartz etc.

Since at the nanoscale particle interactions and factors like van der Waals forces are more dominant than gravity, the ratio of surface area to the mass of the particle becomes important. This behavior can be observed for fine reactive clays such as bentonite which mainly contains of very fine Na-montmorillonite particles (surface area of 800 m^2/kg) (National Research Council, 2006). This high surface area exposes most of the constituent atoms to reactions as the nanoparticles have higher adsorption ability and are more sensitive relatively. The nanoparticles in clay also exhibit different pH, ionic concentration and surface chemistry and understanding of these behaviors on the nanoscale can affect their durability over time.

Clay minerals are structured as two-dimensional sheets of SiO_4 tetrahedrons. Each tetrahedron is linked with another through the sharing of its three corners, forming a hexagonal mesh. The fourth oxygen molecule which is perpendicular to the sheet or is at the top of each tetrahedron connects to the adjacent octahedral layer which has smaller cations as Mg^{2+} (magnesium) or Al^{3+} (aluminum). The shared oxygen atoms (O^{2-}) and the hydroxyls (OH), which are unshared, are found at the same level at the junction of the two layers. The partial substitution of the silicon atoms with aluminum has been commonly observed. However, some cases of ferric iron substitution have also been found to replace the cations. The layer arrangement of the tetra- and octahedral sheets classifies the layer.

The classification of clay minerals can be either as 1:1 or 2:1. The originating parameter behind this nomenclature is as they are formed by octahedral hydroxide and tetrahedral silicate sheets. This has been further elaborated in the following description. Clay with 1:1 arrangement has one octa- and one tetrahedral sheet, such as serpentine and kaolinite. However, for the clay with 2:1 arrangement, two tetrahedral sheets sandwich one octahedral sheet, such as montmorillonite, talc and vermiculite. Montmorillonite is from the smectite family, which is the most dominant clay mineral found in a special category of clays called expansive soils which are commonly found in many arid and semi-arid regions around the world and cause many issues for the projects on such structures (U. Hasan, A. Chegenizadeh, M. Budihardjo, & H. Nikraz, 2015), has a 2:1 layer arrangement. This implies that it has two tetrahedral sheets sandwich the octahedral sheet. Each of the two tetrahedral sheets of the layer have one unshared oxygen vertex. These vertexes from adjacent layers point towards each other but do not have a lot of attraction which provides space for the exchangeable cations and water molecules. Due to the replacement of the silicon by aluminum atoms as well as Al^{3+} by the Mg^{2+} at some parts of the octahedral sheets provides the negative charge that gives montmorillonite a high cation exchange capacity (CEC). This high CEC value is in turns responsible for the sensitivity of the expansive soils to the climatic variations.

The fundamentally "nano-nature" of the clay particles which are the building blocks for rock and soil masses, renders the study and research in the field of nanotechnology imperative for better understanding of the mineralogy, crystallography and the soil surface chemical reactions. In addition, better establishment of the clay nano-behavior can help to produced engineered clay particles which may have the desired strength, texture, plasticity and rheological characteristics. The currently developing nano-probing technologies such as MEMS and FTIR techniques are being used to probe the crystallography and mineralogical changes in the clay particles such as the morphology, surface charges and hydrophobicity (National Research Council, 2006).

Apart from some developments, most of nanotechnology applications of the nanotechnology are still in the theoretical and exploratory phase. Researchers are investigating the possibilities of engineering solutions to geotechnical problems at the nanoscale such as building clay cores and soil bases through self-assembling clay particles to achieve the desired rheology, chemistry, frictional and consolidation control. Nano technology is also applied in the use of reinforcing material like carbon nanotubes (CNTs) for cement based materials. Engineered clay particles may have the desired strength to overcome or counter the expected applied loads and other external factors such as climatic variations, for example, in the case of montmorillonite clay minerals, where the swelling potential is influenced by the surface chemistry. Some of the advancements have already surfaced at the commercial scale, such as the production of precipitated carbonates and kaolin for paint and paper industries.

Further developments may be used to produce engineered nanoparticles that can be induced in the rock and soil mass for the purpose of research and investigation and function as nano-sensors to produce useful and accurate data about the in-situ particle characterization, chemical analysis and the inter-particle interactions. Biogeomembranes are one of the technologies being researched and adapted for geotechnical engineering applications. The technique employs biological organisms for the improvement and stabilization of the soil and groundwater. The nanotechnology applications in geotechnical engineering are essentially:

- **Nanoscale Research:** Soil Characterisation.
- **Numerical Modelling:** Empirical relations and modelling soil and rock behaviour.
- Soil Stabilisation and Pavement Durability.

NANO-SENSORS FOR SOIL CHARACTERISATION

Background

The accuracy of the acquisition of information and data about the behavior of soil and rock masses such as the mineralogy, morphology, crystal structure and chemistry can be greatly enhanced if they can be studied at the basic level trough nano-sensors. The technology can make the geomechanical, spatial, temporal and visual data available about the response of the soil and rock masses about the effect of the external climatic, human and loads with real-time in-situ projection on a computer display through intercommunicating nano-sensors with minimum disturbances to the soil strata and the climate as compared to digging boreholes.

Application of the nanotechnology can be the development of nano-sensors and Micro-Electro-Mechanical Systems (MEMS) to be used for the characterization and monitoring of the clay particles, (Z. L. Wang, Liu, & Zhang, 2002). The MEMS sensors can be employed along with the microdrilling technique which can drill 1.25 inches wide holes with probable 5000 feet depth. The real-time in-situ monitoring may be beneficial for the understanding of the ground strata response to natural hazards and loads from external structures. It can also be used for developing self-healing leach barriers in waste management.

The applications of nano-sensors in the field of geotechnical investigations at the laboratory scale are widely documented. Such systems involve different techniques which vary based on their approaches, amount and type of data needed, cost-effectiveness, accuracy and practicability Table 1.

The applications of nanotechnology for the probing of the texture, composition and chemistry of the materials at the nanometric scale will now be further elaborated.

Atomic Force Microscopy (AFM)

The atomic force microscopy has been among the revolutionizing outcomes of the advancing nanotechnology. It uses the scanning probe technology to provide morphology, grain sizes, texture, nano-mechanical specifications and other rheological characteristics at the nanometric scale. The sharp probe is in form of cantilever tip and works on Hooke's law. It works on three modes based on tip motion; contact mode, tapping mode and non-contact mode. The operation of the device is in ultra-high vacuum to give true atomic resolution of the testing materials. The device does not require specific preparation techniques to be applied during the sample preparation phase such as coatings or vacuum storage. However, the scan area is

Nanotechnology Future and Present in Construction Industry

Table 1. Summary of laboratory nano-sensing systems

Purpose	Benefits	Limitations
Electron Microscopy		
The electron microscopes bombard the surfaces of the test specimen with concentrated beam of high energy electrons, the detectors are furnished with the devices to scan the electron signals from the atoms and provide information about the morphology and mineralogy of the specimen.		
Scanning Electron Microscopy (SEM)		
• Uses scattered electrons to give images of the samples. • Soil Analysis and detection. • Provide particle surface imagery, size and shape of the particles.	• Easy to prepare and place the samples. • Gives accurate live images of the particle surface with 0.4 nm resolution. • The speed of analysis is significantly high. • The data produced has a high accuracy rate and can pick up minute details on the particle surface as it uses the electro-diffraction techniques. • It is able to determine the elemental composition of the test specimens.	• Due to non-conductive nature of most clay particles and geomaterials, the charged bombarded by the electron beam tends to segregate and provides imagery that is difficult to interpret. However, this "particle-charging effect" can be rectified through the use of conductive coatings such as of platinum or carbon, as thin as 5nm. • It can only provide data about the particle surface. • The procurement, operation and maintenance of the devices are very expensive. • It is limited in its scope of operation as it cannot produce atomic scale imagery.
Transmission Electron Microscopy (TEM)		
• Uses transmitted electrons that pass through the samples. • Produces data about minute details on the nanoscale about the elemental composition of the samples.	• High magnification level with image resolution of 0.05nm. • Provides information about the internal particle morphology and crystallography. • The quality and accuracy of the images produced are very high.	• The sample has to be polished to account for the depth of field. • Sample has to prepared and placed in specialized TEM grid, making sample preparation difficult. • Due to the advanced nature of test and production of 2D images, it requires expert user to interpret the procured data. • The high energy and voltage of the electron ray may destroy sensitive samples.
X-Ray Diffraction (XRD)		
• Uses the x-rays to give information about the mineralogy and crystallography of the specimens through the peaks and angle (2-theta) of the reflection spectra.	• The sample does not need to be polished or prepared with specific coatings. • The crystallography and mineralogical data obtained is highly accurate.	• The quality and accuracy of the data is limited by the size of the specimens. • As non-crystalline samples give weak signals, the technique is more suitable for the crystalline samples.

continued on following page

Table 1. Continued

Purpose	Benefits	Limitations
Fourier Transform Infrared (FTIR)		
• Produces infrared absorption spectrum about the test specimens and works on the principle of infrared spectroscopy	• Nano-clays and other objects at the nanometric scale can be studied at high resolution. • Due to Fellget's advantage, the scanning time is very low and has higher signal-noise ratio, making it highly useful for kinetic work. • Second benefit is Jacquinot's advantage giving a high optical throughput. • Conne's advantage provides the control over the wavelength calibration.	• FTIR has to be used in conjunction with the Fourier transform spectroscopy so that the FTIR interferograms can be interpreted. • It uses only single beam which may be problematic with atmospheric variations and longer scanning rates in infrared-absorbing gases can affect the accuracy of the results. • Fellget's disadvantage from noise on one region of the spectrum can affect the rest of the spectrum.

smaller than the SEM and the scanning image is also slower than SEM and may damage the samples due to the extra period of exposure.

Due to the capability of sensing changes in the material microstructure, chemistry and morphology at higher spatial resolutions; the AFM has been used for the investigation of the surface morphology of the cement pastes undergoing hydration, identify the phases of hydration, carbonation and the cohesion (Lesko, Lesniewska, Nonat, Mutin, & Goudonnet, 2001; T. Yang, Keller, & Magyari, 2002). The calcium-silicate-hydrate (C-S-H) formation is commonly observed during hydration of soils stabilized with the cement composites. When the C-S-H was observed under AFM, similar C-S-H cluster formation at the nanometric scaled of (Plassard, Lesniewska, Pochard, & Nonat, 2004). Research and nanoscale investigation have yielded that the antiparticle attractive forces between the particles of cement have an electrostatic nature (Lesko et al., 2001) (Plassard, Lesniewska, Pochard, & Nonat, 2005).

Micro- and Nano-Electro-Mechanical Systems

The Micro-electro-mechanical systems (MEMS) assimilate the mechanical and electronic sensors through the use of the micro-level fabrication engineering. The monitoring of the geotechnical structures like rocks and soil strata, leachates, geo-membranes and groundwater through micro-sensors is accurate, efficient and cost-effective. These sensors have comparatively higher sensitivities than the conventional sensors due to the denser sensing arrays and are able to estimate the moisture content, water table, strain and strain-rate, displacement, temperature, and other parameters of the soil's chemical and mechanical characteristics. However, as the data has to be collected wirelessly through radio signals by methods like magnetic, gravitational, electromagnetic and seismic potential, the frequency depth of penetration must be considered and necessary data filtration has to be performed for better and overtime, automatic, collection of the subsurface data about the spatial rock and soil profile (National Research Council, 2006).

The fiber-optic seismic sensor technology is available for the boreholes that can provide high resolution imagery of the subsurface strata without the boreholes interferences. Another example is the Badger Explorer (Bradbury, 2004) which collects data like porosity, temperature, bulk density and pore pressure from the ground. Further advancements in sensing, data collection and delivery may accelerate the geotechnical applications and faults in the geosystems such as leakages is landfills, seepages, displacements and settlements can be monitored in real-time. The Smardust MEMS system is one example of such advancements where the operation is performed wirelessly (Warneke et al., 2002).

The further miniaturization of the MEMS sensors is the nano-electro-mechanical systems (NEMS). This next generation technology uses materials like carbon nanotubes (CNTs) for nano-probing. The CNTs were discovered in 1991 (ref) and are among the most popularly researched materials. It is able to perform as strain sensor when loads are applied as it can monitor the changes in resistance (Zhao et al., 2010). The parameter of strain is especially important from the geotechnical engineering perspective as it is often necessary to monitor the deformation of the strata at a resolution >10nε, that can be crustal deformation due to oceanic tides. This can be achieved through the use of conventional sensing technology but such sensors have large sizes. Fiber Bragg grating (FBG) has been researched under laboratory and filed conditions as an alternative nano-strain resolution sensor for geoengineering applications (Zuyuan, Qingwen, & Tokunaga, 2012). The researchers performed measurements of the changes in strain as induced through seismic activity of an earthquake of M7.1 activity at Aburatsubo Bay which has a Japan Meteorological Agency seismic intensity scale of 2. The nano-sensors are further being improved to be capable of sensing both dynamic and static strain with a high nano-strain resolution (Qingwen, Zuyuan, Tokunaga, & Hotate, 2011; Zuyuan, Liu, & Tokunaga, 2013).

Nanoindentation

The indentation hardness test methods are used for the determination of the hardness of the materials to deformations (Tabor, 1996). The nanoindentation is a type of the indentation testing that is used for the indentation of smaller volume of materials which have thicknesses around 100nm (Poon, Rittel, & Ravichandran, 2008). Small load is applied through a probing tip that has the size in the nanometric scale, this produces the difficulty of the identification of the contact area as the dimensions of the contact area are very small compared to the large scale indentation techniques. The problem can be mitigated through the use of the AFM or SEM techniques but may introduce further complications in the testing such as sample preparation and or issues related with the operation of the electron microscope. In actual practice, Berkovich tip is used for determining the indentation hardness of the test material. The tip is named after its inventor E.S. Berkovich and is basically geometrical self-similar triple-sided pyramid with a flat profile (Berkovich, 1950). Since the geometry of the indentation tip is already know, the depths of penetrations are recorded to calculate the contact area. The results of the nanoindentation technique involves load and penetration depth and the load-displacement curves can be plotted to give the mechanical characteristics like Young's moduli and hardness (Oliver & Pharr, 2004). The example is the estimation of the material characteristics of cementitious composites (Constantinides, Ulm, & Van Vliet, 2003).

Nano-FTIR Systems

Nano-FTIR is a new technique that uses a combination of scattering-type scanning near-field optical microscopy (s-SNOM) and infrared spectroscopy (FTIR) and can have very sensitive resolution (< 20 nm). The FTIR nano-spectroscopy system Figure 3 records infrared spectrum with heat source much lower than the diffraction limit and characterizes the material.

Figure 3. Schematic setup of thermal-sourced s-SNOM system NeaSNOM
Sources: Huth et al., 2012; and Neaspec GmbH, n.d.

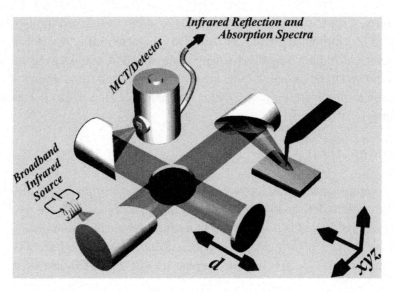

Such high resolution can help to identify chemical contagions by providing localized infrared spectrum accurate to nanometer scales at central points of particle which cannot be achieved by conventional FTIR (Huth et al., 2012). Therefore crystallography, electron paramagnetic and nuclear magnetic resonance, atomic probing, scanning and transmission electron microscopic imagery of the geosystems and geotechnical materials can help in identifying the chemical and mechanical phenomena occurring in the soil mass under in-situ conditions or after soil stabilization (Yao & Wang, 2005).

The technique has been researched for the mapping and characterization of the microstructural and mineralogical properties of the nano-systems through the use of lower intensity broadband source over wider spectral ranges (Hermann et al., 2013).

EMPIRICAL RELATIONS AND MODELLING SOIL AND ROCK BEHAVIOUR

The understanding of mechanical behavior of clays on nanoscale can be used to understand and predict the geomechanical behavior of the soils and rocks (Borja, 2011). In some cases, the fluid-interaction behavior of soils and rocks can be dictated by nanoparticles as observed for seismic dynamic weakening (De Paola et al., 2011). The governance of the proper understanding of the behavior of objects at nanoscale over the better knowledge, modelling and manipulation of the materials at larger scale also exists for the non-living materials. The technological advances in nanotechnology and revolutionizing approach of moving from nano- to the macroscopic level have garnered curiosity of the scientific research community and the industrial sectors toward nanotechnology.

The advancement in the fields of modelling and empirical simulations such as the classical and quantum Monte Carlo techniques, density functional theory, fast multigrid algorithms and the molecular dynamics simulations or MD simulations are also in line with the advancements in the field of nanotechnology. The computation and simulation techniques of various theoretical models and empirical relations has significant importance and provides much insight and control over the computational power during the complex systems simulations when able to model or manipulate the nanostructures to model or predict the behavior of the macroscopic scale structures (Musa, 2012).

Nano to microscale characterization and modelling also facilitate with highly accurate computational capabilities. This high accuracy is beneficial in predicting simulation empirical models of the geomaterial's behavior upon exposure to extreme settings. The ability of nano-mechanics in clarifying frictional response between graphite layers on a nanoscale has been determined as possible by using ab-initio methods (Neitola, Ruuska, & Pakkanen, 2005). The mathematical models for interpretations and or simulations of the nano-, micro- and macro-structural numerical problems can be carried out through the classical Newtonian mechanics and the quantum mechanics. The Newtonian mechanics relations can be applied through dislocation dynamics, classical mechanics and molecular methods (Hirth & Lothe, 1982). The Newtonian dynamic law governs the motion of atoms which for an "i^{th}" particle is (Musa, 2012):

$$F_i = \frac{d^2 r_i}{dt^2} m_i$$

The interatomic characteristics and interactions become more pronounced at the nanoscale, the global optimization models that are categorized as deterministic and stochastic methods can be used to model the microstructures to study the materials at the smaller scale (Liberti & Kucherenko, 2005; Musa, 2012). The quasi Monte Carlo approach is an example of the stochastic approach while the approaches that governs the motion of objects through the deterministic approach that does not leave any room for random occurrences contains static and molecular dynamic (MD) approaches (Rapaport, 2004). Due to the smaller size at the nanoscale, temperature and velocities of the atoms become significant, the MD simulation have occupied an important niche in the nanotechnology research and studies on nanoparticles (Tian, 2008).

The MD simulation techniques function on the basis of different ensembles through observation of the momenta and positions of the atoms. The dynamical and thermodynamic material characteristics are simulated and studies through the statistical relations such as the relation between the velocity and temperature of the atoms (McQuarrie, 2008). If the system has "N" number of particles which has Nf degrees of freedom expressed as $N_f = 3N - 3$, the nanoscale level temperature-velocity relationship of the atoms is (Tian, 2008):

$$T = \left\langle \sum_{i=1}^{N} \frac{m_i v_i^2}{k_B N_f} \right\rangle$$

The angled brackets indicate the ensemble average (Crooks, 1998) which corresponds to the average of the all the atomic velocities for any "i^{th}" particle with mass and velocity as m_i and v_i respectively, whereas k_B = Boltzmann constant (So & Voter, 2000).

The rheological characteristics of soils and rocks can be studied through molecular dynamics simulation of nanoparticles as stated by Sharma, Wang and Gutierrez (2007). They also specified MD simulation techniques using Euler's method or Verlet's method and stress or strain controlled simulations using Monte Carlo method. The classical approach of the molecular dynamic simulation postulates that the attractive and repulsive forces existing between particles play an important role at nanoscale level. These are described by interatomic Lennard-Jones potential energy function, which approximates the relationship between the pair of atoms as influenced by the distance between the two atoms which do not have any boding. If the potential well is expressed by ε, the distances between the atoms is "r" and "σ" can be referred to as the set distance where potential between the atoms can be zero; Lennard-Jones potential function as commonly expressed below (Jones, 1924):

$$V_{LJ} = 4\varepsilon \left[\left(\frac{\sigma}{r}\right)^{12} - \left(\frac{\sigma}{r}\right)^{6} \right]$$

Finite element analysis is another approach utilized for the study of mechanical, morphological, optical and microstructural characteristics of materials. FEM models are well-studied and well-researched in the fields of solid and soil mechanics for the stress analyses by dividing the complex physical systems into smaller elements and analyzing each small element as a unit. Computational techniques can be used for the determination of the mechanical behavior, like flexural and tensile behaviors, of the nanomaterials. The two parameters that placed as inputs for the FEM models are material density and the modulus of elasticity which form agreement with the MD simulation values performed on different scales. The total elastic energy as per the Hookean potential function can be expressed as (Musa, 2012):

$$U_v = \frac{1}{2} \int dr \sum_{\mu,v,\lambda,\sigma}^{3} \varepsilon_{\mu v}(r) C_{\mu,v,\lambda,\sigma} \varepsilon_{\lambda,\sigma}(r)$$

The mass density parameter is expressed through "r" and the Greek indices have been used to denote the directions in the Cartesian coordinate system. The bonded interactions of the atoms is studied through the Morse's potential function which describes the energy for the diatomic system (Lim, 2007). If the atoms are "r" distance apart while the equilibrium bond distance is "r_e", "a" is potential well depth and "D_e" represents depth, the Morse's potential function can be formulated as (Morse, 1929):

$$V(r) = D_e \left(1 - e^{-a(r-r_e)}\right)^2$$

The nanoscale MD simulation is same for all the scales and the parameters remain consistent and a combination of different forces of attraction and repulsion (Coulomb's and van der Waal's forces, Born-Huggins-Meyer potential, van der Waals and Morse terms. The FEM approach can also be used to perform the MD simulation. The MDFEM can be used to perform simulations across nanoscale and macroscale at the same time and complex models can be developed for accurate prediction of the soil and rock materials containing nano-sized objects (Musa, 2012).

In essence, the nanoscale modelling can be used to provide proper understanding of the heterogeneous materials' behavior at the grain level. The properties of the material that can then be interpreted include characteristics such as the soil or rock permeability, porosity and permittivity (Bear, 2013; Lefeuvre, Federova, Gomonova, & Tao, 2010). The application of the automatic meshing techniques at the nanoscale to study the capillaries formed between the grains of the soil or rock strata at the nano-level can be used to study the fluidity of groundwater through the strata and the permeability of the material can be studied by modelling on computational software such as MATLAB and Comsol Multiphysics as has been demonstrated through several simulations that investigated the fluid motion or filtration in a fictitious soil conducted by Serge Lefeuvre and Olga Gomonova (2012).

SOIL STABILIZATION AND DURABILITY OF PAVEMENTS

The stabilization of the soil and pavements is a common practice in most geotechnical engineering applications. The geotechnical engineers are often required to stabilize the underlying soil mass so that the safety, durability and strength of the overlying structure can be ensured (U Hasan et al., 2015). The advancements in the nanotechnology field have been widely investigated to achieve the geotechnical purpose of soil, rock and pavement stability. These have been further discussed in the succeeding pages. For simplification of the nano-stabilization techniques effects, the ratio between the strength of the stabilized soil specimen and the strength of the unstabilized soil specimen, referred to as the ratio of strength, based upon the strength development index introduced by U. Hasan, A. Chegenizadeh, M. A. Budihardjo, and H. Nikraz (2015), has been used. The ratio of the strengths of the stabilized and unstabilized samples quantifies the strength enhancement effect of the nanomaterial in an easily comparable scale. Therefore, the higher the ratio better is the stabilizing nature of the nanomaterial.

Soil Stabilization with Nanoclays and Alternate Nanomaterials

Nanotechnology can also be used to produce organically engineered nanoclays (Azzam, Sayyah, & Taha, 2013). Nanoclays have also been found to improve the mechanical characteristics of asphalt binder (You et al., 2011). The most important raw material for nanoclays is montmorillonite (Nanocor Technical, n.d.). Montmorillonite, sedimentary soil and kaolinite were pulverized by Taha (2009) in a ball-mill to obtain nano-sized soils termed as nanosoils. These nanosoils were mixed with the original soils and were observed to improve soil properties by increasing the specific surfaces and therefore, reducing the plasticity index.

Different nanomaterials such as nano metallic oxides (MgO and CuO) and nanoclays was used by Majeed, Taha, and Jawad (2014) to stabilize weak soils. They used two types of soils, namely organic silt (Soil 1) and highly plastic clay (Soil 2). The effects of the stabilizing admixtures on the Atterberg's limits were investigated and the change in the compaction parameters and the linear shrinkage were also researched. They found that the Atterberg's limits and the linear shrinkage of the treated soil decreased with the induction of the additives. Contrarily, the optimum moisture content decreased when percentage of the additive grew while a maximum dry density rise was observed. The unconfined compressive strength (UCS) of the treated samples also improved with increasing percentage of the nanomaterials. The UCS enhancement effect of the nanomaterials for both soils showed somewhat of a bell-curve, with the Soil 2 showing comparatively sharper increase and subsequent UCS decrement with introduction of the stabilizing nanomaterials. The optimum percentages of all the stabilizing nanomaterials were documented to be less than 1% for both of the weak soils investigated in the laboratory experiments. Figure 4 shows the ratios between the strengths of unstabilized and stabilized Soils 1 and 2, as stabilized with the optimum respective nanomaterial percentage.

CNT FIBER NANO SOIL STABILIZATION

Fiber reinforcement has been investigated to enhance the unconfined compressive strength of weak fine grained soil. Jafari and Esna-ashari (2012) added fiber reinforcements obtained from alternate recycled sources in different percentages along with lime. They observed that by adding fiber, the samples stabilized by 4% lime produced higher compressive strength values than the unstabilized and samples with 8% of lime.

Carbon Nanotubes or widely known as CNTs are nanofibers example which are commonly utilized in the world. CNTs exhibit remarkable mechanical properties; for example higher stiffness, high strengths, high young modulus as well as high

Figure 4. Effects of nanomaterials on unconfined compressive strength
Source: Majeed et al., 2014.

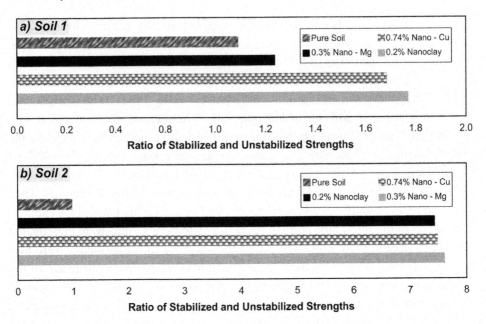

elastic stress-strain behavior. Those remarkable properties suggest CNTs as great materials for reinforcement of soil. Moreover, compare to other nanomaterials CNTs have better properties which suggest its potential as nanofibers for soil stabilization reinforcement materials. Safety precaution during handling CNTs should be followed in order to maintain safety and prevent any harmful effects.

CNTs have a much higher strength when compared with steel, prepared from carbon manipulation and modification. Structurally they are composed of single or more graphene layers, which is a sheet shaped by carbon atoms bonded in hexagonal honeycomb shape, forming a cylinder structure which are named after a "n,m" nomenclature, based upon the rolling of the graphene sheet vector to form the tube-like nanostructure where diameters are usually ≤ 2nm (S. Kumar et al., 2012; NanoScienceWorks, n.d.). It has been further illustrated Figure 5.

The unique characteristics of CNTs have garnered research and commercial attention towards the carbon nanotubes. They have high aspect ratios, such as 2,500,000:1 (Zheng et al., 2004) and lower diameters (Valentini, Biagiotti, Kenny, & Santucci, 2003). The electrical resistance of the CNTs has been observed to be significantly low which makes them good conductors of electricity (Masahito et al., 2011; C. Yang et al., 2010). CNTs thermal conductivities have also been observed to be significantly higher, with value as high as 6,600W/m-K for an isolated CNT at ambient conditions (Berber, Kwon, & Tomanek, 2000; Siddique & Mehta, 2014).

Figure 5. Schematic representation of different CNTs
Sources: Kramer, 2003; Ebbesen, 1996; and Barone, Hod, and Scuseria, 2006.

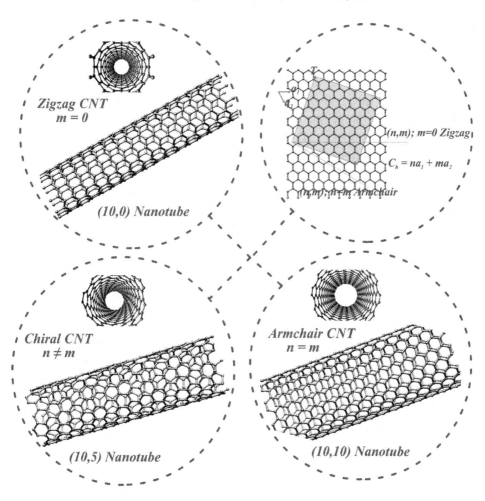

The CNTs have also been researched to be stronger and stiffer in terms of the elastic moduli and strengths when compared with other nanomaterials. The strength of CNT is also higher than nano-silica and also exhibits highly elastic stress-strain behaviors. Research has also established that the mechanical properties, strengths and Young's moduli are higher than steel Past studies found that CNTs Young modulus was higher than steel. The elastic modulus of CNTs has been tested to be ten to hundred times higher than the strongest steel, the elastic moduli values have been documented to be approximately 1TPa (Jose et al., 2007; Zhang et al., 2007).

Due to CNTs high strength and Young's moduli, they have been observed to improve flexural and compression strengths of cement-based mixtures. In an experimental

Nanotechnology Future and Present in Construction Industry

investigations by Nochaiya and Chaipanich (2011), multi-walled carbon nanotubes (MWCNTs) were used to produce cement-water pastes. They observed that MWCNT addition reduced the Portland cement paste's porosity and microstructural analysis showed structural interactions and formation of the cement hydration products, like ettringite, calcium hydroxide and calcium-silicate-hydrate, filled by the MWCNT fibers; thus making denser and stronger microstructure with high strengths.

Mohd Raihan Taha and Ying (2010) performed initial research work on the effect of CNTs induction on the rheological characteristics of the kaolinite clay and noted that the addition of CNTs can increase the Atterberg's limits; such as liquid limit, plastic limit and plasticity index of the clay. The CNTs induction also caused the compressibility of the samples to increase while the hydraulic conductivity or permeability was decreased. Also an inclusion of 1% CNTs fibers in the soil matrix resulted in the rise in the void ratio of the treated mixture. They also observed that the CNTs addition resulted in significant enhancement in the compressive strength of the stabilized mixture Figure 6.

CNTs can also be designed or consolidated inside a soil matrix and aid in resisting overburden stresses or help to control the porosity of the composition. Significant research has been made in identifying the practicality of the CNT usage (Office of the Press Secretary, 2000).

Since carbon nanotubes are considered as new materials, safety factors such as toxicity was one of the concern. It was not much finding regarding the toxicity or safety factors of carbon nanotubes. Available data suggest carbon nanotubes are harmful if they reach the body organs. It could bring harmful effect such as inflammatory and fibrotic reactions(Siddique & Mehta, 2014). Consequently, it is recommended to use personal protective equipment to handle carbon nanotubes. Safety glasses, nitrile gloves, respiratory mask and lab coat are necessary to prevent direct exposure and carbon nanotubes effect inside the body.

Figure 6. Effect of CNTs inclusion on compressive strength of clay
Source: Mohd Raihan Taha and Ying, 2010.

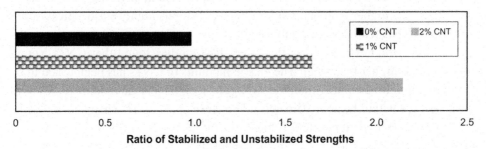

Effect of Nano-Silica on Durability of Concrete Pavements

Nano-silica is another application of nanotechnology for enhancing the engineering characteristics of the soil strata and pavements. The nano-silica has fine highly reactive pozzolanic silica particles and has been commercially utilized in the oilfield and construction industries as cement admixture to enhance the mechanical properties and decrease the permeability of the mixture. As it is with any form of particles, the nano-silica particles have a distinctive behavior when introduced in the Portland cement paste or concrete mixtures, depending upon the size of the particles. The inclusion of nanoparticles can increase the durability, strength, shrinkage and inter-particle bonds. The nano-silica particles can fill the calcium silicate hydrate, which is cement hydration product, inter-particle voids functioning as nano-fillers (Senff, Labrincha, Ferreira, Hotza, & Repette, 2009). Nano-silica has also been evaluated in laboratory to decrease the porosity of the cement matrix by 33.3% while increasing the compressive strength by approximately 155%. Furthermore, the induction of the nano-silica decreased the hydraulic conductivity by 99% (Ershadi, Ebadi, Rabani, Ershadi, & Soltanian, 2011).

This cement compressive strength enhancement characteristic of nano-silica is useful for reinforced concrete pavements. After the completion of the first concrete pavement in 1893 (Snell & Snell, 2002), the concrete pavements are being used for bridge decks, road surfaces, parking lots, streets and airfield runways. Even though concrete pavements have higher durability, due to concrete's characteristic of gaining strength during its lifetime; low cost of maintenance, due to ability of outlasting designed life-expectancy; rigidity and ease of operation; it requires repairs to be done (Kuennen, 2003). The repair operations are sometimes carried out to rectify an initial problem that developed due to poor workmanship during the construction phase or may be required to prolong the service-life or guard against weathering effects, such as frosting. As the presence of the micro-cracks within the cement paste is one of the major factors that result in the reduced durability and the frost-resistance of the cement paste. Pozzolans like pulverized fuel ash, has been investigated to densify the particle structure of the cement which may mitigate or prevent the micro-crack formation (Gao, Lo, & Tam, 2002). It has also been documented that the pozzolans can enhance frost-resistance characteristic of the concrete and improve the microstructure (Pacewska, Bukowska, Wilińska, & Swat, 2002).

The induction of specifically tailored nanoparticles, such as nano-silica, can cause a dramatic improvement in the microstructure and rheological properties of cement paste. Nano-silica induction can also increase the material properties of cement mortars (H. Li et al., 2004). The smaller size of nano-silica produces higher surface area- volume ratios, producing higher nano-silica particles pozzolanic reactivity. The nano-silica pozzolanic behavior results in rise in the amount of the calcium silicate hydrates due to reaction with the calcium hydroxide to produce stronger and durable structure and also act as nano-filler for densification of the microstructure (Senff et al., 2009). The nano-silica has also been observed to enhance the water-penetration resistant of the cement paste (Ji, 2005). In a study conducted by Salemi and Behfarnia (2013), nano-silica was observed to enhance the concrete UCS by 30% when nano-silica percentage by the weight of the cementitious material was set at 5%. The frost-resistance of the concrete mix for concrete pavements was also found to be increased by adding nano-silica. The deterioration of the concrete was noted to be decreased by approximately 84% with the 5% nano-silica inclusion. The effects of different freeze-thaw cycles upon UCS loss are shown in Figure 7.

Effect of Nano-Alumina on Durability of Concrete Pavements

Nano-alumina or nano-Al_2O_3 has also been investigated to produce calcium aluminum silicate gels in the concrete mixture. Salemi and Behfarnia (2013) proposed that due to the reaction of nano-Al_2O_3 with $Ca(OH)_2$, which is affected by the surface area available for the reaction, highly pure nano-Al_2O_3 with higher blain fineness may be used to enhance the mechanical properties, such as tensile and compressive strengths, of concrete (Jo et al., 2007). It has also been investigated to improve the design life or durability of concrete by improving the resistance against chloride penetration and water absorption (Nazari, Sh, Sh, Shamekhi, & Khademno, 2010).

The elasticity modulus of cement mortar can also be increased significantly with the addition of nano-Al_2O_3. It has been demonstrated through experiments that with 5% incorporation of nano-Al_2O_3 in the mortar, the elasticity modulus increased by 143% after 28 days of curing but did not produce much increase in the compressive

Figure 7. Loss in Compressive strength of specimens with freeze-thaw cycles
Source: Salemi and Behfarnia, 2013.

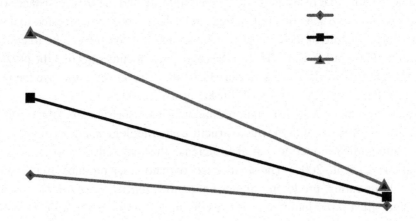

strength (Z. Li, Wang, He, Lu, & Wang, 2006). Similar to nano-silica, nano-alumina also acts as a filler material. Nano-Al2O3 also promotes the pozzolanic reaction in the concrete in addition to improving the microstructure. This results in reduction in the number of voids in the concrete. The water absorption capability of the concrete pavements decreases with the incorporation of nano-alumina in the concrete mixture and render it less susceptible to impacts of climatic changes, such as volume variation due to freezing or thawing of water, thereby significantly increasing the durability and life expectancy of the concrete pavements by enhance resistance against freeze-thaw. The decrease of concrete UCS containing nano-alumina when subjected to freeze-thaw cycles has been observed to be only 18.19% after 300 cycles, while the concrete mixture without any nano-Al_2O_3 showed 100% reduction in the UCS, upon experiencing equal freeze-thaw cycles (Salemi & Behfarnia, 2013).

FUTURE RESEARCH DIRECTIONS

Even though, the present applications of nanotechnology in geotechnical engineering are mainly in the research and pre-design phases, the possibility of controlling the behavior of clay particles on nanoscale may result in controlled consolidation, self-assembling aggregate formation, clay liners, engineered magnetic and polar characteristics designed for wet conditions. Nanoparticles may be designed in future to act as sensors that can measure the in-situ parameters of a soil composition enabling researchers to predict accurate and reliable models.

The advancements in the field of nanotechnology, specifically the impacts regarding cement-based materials' engineering and production are to be established further. Environmental impacts must also be properly understood so as to reduce harmful carbon dioxide emissions to the environment that are associated with the current practices of cementitious materials in concrete pavements. Additional materials such as nano-silica and nano-alumina may also be induced in the mixture to produce durable concrete pavements that may also reduce the overall construction and maintenance costs of reinforced concrete pavements. The new approach to geotechnical engineering may produce smarter materials that may result in the creation of an inter-particle communication and sensing network making the structures greener, effective, efficient and durable. The effective real-time monitoring of geotechnical structures during the occurrence of hazards such as seismic activities, floods and hurricanes etc.; may enable geotechnical engineers to counter problems such as settlements, seepage and dislocation (National Research Council, 2006).

Nanomaterial such as CNTs can also be used for stabilizing soils. Also the usage of atomic force microscopes to study the nano clay particle surface and hydrophobicity or chemical force microscopy for the study of surface reactions and fabric formation of clay soil can help to engineer better clay particles. The integration of numerical modeling, nano-engineering and sensing technologies such as MEMS and NEMS will change the traditional approaches to the understanding of soil profile and ground situations. Many countries are investing in the nanotechnology research contributing to the global nanotechnology research. Countries such as Japan and US alone have contributed more than 3 billion US dollars in the nano-research in 2003 (Nordan, 2005).

In summary, the applications of nanotechnology in the field of geotechnical and pavement engineering hold great potential. Nano-modification of ground under foundations or subgrades may increase the mechanical and chemical characteristics and produce economical, greener and safer structures that are more durable. After the passage of almost half a century after the famous lecture by Richard Feynman, geotechnical engineers are working to bring the applications of the nanotechnology within the commercial and civil infrastructure spheres.

CONCLUSION

The most commonly encountered material during the construction of any civil engineering project is the underlying soil and rock strata, which is tackled by the geotechnical engineers. These soil and rock masses may exist in many states such as loose, dense, wet, dry, cohesive, cohesionless and weak or soft conditions. Fine grained soils or clays are usually found on project sites and the improved knowledge of nanotechnology may help in understanding the clay behavior at the nanoscale. The nano-probing technologies such as atomic force microscopy, nano-FTIR, nanoindentation, MEMS and NEMS may help to localize issues in the mineralogical and morphological studies of the clay particles and understand characteristics such as the texture, surface chemistry, inter-particle attraction and mechanical properties under in-situ conditions.

Although, the understanding of the soil and rock masses at the nanoscale may be immensely helpful, the study of nanostructures produces additional problems. Therefore, improved computational techniques at the nanoscale must also be established. The finite element molecular dynamics simulations may also be conducted for modelling of the materials as it permits the concurrent modelling at different scales, i.e. the nano and the macroscopic scales. It can be used for the analysis of the variations in the particle size distributions and inter-particle interactions within the soil and rock matrices.

The fabrication of nanoclays composites has also been conducted by researchers to produce engineered nanoclays which have mineralogical materials such as montmorillonite. It has been used by researchers along with other sedimentary clays to produce the nanosoils. These nanoclays can then be used as an additive that may enhance the asphalt binders' mechanical characteristics. Researchers have observed that the addition of nanoclays may increase the complex shear modulus and the asphalt binders' viscosities and may also produce significant effects on the lower temperature resistance against cracking depending upon the percentage, type and mixing procedure of the nanoclays and the asphalt binders.

Nanoclays and nano-metallic oxides have also been research to increase the dry densities and decrease the plasticity, linear shrinkages, optimum moisture contents and Atterberg's limits of the weak soils when added in optimum percentages. In addition, the compressive strengths of the weak soils may also been enhanced with the induction of nanomaterials such as nanoclays when added in optimum percentages (i.e. $\leq 1\%$).

Nanomaterials such as nano-alumina and nano-silica may be used for the stabilization of the concrete mixes in reinforced concrete pavement. Nano-silica has been observed to improve the rheological characteristics and microstructural properties of cement paste. It has also been observed under laboratory conditions by some researchers to increase the material properties of cement mortars. The smaller size of nano-silica produces higher surface area- volume ratios, producing higher nano-silica particles pozzolanic reactivity. On the other hand, the effect of nano-alumina has also been investigated under the laboratory conditions. Researchers have proposed that the inclusion of nano-Al2O3 in the concrete pavement mortar may be able to produce calcium aluminum silicate gels in the concrete mixtures that may improve the overall strength and durability of the pavements.

REFERENCES

Abdel Hameed, R. S., Abu-Nawwasb, A. H., & Shehataa, H. A. (2013). Nano-composite as corrosion inhibitors for steel alloys in different corrosive media. *Advances in Applied Science*, *4*, 126–129.

Azzam, E. M. S., Sayyah, S. M., & Taha, A. S. (2013). Fabrication and characterization of nanoclay composites using synthesized polymeric thiol surfactants assembled on gold nanoparticles. *Egyptian Journal of Petroleum*, *22*(4), 493–499. doi:10.1016/j.ejpe.2013.11.011

Barone, V., Hod, O., & Scuseria, G. E. (2006). Electronic structure and stability of semiconducting graphene nanoribbons. *Nano Letters*, *6*(12), 2748–2754. doi:10.1021/nl0617033 PMID:17163699

Bear, J. (2013). *Dynamics of fluids in porous media*. Courier Corporation.

Berber, S., Kwon, Y.-K., & Tomanek, D. (2000). Unusually high thermal conductivity of carbon nanotubes. *Physical Review Letters*, *84*(20), 4613–4616. doi:10.1103/PhysRevLett.84.4613 PMID:10990753

Berkovich, E. (1950). Three-faceted diamond pyramid for studying microhardness by indentation. *Zavodskaya Laboratoria*, *13*(3), 345–347.

Borja, R. I. (2011). *Multiscale and Multiphysics Processes in Geomechanics: Results of the Workshop on Multiscale and Multiphysics Processes in Geomechanics, Stanford, June 23-25, 2010*. Springer.

Chang, T.-P., Shih, J.-Y., Yang, K.-M., & Hsiao, T.-C. (2007). Material properties of Portland cement paste with nano-montmorillonite. *Journal of Materials Science*, *42*(17), 7478–7487. doi:10.1007/s10853-006-1462-0

Constantinides, G., Ulm, F.-J., & Van Vliet, K. (2003). On the use of nanoindentation for cementitious materials. *Materials and Structures*, *36*(3), 191–196. doi:10.1007/BF02479557

Crooks, G. E. (1998). Nonequilibrium measurements of free energy differences for microscopically reversible Markovian systems. *Journal of Statistical Physics*, *90*(5-6), 1481–1487. doi:10.1023/A:1023208217925

De Paola, N., Hirose, T., Mitchell, T., Di Toro, G., Viti, C., & Shimamoto, T. (2011). Fault lubrication and earthquake propagation in carbonate rocks. In R. Borja (Ed.), *Multiscale and Multiphysics Processes in Geomechanics* (pp. 153–156). Springer Berlin Heidelberg. doi:10.1007/978-3-642-19630-0_39

Drexler, K. E. (1996). *Engines of Creation*. Fourth Estate.

Ebbesen, T. W. (1996). Carbon nanotubes. *Physics Today*, *49*(6), 26–35. doi:10.1063/1.881603

Ershadi, V., Ebadi, T., Rabani, A., Ershadi, L., & Soltanian, H. (2011). The effect of nanosilica on cement matrix permeability in oil well to decrease the pollution of receptive environment. *Int. J. Environ. Sci. Develop*, *2*, 128–132. doi:10.7763/IJESD.2011.V2.109

Feynman, R. P. (1961). *There's plenty of room at the bottom*. New York: Reinhold.

Gao, X. F., Lo, Y. T., & Tam, C. M. (2002). Investigation of micro-cracks and microstructure of high performance lightweight aggregate concrete. *Building and Environment*, *37*(5), 485–489. doi:10.1016/S0360-1323(01)00051-8

Hasan, U., Chegenizadeh, A., Budihardjo, M., & Nikraz, H. (2015). A review of the stabilisation techniques on expansive soils. *Australian Journal of Basic & Applied Sciences*, *9*(7), 541–548.

Hasan, U., Chegenizadeh, A., Budihardjo, M. A., & Nikraz, H. (2015). *Experimental evaluation of construction waste and ground granulated blast furnace slag as alternative soil stabilisers*. Manuscript submitted for publication.

Hashmi, S. (2014). *Comprehensive Materials Processing*. Elsevier Science.

Hermann, P., Hoehl, A., Patoka, P., Huth, F., Rühl, E., & Ulm, G. (2013). Near-field imaging and nano-Fourier-transform infrared spectroscopy using broadband synchrotron radiation. *Optics Express*, *21*(3), 2913–2919. doi:10.1364/OE.21.002913 PMID:23481749

Hirth, J. P., & Lothe, J. (1982). *Theory of Dislocations*. Krieger Publishing Company.

Hull, M. S., Quadros, M. E., Born, R., Provo, J., Lohani, V. K., & Mahajan, R. L. (2014). Sustainable Nanotechnology: A Regional Perspective. In M. S. Hull & D. M. Bowman (Eds.), *Nanotechnology Environmental Health and Safety* (2nd ed.; pp. 395–424). Oxford, UK: William Andrew Publishing.

Huth, F., Govyadinov, A., Amarie, S., Nuansing, W., Keilmann, F., & Hillenbrand, R. (2012). Nano-FTIR Absorption Spectroscopy of Molecular Fingerprints at 20 nm Spatial Resolution. *Nano Letters*, *12*(8), 3973–3978. doi:10.1021/nl301159v PMID:22703339

Jafari, M., & Esna-ashari, M. (2012). Effect of waste tire cord reinforcement on unconfined compressive strength of lime stabilized clayey soil under freeze–thaw condition. *Cold Regions Science and Technology*, *82*(0), 21–29. doi:10.1016/j.coldregions.2012.05.012

Ji, T. (2005). Preliminary study on the water permeability and microstructure of concrete incorporating nano-SiO2. *Cement and Concrete Research*, *35*(10), 1943–1947. doi:10.1016/j.cemconres.2005.07.004

Jo, B.-W., Kim, C.-H., Tae, G.-, & Park, J.-B. (2007). Characteristics of cement mortar with nano-SiO2 particles. *Construction & Building Materials*, *21*(6), 1351–1355. doi:10.1016/j.conbuildmat.2005.12.020

Jones, J. E. (1924, October 1). *On the determination of molecular fields. II. From the equation of state of a gas*. Paper presented at the Royal Society of London. Series A: Mathematical, Physical and Engineering Sciences, London, UK.

Jose, M. V., Steinert, B. W., Thomas, V., Dean, D. R., Abdalla, M. A., Price, G., & Janowski, G. M. (2007). Morphology and mechanical properties of Nylon 6/MWNT nanofibers. *Polymer*, *48*(4), 1096–1104. doi:10.1016/j.polymer.2006.12.023

Kadivar, Barkhordari, & Kadivar. (2011). Nanotechnology in Geotechnical Engineering. *Advanced Materials Research*, *261-263*, 524-528. doi: 10.4028/www.scientific.net/AMR.261-263.524

Kramer, B. (2003). *Advances in Solid State Physics* (Vol. 43). Springer. doi:10.1007/b12017

Kuennen, T. (2003). Correcting Problem Concrete Pavements. *Better Roads*, *73*, 32–38.

Kuhlbusch, T., Asbach, C., Fissan, H., Gohler, D., & Stintz, M. (2011). Nanoparticle exposure at nanotechnology workplaces: A review. *Particle and Fibre Toxicology*, *8*(1), 22. doi:10.1186/1743-8977-8-22 PMID:21794132

Kumar, A. P., Depan, D., Singh Tomer, N., & Singh, R. P. (2009). Nanoscale particles for polymer degradation and stabilization—Trends and future perspectives. *Progress in Polymer Science*, *34*(6), 479–515. doi:10.1016/j.progpolymsci.2009.01.002

Kumar, S., Kolay, P., Malla, S., & Mishra, S. (2012). Effect of Multiwalled Carbon Nanotubes on Mechanical Strength of Cement Paste. *Journal of Materials in Civil Engineering*, *24*(1), 84–91. doi:10.1061/(ASCE)MT.1943-5533.0000350

Lefeuvre, S., Federova, E., Gomonova, O., & Tao, J. (2010). Microwave Sintering of Micro-and Nano-Sized Alumina Powder. *Advances in Modeling of Microwave sintering*, 8-9.

Lefeuvre, S., & Gomonova, O. (2012). Modeling at the Nano Level: Application to Physical Processes. In S. M. Musa (Ed.), *Computational Finite Element Methods in Nanotechnology* (pp. 559–583). Taylor & Francis. doi:10.1201/b13002-17

Lesko, S., Lesniewska, E., Nonat, A., Mutin, J.-C., & Goudonnet, J.-P. (2001). Investigation by atomic force microscopy of forces at the origin of cement cohesion. *Ultramicroscopy*, *86*(1), 11–21. doi:10.1016/S0304-3991(00)00091-7 PMID:11215612

Li, G. Y., Wang, P. M., & Zhao, X. (2005). Mechanical behavior and microstructure of cement composites incorporating surface-treated multi-walled carbon nanotubes. *Carbon*, *43*(6), 1239–1245. doi:10.1016/j.carbon.2004.12.017

Li, H., Xiao, H.-, Yuan, J., & Ou, J. (2004). Microstructure of cement mortar with nano-particles. *Composites. Part B, Engineering*, *35*(2), 185–189. doi:10.1016/S1359-8368(03)00052-0

Li, W., Li, Y., & Wang, X. (2010). *Notice of retraction nano-TiO$_2$ modified super-hydrophobic surface coating for anti-ice and anti-flashover of insulating layer.* Paper presented at the Power and Energy Engineering Conference (APPEEC).

Li, Z., Wang, H., He, S., Lu, Y., & Wang, M. (2006). Investigations on the preparation and mechanical properties of the nano-alumina reinforced cement composite. *Materials Letters, 60*(3), 356–359. doi:10.1016/j.matlet.2005.08.061

Liberti, L., & Kucherenko, S. (2005). Comparison of deterministic and stochastic approaches to global optimization. *International Transactions in Operational Research, 12*(3), 263–285. doi:10.1111/j.1475-3995.2005.00503.x

Lim, T.-C. (2007). Long range relationship between Morse and Lennard–Jones potential energy functions. *Molecular Physics, 105*(8), 1013–1018. doi:10.1080/00268970701261449

Little, D. N., Males, E. H., Prusinski, J. R., & Stewart, B. (2000). *Cementitious stabilization.* Transportation in the New Millennium.

Majeed, Z. H., Taha, M. R., & Jawad, I. T. (2014). Stabilization of Soft Soil Using Nanomaterials. *Research Journal of Applied Sciences. Engineering and Technology, 8*(4), 503–509.

Makar, J., Margeson, J., & Luh, J. (2005). *Carbon nanotube/cement composites-early results and potential applications.* Paper presented at the 3rd International Conference on Construction Materials: Performance, Innovations and Structural Implications, Vancouver, Canada.

Masahito, S., Kashiwagi, Y., Li, Y., Arstila, K., Richard, O., Cott, D. J., & Vereecken, P. M. et al. (2011). Measuring the electrical resistivity and contact resistance of vertical carbon nanotube bundles for application as interconnects. *Nanotechnology, 22*(8), 085302. doi:10.1088/0957-4484/22/8/085302 PMID:21242623

McQuarrie, D. A. (2008). *Quantum chemistry.* University Science Books.

Morse, P. M. (1929). Diatomic Molecules According to the Wave Mechanics. II. Vibrational Levels. *Physical Review, 34*(1), 57–64. doi:10.1103/PhysRev.34.57

Musa, S. M. (2012). *Computational Finite Element Methods in Nanotechnology.* Taylor & Francis. doi:10.1201/b13002

Nanocor Technical. (n.d.). *Nanoclay Structures.* Retrieved 12 November 2014, from http://www.nanocor.com/nano_struct.asp

NanoScienceWorks. (n.d.). *Types of Carbon Nanotubes*. Retrieved 04/03/2015, 2014, from http://www.nanoscienceworks.org/Members/siebo/657px-types_of_carbon_nanotubes.png/view

National Research Council. (2006). *Geological and geotechnical engineering in the new millennium: Opportunities for research and technological innovation.* Washington, DC: The National Academies Press.

Nazari, A., Sh, R., Sh, R., Shamekhi, S. F., & Khademno, A. (2010). Influence of Al2O3 nanoparticles on the compressive strength and workability of blended concrete. *Journal of American Science, 6*(5), 6–9.

Neaspec Gmb, H. (n.d.). *Nano-FTIR – Nanoscale Infrared Spectroscopy with a thermal source*. Retrieved 03/03/2015, 2015, from http://www.neaspec.com/application/nano-ftir-nanoscale-infrared-spectroscopy-with-a-thermal-source/

Neitola, R., Ruuska, H., & Pakkanen, T. A. (2005). Ab Initio Studies on Nanoscale Friction between Graphite Layers: Effect of Model Size and Level of Theory. *The Journal of Physical Chemistry B, 109*(20), 10348–10354. doi:10.1021/jp044065q PMID:16852254

Nochaiya, T., & Chaipanich, A. (2011). Behavior of multi-walled carbon nanotubes on the porosity and microstructure of cement-based materials. *Applied Surface Science, 257*(6), 1941–1945. doi:10.1016/j.apsusc.2010.09.030

Nordan, M. M. (2005). *Nanotechnology: where does the US stand*. Lux Research Incorporated.

Oberdürster, G. (2000). Toxicology of ultrafine particles: in vivo studies. *Philosophical Transactions of the Royal Society of London. Series A: Mathematical, Physical and Engineering Sciences, 358*(1775), 2719-2740.

Office of the Press Secretary. (2000). National Nanotechnology Initiative - Leading To The Next Industrial Revolution. *Microscale Thermophysical Engineering, 4*(3), 205–212. doi:10.1080/10893950050148160

Oliver, W. C., & Pharr, G. M. (2004). Measurement of hardness and elastic modulus by instrumented indentation: Advances in understanding and refinements to methodology. *Journal of Materials Research, 19*(01), 3–20. doi:10.1557/jmr.2004.19.1.3

Pacewska, B., Bukowska, M., Wilińska, I., & Swat, M. (2002). Modification of the properties of concrete by a new pozzolan—a waste catalyst from the catalytic process in a fluidized bed. *Cement and Concrete Research, 32*(1), 145–152. doi:10.1016/S0008-8846(01)00646-9

Picraux, S. T. (2014). *Nanotechnology*. Retrieved 9 November 2014, from http://www.britannica.com/EBchecked/topic/962484/nanotechnology/236436/Properties-at-the-nanoscale

Plassard, C., Lesniewska, E., Pochard, I., & Nonat, A. (2004). Investigation of the surface structure and elastic properties of calcium silicate hydrates at the nanoscale. *Ultramicroscopy, 100*(3), 331–338. doi:10.1016/j.ultramic.2003.11.012 PMID:15231326

Plassard, C., Lesniewska, E., Pochard, I., & Nonat, A. (2005). Nanoscale experimental investigation of particle interactions at the origin of the cohesion of cement. *Langmuir, 21*(16), 7263–7270. doi:10.1021/la050440+ PMID:16042451

Poon, B., Rittel, D., & Ravichandran, G. (2008). An analysis of nanoindentation in linearly elastic solids. *International Journal of Solids and Structures, 45*(24), 6018–6033. doi:10.1016/j.ijsolstr.2008.07.021

Portland Cement Association. (n.d.). *Concrete Pavement*. Retrieved 13/03/2015, 2015, from http://www.cement.org/cement-concrete-basics/products/concrete-pavement

Pospíchal, O., Kucharczyková, B., Misák, P., & Vymazal, T. (2010). Freeze-thaw resistance of concrete with porous aggregate. *Procedia Engineering, 2*(1), 521–529. doi:10.1016/j.proeng.2010.03.056

Qingwen, L., Zuyuan, H., Tokunaga, T., & Hotate, K. (2011, Aug. 28). *An ultra-high-resolution large-dynamic-range fiber optic static strain sensor using Pound-Drever-Hall technique.* Paper presented at the Quantum Electronics Conference & Lasers and Electro-Optics (CLEO/IQEC/PACIFIC RIM).

Rapaport, D. C. (2004). *The art of molecular dynamics simulation.* Cambridge university press. doi:10.1017/CBO9780511816581

Sajanlal, P. R., Sreeprasad, T. S., Samal, A. K., & Pradeep, T. (2011). *Anisotropic nanomaterials: structure, growth, assembly, and functions.* doi: 10.3402/nr.v2i0.5883

Salemi, N., & Behfarnia, K. (2013). Effect of nano-particles on durability of fiber-reinforced concrete pavement. *Construction & Building Materials, 48*(0), 934–941. doi:10.1016/j.conbuildmat.2013.07.037

Sanchez, F., & Sobolev, K. (2010). Nanotechnology in concrete – A review. *Construction & Building Materials, 24*(11), 2060–2071. doi:10.1016/j.conbuildmat.2010.03.014

Schodek, D. L., Ferreira, P., & Ashby, M. F. (2009). *Nanomaterials, nanotechnologies and design: an introduction for engineers and architects.* Butterworth-Heinemann.

Senff, L., Labrincha, J. A., Ferreira, V. M., Hotza, D., & Repette, W. L. (2009). Effect of nano-silica on rheology and fresh properties of cement pastes and mortars. *Construction & Building Materials, 23*(7), 2487–2491. doi:10.1016/j.conbuildmat.2009.02.005

Sharma, A., Wang, J., & Gutierrez, M. S. (2007). *Nanoscale Simulations of Rock and Clay Minerals*. Advances in Measurement and Modeling of Soil Behavior.

Shi, T., Schins, R., Knaapen, A., Kuhlbusch, T., Pitz, M., Heinrich, J., & Borm, P. (2003). Hydroxyl radical generation by electron paramagnetic resonance as a new method to monitor ambient particulate matter composition. *Journal of Environmental Monitoring, 5*(4), 550–556. doi:10.1039/b303928p PMID:12948226

Siddique, R., & Mehta, A. (2014). Effect of carbon nanotubes on properties of cement mortars. *Construction & Building Materials, 50*(0), 116–129. doi:10.1016/j.conbuildmat.2013.09.019

Snell, L. M., & Snell, B. G. (2002). Oldest concrete street in the United States. *Concrete International-Detroit, 24*(3), 72–74.

So, M. R., & Voter, A. F. (2000). Temperature-accelerated dynamics for simulation of infrequent events. *The Journal of Chemical Physics, 112*(21), 9599–9606. doi:10.1063/1.481576

Stankiewicz, A., Szczygieł, I., & Szczygieł, B. (2013). Self-healing coatings in anti-corrosion applications. *Journal of Materials Science, 48*(23), 8041–8051. doi:10.1007/s10853-013-7616-y

Tabor, D. (1996). Indentation hardness: Fifty years on a personal view. *Philosophical Magazine A, 74*(5), 1207–1212. doi:10.1080/01418619608239720

Taha, M. R. (2009). Geotechnical Properties of Soil-Ball Milled Soil Mixtures. In Z. Bittnar, P. M. Bartos, J. Němeček, V. Šmilauer, & J. Zeman (Eds.), *Nanotechnology in Construction 3* (pp. 377–382). Springer Berlin Heidelberg. doi:10.1007/978-3-642-00980-8_51

Taha, M. R., & Ying, T. (2010). Effects of carbon nanotube on kaolinite: Basic geotechnical behavior. *World Journal Of Engineering, 7*(2), 472–473.

Tian, P. (2008). Molecular dynamics simulations of nanoparticles. *Annual Reports Section "C"(Physical Chemistry), 104*, 142-164.

Valentini, L., Biagiotti, J., Kenny, J. M., & Santucci, S. (2003). Morphological characterization of single-walled carbon nanotubes-PP composites. *Composites Science and Technology, 63*(8), 1149–1153. doi:10.1016/S0266-3538(03)00036-8

Virji, M. A., & Stefaniak, A. B. (2014). 8.06 - A Review of Engineered Nanomaterial Manufacturing Processes and Associated Exposures. In *Comprehensive Materials Processing* (pp. 103–125). Oxford, UK: Elsevier. doi:10.1016/B978-0-08-096532-1.00811-6

Wang, Z., Han, E., & Ke, W. (2006). Effect of nanoparticles on the improvement in fire-resistant and anti-ageing properties of flame-retardant coating. *Surface and Coatings Technology, 200*(20–21), 5706–5716. doi:10.1016/j.surfcoat.2005.08.102

Wang, Z. L., Liu, Y., & Zhang, Z. (2002). *Handbook of nanophase and nanostructured materials: Characterization* (Vol. 2). Kluwer Academic/Plenum.

Warneke, B. A., Scott, M. D., Leibowitz, B. S., Zhou, L., Bellew, C. L., Chediak, J. A., . . . Pister, K. S. (2002). *An autonomous 16 mm 3 solar-powered node for distributed wireless sensor networks.* Paper presented at the Sensors. doi:10.1109/ICSENS.2002.1037346

Yang, C., Hazeghi, A., Takei, K., Hong-Yu, C., Chan, P. C. H., Javey, A., & Wong, H. S. P. (2010, 6-8 Dec. 2010). *Graphitic interfacial layer to carbon nanotube for low electrical contact resistance.* Paper presented at the Electron Devices Meeting (IEDM), 2010 IEEE International.

Yang, T., Keller, B., & Magyari, E. (2002). AFM investigation of cement paste in humid air at different relative humidities. *Journal of Physics-London-D Applied Physics, 35*(8), L25–L28. doi:10.1088/0022-3727/35/8/101

Yao, N., & Wang, Z. L. (2005). *Handbook of Microscopy for Nanotechnology.* Springer. doi:10.1007/1-4020-8006-9

Yilbas, B. S. (2014). Introduction to Nano- and Microscale Processing – Modeling. In M. S. J. Hashmi (Ed.), *Comprehensive Materials Processing* (pp. 1–2). Oxford, UK: Elsevier. doi:10.1016/B978-0-08-096532-1.00700-7

You, Z., Mills-Beale, J., Foley, J. M., Roy, S., Odegard, G. M., Dai, Q., & Goh, S. W. (2011). Nanoclay-modified asphalt materials: Preparation and characterization. *Construction & Building Materials, 25*(2), 1072–1078. doi:10.1016/j.conbuildmat.2010.06.070

Zhang, X., Li, Q., Holesinger, T. G., Arendt, P. N., Huang, J., Kirven, P. D., & Zhao, Y. et al. (2007). Ultrastrong, stiff, and lightweight Carbon-Nanotube fibers. *Advanced Materials, 19*(23), 4198–4201.

Zhao, H., Zhang, Y., Bradford, P. D., Zhou, Q., Jia, Q., Yuan, F.-G., & Zhu, Y. (2010). Carbon nanotube yarn strain sensors. *Nanotechnology, 21*(30), 305502. doi:10.1088/0957-4484/21/30/305502 PMID:20610871

Zheng, L., O'connell, M., Doorn, S., Liao, X., Zhao, Y., Akhadov, E., & Dye, R. et al. (2004). Ultralong single-wall carbon nanotubes. *Nature Materials, 3*(10), 673–676. doi:10.1038/nmat1216 PMID:15359345

Zuyuan, H., Liu, Q., & Tokunaga, T. (2013). *Sensing the Earth Crustal Deformation with Fiber Optics.* Paper presented at the Frontiers in Optics.

Zuyuan, H., Qingwen, L., & Tokunaga, T. (2012, 6-11 May 2012). *Realization of nano-strain-resolution fiber optic static strain sensor for geo-science applications.* Paper presented at the Conference on Lasers and Electro-Optics (CLEO).

KEY TERMS AND DEFINITIONS

Anthropogenic: The objects, effects and/or materials produced either as direct or indirect consequence of human activities.

Asphalt: The thick, viscous, gluey and dark-colored petroleum-based material that is commonly used as a binding agent for aggregates in roads and pavement engineering.

Cation Exchange Capacity: The ability/capacity of the soil or rock mass for holding the cations that are exchangeable. It mainly effects the capacity of the soil mass to provide a buffer against the acidification of the soil.

Dowel: The cylindrical solid road often constructed from metals, plastics or woods used for support at joints or as tensile reinforcement.

Equiaxed: Often used in the crystallography nomenclature to convey that the axes of the crystals have almost same lengths.

Expansive Soils: The soils that show high reactivity to moisture variations, where the climatic changes produce pronounced shrinkage or swelling in the swell mass depending upon the wetting-drying seasonal or environmental conditions.

Fullerenes: The carbon molecules that exist in the shape of hollow spheres, tubes, ellipsoids and other similar spherical or "bulkyball- like" shapes.

Fullerenes: The carbon molecules that exist in the shape of hollow spheres, tubes, ellipsoids and other similar spherical or "bulkyball- like" shapes.

Magnetotactic Bacteria: These are microorganisms that show the behavior of migrating and orienting along the magnetic field of the earth.

Nanoclay: A nanoclay contains nanoparticles that have members of the smectite silica family in stacked in layers.

Subgrade: The in-situ material which is underlying the pavements or roads and other transportation structures.

Van der Waals Forces: These are the inter-particle or inter-molecular forces of repulsion/attraction that exists between different molecules.

Chapter 6
Applications of Nanotechnology in Transportation Engineering

Imtiaz Ahmed
Mirpur University of Science and Technology, Pakistan

Israr Ul Haq
Mirpur University of Science and Technology, Pakistan

Naveed Ahmad
University of Engineering and Technology Taxila, Pakistan

Muhammad Hassan
Mirpur University of Science and Technology, Pakistan

Imran Mehmood
Mirpur University of Science and Technology, Pakistan

Muhammad Umer Arif Khan
Mirpur University of Science and Technology, Pakistan

ABSTRACT

Nanotechnology is the latest development in science, where design, construction and applications of various particles involve at least one dimension in nanometers. The nanotechnology has been utilized in many of the scientific and societal disciplines including electronics, medicine, materials science and many more. It has also influenced the broader fields like civil engineering as well as the sub-disciplines including transportation, structural, geotechnical, water resources and environmental engineering. The current focus of the researchers in transportation field is to develop the materials for sustainable transportation facilities, by using the concepts of nanotechnology. The chapter is concerned with the literature review of potential applications of the nanotechnology in transportation engineering including safety, durability, sustainability and economy. The practical applications of the nanotechnology and nanomaterials shall prove to be an asset in transportation engineering.

DOI: 10.4018/978-1-5225-0344-6.ch006

Applications of Nanotechnology in Transportation Engineering

HISTORICAL BACKGROUND

In recent times Nanotechnology had started to attract significant factions of media and investment industry. (Zhu et al, 2004). It is basically concerned with the development of new materials with the help of better understanding of the basic building units of all the materials i.e. atoms and molecules. With the backing of unprecedented funding, nanotechnology is fast emerging as the industrial revolution of the 21st century (Siegel et al, 1999). According to Morse, 2004 interpretation of Nano Technology, it can be considered as the future of industries. It would play a vital role in transformation and creation of whole new industries.

Since 1990s, implementation of various areas of nanotechnology has rapidly grown such as science and education, construction and manufacturing, Nano-electronics and information technology, healthcare, aeronautics, environment, biotechnology, agriculture, national security and many more.(Sahoo et al. 2007. Tegart 2009, Salerno et al. 2008, Sobolev et al. 2006).

In spite of very huge research and funding in nanotechnology, it is yet much less well-defined and well-structured discipline as compared to some of the other scientific disciplines. A lot of it is still needed to be explored by the researchers in order to better understand and utilize this already very beneficial discipline for the betterment of human beings.

The word "Nano" has a Greek root from the word "Dwarf" representing a billionth (Zhu et al, 2004). Considering this descent, a nanometer is a billionth part of a meter; as small as 1.25×10^{-6} of the diameter of a human hair.

A Japanese engineer, Norio Taniguchi can be regarded as the individual who introduced the term nanotechnology (Taniguchi, 1974). His description encircled precise manufacturing of parts with finishes and tolerances ranging 0.1 to 100nm. It was a major breakthrough as it intended to control materials and engineering measurements beyond the micro scale. Later in 1981, Drexler (Drexler, 1981) pointed out a new approach that involved atom-by-atom manipulative which is more related to the meaning and application today. (Sahoo et al. 2007, Zhu et al. 2004, Salerno et al. 2008, Rodunar 2006, Sanchez et al. 2010, Cao 2006, Islam et al. 2010, Pacheco-Torgal et al. 2011 and Steyn 2008).

Recent interest in nanostructured materials can be attributed to more refined knowledge about creative manipulations of materials on the nanoscale to perform functions, that otherwise would not have been possible. According to Zhu et al. 2004 and Uskokovic 2007 development in this field has allowed more precision enabling a magnified view; hence surfacing various unexpected and unusual features.

INTRODUCTION TO NANOTECHNOLOGY

Many researchers have attempted to define and explain nanotechnology in their own way. Table 1 summarizes some of the attempts to define this emerging scientific discipline.

In a very general prospect, developments in basic physics and chemistry are leading the way for research in the field of nanotechnology. It is because of the fact that these basic branches of science deal with particles on atomic and molecular level and enable materials and structures to perform tasks that are impossible in their original macroscopic form. This drives the fact that evolution in technology and other related scientific areas e.g. physics and chemistry is making a significant contribution in fast and aggressive developments in its research (Chong, 2002).

Considering the benefits and its future expectances, a significant amount of fortune is currently being invested in nanotechnology. U.S. National Nanotechnology Initiative (NNI) spends more than $1 billion annually plus President's 2008 budget for NNI at 1$.5 billion. But, it needs to be said that such spending is mostly motivated by short term return on investment generated via high value commercial products (Dhir et al. 2005). ARI News has reported that the involvement of nanotechnology in various construction projects carried in developing world is placed at number 8 out of 10 applications that can have a high impact on these countries (ARI News, 2005).

Table 1. Definition of nanotechnology according to researchers

Sr. No.	Reference	Definition
1	(Sahoo et al. 2007).	Nanotechnology is the science that involves design, synthesis, characterization and material application represented in the smallest functional organization with at least one dimension on the Nano scale (10-9 m).
2	(Kelsall et al, 2004)	Nanotechnology covers the design, construction and utilization of functional structures with at least one characteristic Nano scale dimension.
3	(Pieter et al, 2011)	Nanotechnology is science that creates opportunities to harvest construction materials with innovative functionalities and improved properties.
4	(Chong et al, 2004)	Nanotechnology is concerned with the formation of fresh materials and systems at the molecular level because of the fact that the macroscopic properties of the materials are greatly influenced by the atomic and molecular level interactions of the materials.
5	(Ashwani K. Rana et al, 2009)	Nanotechnology is usually dominated by the advancements in physics and chemistry research to produce the materials at micro (atomic and molecular) level which can perform the functions that cannot be accomplished by the conventional materials in their macroscopic form.

Applications of Nanotechnology in Transportation Engineering

Nanotechnology is the field that is involved with the design, erection and operation of various structural elements when one of the characteristic dimensions is measured in nonmetric scale (Goddard III et al. 2007). In last 10 years, considerable development has been achieved in the field of nanotechnology. These developments are mostly the result of dedicated initiatives e.g. National Nanotechnology Initiative, improvements in technology and apparatus and new research into chemistry and physics with respect to nano-scale. Nano-science can be categorized into three broad types, i.e. nanofabrication, nanostructures and nano-electronics with typical applications in life sciences & energy (Goddard III et al. 2007).

In general, nanomaterials may have globular, plate-like, rod like or more complex geometries. Typically, near-spherical particles which are smaller than 10 nm are called clusters. The number of atoms in a cluster increases greatly with its diameter. At 1 nm diameter, there are 13 atoms in a cluster and at 100 nm diameter the cluster can accommodate more than 107 atoms. Clusters may have a symmetrical structure which is, however, often different in symmetry from that of the bulk (Abdullah et al., 2015).

Clusters may also have an irregular or amorphous shape. As the number of atoms in a cluster increases, there is a critical size which a particular bond geometry that is characteristic of the extended bulk (Glenn, 2006 and Rodunar, 2006).

Since its birth, nanotechnology has proved to be very beneficial for human beings in almost all professions of life. From medicine to healthcare, from science and education to application; from materials and construction; manufacturing; nano-electronics and computer information technology; aeronautics; energy and environment; agriculture; biotechnology; and national security, nanotechnology has proved its worth.

NANOTECHNOLOGY AND CIVIL ENGINEERING

Like all the other scientific fields, nanotechnology has also influenced the civil engineering to a great deal. Many researchers have been working on this discipline to form the nanomaterials for construction. A few of such attempts are listed in Table 2.

Like all other fields of life, nanotechnology is establishing its importance and significance in civil engineering with every passing day. The latest researches, some of those discussed in above table, have been very beneficial to produce the materials using nanotechnology concepts for civil engineering applications. It plays its part in enabling production of construction material with unforeseen functionalities and improved properties (Broekhuizen et al, 2010).

It is said that Nanoparticles (NPs) use silica fume to reduce weight of concrete with increased strength and elasticity. Moreover, it reduces energy consumption of

Table 2. Applications of nanotechnology in civil engineering: a brief history

Sr. No.	Researcher	Title	Year
1	Abdullah et al.	A Review on The Exploration of Nanomaterials Application in Pavement Engineering	2015
2	Das and Mitra	Nanomaterials for Construction Engineering-A Review	2014
3	Ugwu et al.	Nanotechnology as a Preventive Engineering Solution to Highway Infrastructure Failures	2013
4	Onuegbu O Ugwu	Nanotechnology for Sustainable Infrastructure in 21st Century Civil Engineering	2013
5	Yao et al.	Rheological properties and chemical analysis of nanoclay and carbon microfiber modified asphalt with Fourier transform infrared spectroscopy.	2013
6	Broekhuizen et al.	Use of nanomaterials in the European construction industry and some occupational health aspects thereof	2011
7	You et al.	Nanoclay-modified asphalt materials: Preparation and characterization	2011
8	Pacheco-Torgal et al.	Nanotechnology: Advantages and drawbacks in the field of construction and building materials	2011
9	Sanchez et al.	Nanotechnology in concrete - A review	2010
10	Jahromi et al.	Engineering Properties of Nanoclay Modified Asphalt Concrete Mixtures	2010
11	Rana et al.	Significance of Nanotechnology in Construction Engineering	2009
12	Wynand JvdM Steyn	Potential Applications of Nanotechnology in Pavement Engineering	2009
13	Jahromi et al.	Effects of nanoclay on rheological properties of bitumen binder.	2009
14	WJvdM Steyn	Research and application of nanotechnology in Transportation	2008
15	Sobolev et al.	Nanomaterials and Nanotechnology for High-Performance Cement Composites	2006
16	Ghile, D.	Effects of nanoclay modification on rheology of bitumen and on performance of asphalt mixtures	2006
17	Cassar, L.	Nanotechnology and photocatalysis in Cementitious Materials	2005
18	Zhu et al.	Application of nanotechnology in construction	2004
19	Bhuvaneshwari et al.	Nanoscience to Nanotechnology for Civil Engineering	
20	Harsh et al.	Applications of nanotechnology in cement and concrete for greater sustainability	

houses via improved isolation material characteristics; improves self-cleaning properties of interior, exterior and glass surfaces to facilitate weathering; improves crack resistance of polymer materials and improves fire resistance of various materials. In addition, it also plays its part in traffic exhaust purification coatings for infrastructure and biodical surfaces for surgery room walls. NanoTiO$_2$ is being explored to increase durability of concrete and maintain its whiteness through lifetime of construct. It can also be used to breakdown organic pollutants, and microorganisms. However, contribution of nano-TiO$_2$ towards coating of roads and auditory barrier along highways has not been practically substantial (Van Ganswijk 2009).

In recent times more innovative and sophisticated progresses have been described, such as building materials containing nano-sensors and nano-particulate self-repairing materials (Koleva 2008; Yang et al. 2009). While many of these applications of nanotechnology are still under development, some of them are currently available in market. (Broekhuizen et al, 2010). Leydekker et al). Leydekker also shares a detailed description of their application in building industry. (Leydekker 2008).

Nanotechnology has widely been used and has massively influenced the Civil Engineering mega projects throughout the world. A lot is still being explored to benefit the human beings by utilizing the basic principles and concepts of this emerging field. The most important benefit of the technology lies in the fact that the materials and hence structure developed possess many rare properties. These products can be used in construction of a variety of structures; lighter structures, stronger composite structures, low maintenance coating; pipe joining materials; fire retarders, insulators and acoustic absorbers. They also help in improving reflectivity of glass (Rana et al. 2009).

There have been many utilizations of nanotechnology in civil engineering and sub disciplines like transportation engineering, structural engineering, construction engineering, geotechnical engineering, hydraulics and water resources engineering and environmental engineering.

OVERVIEW OF TRANSPORTATION ENGINEERING

The study of the application of scientific and technological principles for the planning, operative strategy, working and supervision of the amenities for transportation modes is known as transportation engineering. This is provided for the purpose of safe, effective, express, contented, suitable, cost-effective and environmentally harmonious movement of people and goods. It is a scientific discipline in which materials at macroscopic level are utilized in order to provide the safe, smooth and comfortable way of moving from one place to another. On a contrast to transportation engineering, nanotechnology is a discipline which is characterized by the use of the materials at beyond-micro level. These two fields of science, therefore, differ from each other on the basis of the operational scale i.e. 10^{-9} m for nanotechnology to 10^3 m of transportation infrastructural materials. As the infrastructure is developed from macro level materials conventionally, the research must be focused to assess the possible developments of nanomaterial for infrastructural construction to reduce the burden on the conventional materials. (Steyn, 2008)

As the science of transportation engineering is associated with the efficient movement of people and goods with the help of road, rail and air and providing the safe, smooth and cost effective means of movement for people and goods, the major areas of research focus in the sub-domain of transportation engineering currently and in the past have been safety, durability, economy and sustainability etc. Many researchers have been working in these disciplines to provide the safe smooth ways of movement. Some of the examples to develop the latest materials by utilizing the nanotechnology concepts for the improvements of transportation infrastructure have been described in the following paragraphs.

NANOTECHNOLOGY AND PAVEMENT ENGINEERING

The technology has also offered a handful of benefits to the pavement engineering. While studying the beneficial effects of nanotechnology in pavement engineering, it is a major concern for any researchers to understand and manage the potential instruments that are capable of studying the materials and structures at nano-level in a pavement.

One of the hindrances faced by the researchers to evaluate the consequences of nanotechnology in pavement engineering is the formation of the equipment that can be used effectively to study the nanomaterials and structures at individual level. For the purpose, surface analysis techniques and other conventional techniques can be effectively used. Some of the techniques for this objective may include X-Ray Diffraction (XRD), Scanning Electron microscopy (SEM) and Field Emission Scanning Electron Microscope (FE-SEM) (Cao, 2006 and Kostoff et al., 2007, Abdullah et al. 2015).

X-Ray Diffraction (XRD) is an important, efficient, reliable and most widely used technique to understand the crystal structure of the solids, orientation of individual crystal and poly crystals, stresses, defects, geometry and identification of unknown materials in it. XRD is non-destructive testing and also a powerful technique for investigating the following (Akrema, 2011)

- Crystalline phase characterization (crystallinity, lattice parameters).
- Polymorphism.

- Identification of different modifiers and pigments etc.
- Favored alignment.
- Residual stress and strain.
- Detection of preferred orientation in polymers.
- Characterization of thin films (inorganic and organic) on substrates by grazing angle.
- Temperature characterization of phase transformation (-180°C to +300°C).
- Calculation of crystallite size and strain in crystalline phases (inorganic compounds).

High energy electrons based focused beam is used in scanning electron microscope (SEM) to cause a diversity of the signals at surface of the solid samples. Many information including external morphology (texture), chemical composition, crystalline structure and orientation of materials forming the sample is revealed by the signals that are consequent from electron-sample communications. The types of signals produced by SEM include secondary electron images, back-scattered electron images and elemental Xray maps (Abdullah et al., 2015). Like optical microscopes, SEM also produces the topographical information. Further, it also provides the information about the chemical composition of the materials near the surface (Cao 2006, Joy 1997, Kalaitzidis and Christanis 2003, Lim et al 2005).

The field emission scanning electron microscope (FE-SEM) throws a high-energy electron beam on a sample to understand its characteristics by scanning over the sample. The atoms of the sample are interacted with the fired electrons to produce the signals containing the information about surface topography, composition and other properties, such as electrical conductivity. The obtained characteristics by these elctrons beams can vary from a few millimeters to approximately 10 nanometers. FE-SEM can be helpful in many ways which may include;

- Measuring the thickness of coatings etc.
- Finding the relationship between the surface look and surface morphology.
- Measuring the sizes (height, width, depth etc.) of the various nano-scaled substances.
- Investigation of the nano-scaled substances at individual atomic level.
- Investigating the criteria for various failure mechanisms in materials.
- Defect analysis.

The research and application of nanotechnology in flexible pavement engineering focuses mainly on improvement of asphalt binders which can be achieved by nanoclay, nano hydrated lime, carbon nanoparticles and nano-silica whereas nano titanium dioxide (TiO$_2$) can be sued in rigid pavements for the modifications of concrete properties.

Nanoclays

Nanoclays are nanoparticles of layered mineral silicates. Nanoclays can be classified into various different classes such as montmorillonite, bentonite, kaolinite, hectorite, and halloysite on the basis of their chemical composition and nanoparticle structure [Niroumand et al 2013]. Organoclays which are the nanoclays modified with organic materials are an important group of the hybrid organic-inorganic nanomaterials. They can be possibly used in polymer nanocomposites, as modifiiers for rheological properties, gas absorbents and drug delivery carriers. Binder modification can be best achieved by the addition of montmorillonite out of all of these classes.

Currently, many studies have been conducted on the performance of asphalt binder modified with nanoclay because it consists of high purity and compatible montmorillonite (MMT) particles. Most researchers have been using nanoclay at the range of 3 to 6% by weight of the asphalt binder (You et al 2011, Ghile 2006, Jahromi and Khodaii 2009, Yu et al 2009, Galooyak et al 2010, Jahromi et al 2010, Yao et al 2014, Zare-Shahabadi et al 2010, Zhang et al 2009).

Yang and Tighe (2013) reported that the many physical properties (like tensile strength, tensile modulus, stiffness, modulus of thermal stability and flexural strength) of the bitumen can be improved by amending it with the minor quantities of the nano-clay. They suggested the condition for nano-clay modification that the clay is spread at nano-scopic level. It has also been observed that generally nanoclay modified bituminous materials show higher elasticity (Jahromi and Khodaii 2009). These materials also show lower dissipation of mechanical energy as compared to the virgin/unmodified bitumen (Jahromi and Khodaii 2009). There have been many attempts to study the effect of nanotechnology on pavement engineering. Some of the significant projects are discussed here.

Yu et al. (2007) have studied the effects of different proportions of MMT and organo-modified montmorillonite (OMMT) on different properties of modified binders. The results showed an increase in the softening point and viscosity of the modified binders at high temperatures. Moreover, the amended binders showed higher complex modulus and had lower phase angle. The researchers also concluded that the MMT and OMMT enhanced the viscoelastic properties of modified binders which increase its resistance to rutting at high temperatures. Findings stated by You et al. (2011) describe that nanoclay can improve the complex shear modulus, viscosity and has better low-temperature cracking resistance. Meanwhile, Yao et al. (2013) showed the morphological images of asphalt binder modified with nanoclay

Figure 1. FE-SEM microstructure images of nanoclay modified binder
Yao et al, 2013.

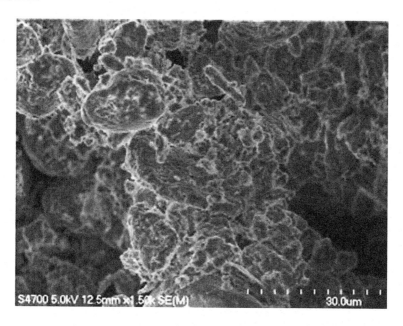

as shown in Figure 1. The rheological findings indicated similar trend as from previous researchers. In addition, they also suggested that the nanoclay has a potential effect as an anti-oxidation, which can reduce oxidation reaction when the modified binders are exposed to heat and daylight.

Similarly, Galooyak et al. (2010) had studied the effect of styrene–butadiene–styrene/organomodified montmorillonite (SBS/OMMT) modified asphalt mixtures. The results exposed that the presence of nanoclay improves the storage stability of Polymer Modified Bitumen (PMB) considerably. The effect of nanoclay (nanofil-15 and cloisite-15A) on rheological properties of binders by using Dynamic Shear Rheometer (DSR) has also been studied by Jahromi and Khodaii (2009). They concluded that the nanoclay changed the rheological properties, increased stiffness, decreased phase angle and increased the ageing resistance in the modified bitumen.

Nano Hydrated Lime

Hydrated lime has widely been used in construction projects. It takes its importance being a well-known and efficient material for modifications of asphalt binders. On addition to the mix, nano-scaled lime undergoes reaction with the aggregates and reinforces the interface bond of bitumen and aggregates. The lime added to the mix reacts with the aggregate molecules to prevent the development of the water soluble soaps which usually endorses the stripping failures in the highways. Water repellent insoluble salts are formed on reaction of the lime with these types of molecules reducing the damaging effects of water (Aragão et al 2010, Cheng et al, 2011, Gorkem and Sengoz 2009, Lee et al 2010, Özen 2011).

The addition of the hydrated lime into an asphalt mix makes the mix more stiff, tough and resistant to rutting. This capability of the lime reflects its greater performance due to the active and helpful effects of mineral filler (Aragão et al 2010, Cheng et al, 2011, Gorkem and Sengoz 2009, Lee et al 2010, Özen 2011). Recently, Lee et al (2010) studied the effect of hydrated lime on dynamic modulus of asphalt-aggregate mixtures. Their findings suggested that the addition of hydrated lime make Hot Mix Asphalt (HMAs) stiff especially at high frequencies and lower temperatures. They also reported that the lime modified HMAs showed smaller decrease of dynamic modulus due to the moisture as compared to the un-modified HMA.

Carbon Nano Particles

Carbon Nanoparticles are sphere shaped blackish high surface area graphitic carbons. The usual sizes of the Nanoscale Carbon Particles are approximately 10 - 45 nanometers (nm) with specific surface area 30-50 m^2/g. They are also termed as Carbon Nanotubes which are made from sheets of graphite that have rolled up to form a tubular structure (Abdullah et al 2015). Carbon nanotubes can be categorized by its particle structure and currently, there are three types of nanotubes: single-walled nanotubes (SWNTs), multi-walled nanotubes (MWNTs) and double-wall nanotubes (DWNT). However, researchers had stated that MWNTs are easier and cheaper to produce compared with the other nanotubes. In addition, MWNTs can efficiently be dispersed when modified with virgin asphalt binder due to its individual molecular structure (Amirkhanian et al 2011, Li et al 2011, Mamalis et al 2004, Xiao et al 2011). There have been many studies in which researchers have tried to evaluate the effect of carbon nanoparticles in improving the properties of bitumen. For example, Amirkhanian et al (2011) and Xiao et al (2011) concluded that generally the MWNT modified asphalt binders increase the rutting resistance at pavement surface temperature. It has been claimed that the paint prepared by the carbon nanotubes and used on the Forth Bridge in Scotland would require renovation after 25 years (Dresselhaus et al (2001) and Saito et al. 1998). UNESCO has announced it as a heritage of the world. After its opening in 1890, the painting of the bridge has been a never ending task which has been addressed with the application of nanotechnology. The nanotechnology based paint has allowed the administration of the bridge to repaint it after a healthy period of 25 years.

Nano Silica

Being one of the most abundantly available compound on land, silica has largely been used in many industries to produce various products including silica gels, colloidal silica and fumed silica etc. Silica nano-composites have been drawing some scientific attention as well. These materials take their advantages from the fact of having less production cost as well as high performance features (Yang and Tighe 2013, Lazzara et al. 2010).

Yao et al (2012) considered the effects of the nanosilica on the rheological properties and chemical bonding of the asphalt binders by adding it at contents of 4% and 6% by weight of the asphalt binder. They studied the effects of nanosilica on asphalt binder as well as the mixture through various laboratory testing procedures. The results suggested a decrease in viscosity of the binder by the addition of nanosilica. Also, the dynamic modulus and flow number of the modified asphalt mixture was improved. The results showed a reduction in rutting susceptibility of the mixture. Generally, it was concluded that the addition of the nanosilica improved the anti-ageing and rutting resistant properties of the asphalt (Yao et al. 2012).

The addition of the 1% nano pulverized rubber improved the rutting resistance and low temperature crack resistant properties of the asphalt binders as compared to the conventional mixtures (Chen et al. 2012). The physical and mechanical properties of asphalt binders as well as mixtures can be enhanced by the addition of 5% SBS plus 2% nano-SiO_2 powder for asphalt modification (Ghasemia et al 2012). Also, the researchers reported the improvement in the moisture susceptibility of the asphalt mixture by incorporation of the nanoclay and carbon microfiber in it (Goh, et al 2010).

Titanium Dioxide (TiO_2)

Titanium dioxide commonly called as titania has been known for its useful photocatalytic qualities. In addition, it provides numerous unique characteristics. Titania is:

1. Comparatively economical, nontoxic material;
2. Has more photocatalytic activity than other materials;
3. Harmonious with conventional building materials, such as cement, in such a way that it may not change unique performance of that material;
4. Effective in any atmospheric environment (Abdullah et al 2015).

Many researchers have attempted to improve the properties of the concrete in rigid pavements by using titanium dioxide. It was suggested by Li et al. (2006, 2007) and Zhang et al. (2011) that any increase in the compressive strength of nano-modified concrete would also increase the resistance to the abrasion thus improving the performance under fatigue. The sensitivity of their fatigue lives to the change of stress is also increased. Besides that, the addition of nano-TiO_2 also polishes the pore structure of concrete and augments the resistance to chloride penetration inside concrete. Hassan et al. (2010) investigated the use of TiO_2 particles as coating for concrete pavements, where these particles can be trapped and decompose the organic and inorganic air pollutants by a photo-catalytic process. They stated that the reinforcement of the samples with 3% TiO_2 slightly improves the nitrogen oxides, NOx ($NO+NO_2$) removal efficiency. They also claimed that the use of TiO_2 coating as a photocatalytic compound would provide satisfactory durability and wear resistance. Fujishima et al. (2008) had coated several road surfaces with cement mixtures containing TiO_2 colloidal solutions. The results obtained in an area of 300 m^2 showed 50 to 60 mg/day NOx degradation. The process of degradation of environmental pollutants is shown in Figure 2 (Khitab et al. 2014).

APPLICATIONS OF NANOTECHNOLOGY IN TRANSPORTATION ENGINEERING

In the recent past, it has been attempted to utilize the principles of nanotechnology in the major areas of research in transportation engineering i.e. safety, economy, sustainability and durability etc. The following discussion provides detailed literature

Figure 2.

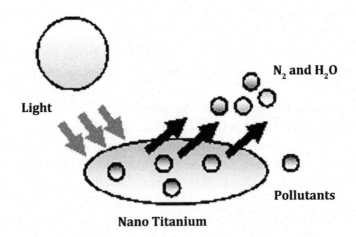

survey regarding the utilization of nanotechnology in Transportation engineering. First of all, the major concerns of any Transport Engineer are listed as follows:

- Economy.
- Safety.
- Durability.
- Sustainability.

Economy

The transportation infrastructures are built by utilizing the public money; thus making the economy the most important factor to be considered and worried about. The economy in terms of long term, short term, maintenance requirements and total life period of the facility should be considered.

It is very important to realize that for transportation infrastructure, nanotechnology can best benefit if its potential effects on maintenance purposes are studied as most of the infrastructure has already been developed in most part of the world and the engineers are concerned / interacted with the maintenance of the highways. On the similar lines, the major portion of highway infrastructure in country like Pakistan has also been laid already. So, the focus of the research should also be to improve / enhance the service life of these facilities as the replacement of the facilities can never be an economical solution. Lastly, the effect of the possible solutions when the infrastructure is discarded is also very important factor to be considered. This becomes very significant when the effects of nano materials for potential solution after discard of the materials and their effect on the environment is considered.

The technology is also concerned with the economy as it may increase the service life of the structure up to many years. This may also be augmented by the enhancement of the infrastructural opposition to the environmental degradational effects. For example, the various types of coatings of nano-composite materials reported by Arafa et al, (2005) and Papakonstantinou, (2005) applied to the concrete faces of bridge abutments and pillars may prolong the service life of these bridges and pillars. The main difference of these specially provided coatings from the conventional coatings is the way of making bond with the substrate materials. They provide a vigorous layer which usually binds chemically with the materials of substrate.

Safety

The most important function to be served by a transportation facility is the safety for the users. Safety is a parameter, which is directly linked to the materials, a facility is constructed of. This is due to the direct contact of the vehicles with the materials.

The research involving the safety of the transportation facilities was reported by Wang et al (2002). They studied the improvement of skid resistance in wet conditions by combining the nano-scale silicon carbide in manufacturing tires. They reported up to 50% decreased abrasion of the tires due to addition of the nano-scale silicon carbide. Many researchers including Degussa (2008) and Hou et al. (2007) have also attempted to apply ZnO_2 to highway infrastructure for introducing the hydrophobic properties in these facilities. The developments in these research fields may lead to hydrophobic facilities with faster water run-off from the surface the reducing the probabilities of hydroplaning (Degussa, 2008; Hou et al, 2007).

Durability

One of the most important features of a highway facility in its durability. The prerequisites for the durable highways can be covered in two main points. One is its ability of provide the expected service with minimum expenditures on the maintenance while the second one is its ability to be termed as a sustainable transportation facility. With respect to the durability, there have been many studies including Munoz et al., 2008. They reported up to 50% increase in shear strength of concrete due to the addition of the nanomaterials.

Sustainability

The basic requirements for sustainable facilities reported by Maher et al (2006) can be termed as follows;

- It should reduce the expenditure of naturally occurring resources;
- It should reduce the usage of energy;
- It should not increase the release of the greenhouse gases;
- It should prevent all types of pollution including air, water, earth, noise etc;
- It should be environmental friendly in such a way to have a positive effect of human health and safety
- It should prevent the risk chances while traveling
- It should also ensure very high level of the user comfort and safety

The potential effect of nanotechnology on the sustainability of the transport infrastructure would again focus on the modification of the current resources that may conventionally prove to be damaging to the environment either through their common application in the infrastructure, or through their production or extracting them from the environment in the first place. In addition to this, re-use of the existing materials may be another option. This may be achieved by modification of these materials and ensuring the structural changes for better and efficiently prolonged service life of the facility. The modification of the materials can also be carried out to make possible the construction at very low energy level i.e. to lower the requirements of the energy during the construction procedure. It can be also attempted to investigate the amendment in the currently used materials to allow the formation of the inert material which will be less damaging to the environment. An example of this can be the modification of the asbestos which would reduce the release of the fibres into the airways (Cassar, 2005).

DRAWBACKS OF NANOTECHNOLOGY

There are many drawbacks of the nanotechnology along with its countless advantages and uses. The scientific developments up till now have warned us about the possible severe health as well as environmental issues related with nanotechnology as there is no complete eco-toxicological outline for any of the nano-materials existing today. As these materials are air-borne and water-borne, the large scale production and release of these materials must be controlled and checked before using these for various construction projects. A lot of research exploration in this regard is still need of the time as an attempt to address the severe health problems associated with the permanent and temporary exposure of the nano-particles. Nanotechnology is likely to carry practical water treatment in some areas (Meridian Institute 2006). It is also to be considered that even leaving aside the potential human health issues, a danger for other conventional water treatment methodologies still exists because these current, public controlled methods may be suspended as the priority may be given to corporate controlled nano-water treatment applications.

In addition to many benefits, this technology exhibits many disadvantages/drawbacks as well: Which can be classified as follows:

1. **Size Differences:** There exists greater difference between the sizes of transportation materials and nano particles. It is one of the major hindrances of the technology as the researcher has to replace macro materials with the nanomaterials without sacrificing the strength of the structure.
2. **Health:** The excessive exposure of the human beings to the nano-particles may have harmful effects on the health. There have been many research attempts which have reported the harmful and dangerous effects of nano-particles to the human beings especially if occupational exposure is encountered. A few of the health related issues may include oxidative stress, fibrosis, cardiovascular effects, cytotoxicity, and possibly carcinogenicity as effects of nano-particle

exposure (Renwick et al.2004; Borm et al. 2006; 2007; Schulte et al. 2008; Trouiller et al. 2009; Knol et al. 2009; Stone et al. 2010, b).

3. **Cost:** Cost is also one of the drawbacks of the nanotechnology. Nanotechnology requires much of the cost for the establishment of the laboratories. The nano materials preparation may also prove to be very expensive. The supporters of the nanotechnology argue that the nanomaterials usage will certainly reduce the energy and natural conventional resources utilization. They argue that the small quantities of the strong nanomaterials may theoretically perform the task as performed by the larger amounts of the macro scale conventional materials. They also support their statement by saying that lighter industrial units constructed by using the light nanomaterials like carbon nanotubes may consume less energy for its proper functioning. However production of nanomaterials has suddenly huge ecological footstep as its production requires:

a. Highly sophisticated production environment,
b. Processing requires high energy and water,
c. Lower levels of productions,
d. An increase in the waste products,
e. The emission of various harmful greenhouse gases like methane, and
f. The utilization of various health harming and toxic materials like benzene for production procedures (Şengül et al 2008).

A separate study of the life cycle of the manufacturing of the carbon nanofibre may reveal that their potential to add to global warming, ozone layer depletion, environmental and human toxicity is as much as 100 times greater per unit mass as compared to the other orthodox materials including aluminum, steel and polypropylene (Khanna et al 2008).

It is expected that for the same product, the amount of the nanomaterial to be used is far less as compared to the amount of traditional materials. So, the impacts of the nanomaterials on global energy and environment can be better understood by the careful evaluation of the life cycle of the products made by these nanomaterials. However, the scientist may conclude based on the available results that the benefits of nanomaterials may be outcome by the cost of their production.

4. **Environmental Issues:** The chemical nature of many of the nano materials may cause serious damage to the environment as well as human beings.

Nanomaterials may themselves be regarded as the new group of the toxicity causing chemicals. With the decrease in the particle size of the nanomaterials, the free radicals production and hence toxicity increases. The research on test tubes have shown that the commercially available nanomaterials can cause harmful effects of the human DNA, depressingly effect cellular function and even cause cell death. Additionally, there have been many scientific attempts to show the toxic effect of these materials on different environmental indicators including algae, invertebrate and fish species (Hund-Rinke and Simon 2006, Lovern and Klaper 2006, Templeton et al 2006, Federici et al 2007, Lovern et al 2007). A confirmation has also been reported about the effect of some nanomaterials to weaken the function of earthworms (Scott-Fordsmand et al 2008). The most alarming and recent development in this scientific discipline has shown the evidences that the nanomaterials can be transferred across generations in both animals as well as humans (Takeda et al 2009, Tsuchiya et al 1996, Lin et al 2009).

Moreover, in-spite of being used in low quantities than the ordinary chemicals, these nano-materials usually has more toxic effects to the environment. It has been estimated by Woodrow Wilson International Center for Scholars' Project on Emerging Nanotechnologies (PEN) in 2006 that during the ten year span between 2011 and 2020; approximately 58,000 metric tons of these materials will be globally produced. They also confirmed that on the basis of the power of nanomaterials, this much production of these materials may possess as much ecological impact as 5 million tons or may be 50 billion tons of traditional materials (Maynard 2006).

CONCLUSION AND RECOMMENDATIONS

Nanotechnology is one of the fastest growing fields of the recent times having a great potential in nanotechnology to improve the transportation engineering. A number of promising developments exist that can potentially change the service life and life-cycle cost of transport infrastructure. In asphalt pavements, the applications of nanotechnology have improved both physical and rheological properties. In terms of physical properties, nanomaterials can decrease penetration and ductility, and increase softening point value. Meanwhile, the rheological properties such as performance grade, rutting resistance, low-temperature cracking resistance and aging resistance also showed healthier improvements. In addition, the engineering properties of mixtures incorporating asphalt binders modified with nanomaterials were significantly improved particularly in the areas of stiffness, rutting resistance, indirect tensile strength and resilient modulus. Additionally, it can also contribute towards safety, economy, durability and sustainability of these infrastuctures.

It is recommended that the researchers can explore more on the production process of nanoclay with optimum cost besides investigate the effectiveness of nanoclay as a modifier for WMA mixtures. With the advances in instrumentation and computational science, it could say that nanotechnology will exploit the improvement of pavement material properties and construction process in the future.

REFERENCES

Abdullah, M. E., Zamhari, K. A., Buhari, R., Kamaruddin, N. H. M., Nayan, N., Hainin, M. R., ... & Yusoff, N. I. M. (2015). A review on the exploration of nano-materials application in pavement engineering. *Jurnal Teknologi, 73*(4).

Amirkhanian, A. N., Xiao, F., & Amirkhanian, S. N. (2011). Characterization of unaged asphalt binder modified with carbon nano particles. *International Journal of Pavement Research and Technology, 4*(5), 281.

Arafa, M. D., DeFazio, C., & Balaguru, B. (2005). Nano-composite coatings for transportation infrastructures: Demonstration projects. *2nd International Symposium on Nanotechnology in Construction*, Bilbao, Spain.

Aragão, F. T. S., Lee, J., Kim, Y. R., & Karki, P. (2010). Material-specific effects of hydrated lime on the properties and performance behavior of asphalt mixtures and asphaltic pavements. *Construction & Building Materials, 24*(4), 538–544. doi:10.1016/j.conbuildmat.2009.10.005

ARI News. (2005). *Nanotechnology in Construction – One of the Top Ten Answers to World's Biggest Problems*. Retrieved July 7, 2015, from www.aggregateresearch.com/article.asp?id=6279

Arkema. (2011). *Analytical Techniques of Structural Characterization* [Online]. Retrieved on October 2011, from http://www.arkema-inc.com/

Bhuvaneshwari, B., Sasmal, S., & Iyer, N. R. (2011, July). Nanoscience to nanotechnology for civil engineering: proof of concepts. In *Proceedings of the 4th WSEAS International Conference on Recent Researches in Geography, Geology, Energy, Environment and Biomedicine (GEMESED'11)* (pp. 230-235).

Borm, P. J., Robbins, D., Haubold, S., Kuhlbusch, T., Fissan, H., Donaldson, K., & Krutmann, J. et al. (2006). The potential risks of nanomaterials: A review carried out for ECETOC. *Particle and Fibre Toxicology, 3*(1), 1. doi:10.1186/1743-8977-3-11 PMID:16907977

Broekhuizen, P. V., Broekhuizen, F. V., Cornelissen, R., & Reijnders, L. Use of nanomaterials in the European construction industry and some occupational health aspects there of. *Journal of Nanoparticle Research*. doi:10.1007/s11051-010-0195-9

Cao, G. (2006). *Nanostructures & Nanomaterials: Synthesis, Properties & Applications* (1st ed.). London: Imperial College Press.

Cassar, L. (2005). Nanotechnology and photocatalysis in cementitious materials. *2nd International Symposium on Nanotechnology in Construction*, Bilbao, Spain.

Chen, S. J., & Zhang, X. N. (2012, October). Mechanics and Pavement Properties Research of Nanomaterial Modified Asphalt. In *Advanced Engineering Forum* (Vol. 5, pp. 259-264). doi:10.4028/www.scientific.net/AEF.5.259

Cheng, J., Shen, J., & Xiao, F. (2011). Moisture susceptibility of warm-mix asphalt mixtures containing nanosized hydrated lime. *Journal of Materials in Civil Engineering*, *23*(11), 1552–1559. doi:10.1061/(ASCE)MT.1943-5533.0000308

Chong, K. P. (2002). Research and Challenges in Nanomechanics, 90-minute Nanotechnology Webcast. *ASME*.

Das, B. B., & Mitra, A. (2014). Nanomaterials for Construction Engineering-A Review. *International Journal of Materials. Mechanics and Manufacturing*, *2*(1), 41–46. doi:10.7763/IJMMM.2014.V2.96

Degussa. (2008). *Zinc Oxide, Cerium Oxide, Indium TinOxide, just part of the growing nano product range from Degussa*. Retrieved April 20, 2015 from http://www.azonano.com/Details.asp?ArticleID=1603

Dhir, R. K., Newlands, M. D., & Csetenyi, L. J. (2005). Introduction. In *Proceedings of the International Conference – Application of Technology in Concrete Design*.

Dresselhaus, M. S., & Dresselhaus, G. (2001). Ph. in Carbon Nanotubes. Topics Appl. Phys, 80, 287–327.

Drexler, K. E. (1981). Molecular engineering: An approach to the development of general capabilities for molecular manipulation. *Proceedings of the National Academy of Sciences of the United States of America*, *78*(9), 5275–5278. doi:10.1073/pnas.78.9.5275 PMID:16593078

Federici, G., Shaw, B. J., & Handy, R. D. (2007). Toxicity of titanium dioxide nanoparticles to rainbow trout (Oncorhynchus mykiss): Gill injury, oxidative stress, and other physiological effects. *Aquatic Toxicology (Amsterdam, Netherlands)*, *84*(4), 415–430. doi:10.1016/j.aquatox.2007.07.009 PMID:17727975

Fujishima, A., Zhang, X., & Tryk, D. A. (2008). TiO 2 photocatalysis and related surface phenomena. *Surface Science Reports*, *63*(12), 515–582. doi:10.1016/j.surfrep.2008.10.001

Galooyak, S. S., Dabir, B., Nazarbeygi, A. E., & Moeini, A. (2010). Rheological properties and storage stability of bitumen/SBS/montmorillonite composites. *Construction & Building Materials*, *24*(3), 300–307. doi:10.1016/j.conbuildmat.2009.08.032

Ghile, D. B. (2006). *Effects of nanoclay modification on rheology of bitumen and on performance of asphalt mixtures*. Delft, The Netherlands: Delft University of Technology.

Glenn, J. C. (2006). Nanotechnology: Future military environmental health considerations. *Technological Forecasting and Social Change, 73*(2), 128–137. doi:10.1016/j.techfore.2005.06.010

Goddard, W. A. III, Brenner, D., Lyshevski, S. E., & Iafrate, G. J. (Eds.). (2007). *Handbook of nanoscience, engineering, and technology*. CRC Press.

Goh, S. W., Akin, M., You, Z., & Shi, X. (2011). Effect of deicing solutions on the tensile strength of micro- or nano-modified asphalt mixture. *Construction & Building Materials, 25*(1), 195–200. doi:10.1016/j.conbuildmat.2010.06.038

Gorkem, C., & Sengoz, B. (2009). Predicting stripping and moisture induced damage of asphalt concrete prepared with polymer modified bitumen and hydrated lime. *Construction & Building Materials, 23*(6), 2227–2236. doi:10.1016/j.conbuildmat.2008.12.001

Hassan, M. M., Dylla, H., Mohammad, L. N., & Rupnow, T. (2010). Evaluation of the durability of titanium dioxide photocatalyst coating for concrete pavement. *Construction & Building Materials, 24*(8), 1456–1461. doi:10.1016/j.conbuildmat.2010.01.009

Hou, X., Zhou, F., Yu, B., & Liu, W. (2007). Superhydrophobic zinc oxide surface by differential etching and hydrophobic modification. *Materials Science and Engineering A, 452*, 732–736. doi:10.1016/j.msea.2006.11.057

Hui, Y., Zhanping, Y., Liang, L., Huei, L. C., David, W., Khin, Y. Y., & Wei, G. S. et al. (2012). Properties and Chemical Bonding of Asphalt and Asphalt Mixtures Modified with Nanosilica. *Journal of Materials in Civil Engineering*. doi:10.1061/(ASCE)MT.1943-5533.0000690

Hund-Rinke, K., & Simon, M. (2006). Ecotoxic effect of photocatalytic active nanoparticles (TiO2) on algae and daphnids (8 pp). *Environmental Science and Pollution Research International, 13*(4), 225–232. doi:10.1065/espr2006.06.311 PMID:16910119

Islam, N., & Miyazaki, K. (2010). An empirical analysis of nanotechnology research domains. *Technovation, 30*(4), 229–237. doi:10.1016/j.technovation.2009.10.002

Jahromi, S. G., Andalibizade, B., & Vossough, S. (2010). Engineering properties of nanoclay modified asphalt concrete mixtures. *The Arabian Journal for Science and Engineering, 35*(1B), 89–103.

Jahromi, S. G., & Khodaii, A. (2009). Effects of nanoclay on rheological properties of bitumen binder. *Construction & Building Materials, 23*(8), 2894–2904. doi:10.1016/j.conbuildmat.2009.02.027

Joy, D. C. (1997). Scanning electron microscopy for materials characterization. *Current Opinion in Solid State and Materials Science, 2*(4), 465–468. doi:10.1016/S1359-0286(97)80091-5

Kalaitzidis, S., & Christanis, K. (2003). Scanning electron microscope studies of the Philippi peat (NE Greece): Initial aspects. *International Journal of Coal Geology, 54*(1), 69–77. doi:10.1016/S0166-5162(03)00020-X

Kelsall, R. W., Hamley, I. W., & Geoghegan, M. (2004). *Nanoscale science and technology*. Chisester, UK: John Wiley and Sons, Ltd.

Khanna, V., Bakshi, B. R., & Lee, L. J. (2008). Carbon nanofiber production. *Journal of Industrial Ecology, 12*(3), 394–410. doi:10.1111/j.1530-9290.2008.00052.x

Khitab, A., & Arshad, M. T. (2014). Nano Construction Materials, *Rev. Adv. Mat. Sci., 3*, 131–138.

Knol, A. B., de Hartog, J. J., Boogaard, H., Slottje, P., van der Sluijs, J. P., Lebret, E., & Brunekreef, B. et al. (2009). Expert elicitation on ultrafine particles: Likelihood of health effects and causal pathways. *Particle and Fibre Toxicology, 6*(1), 1. doi:10.1186/1743-8977-6-19 PMID:19630955

Koleva, D. A. (2008). *Nano-materials with tailored properties for self healing of corrosion damages in reinforced concrete, IOP self healing materials*. SenterNovem, TheNetherlands.

Kostoff, R. N., Koytcheff, R. G., & Lau, C. G. (2007). Global nanotechnology research literature overview. *Technological Forecasting and Social Change, 74*(9), 1733–1747. doi:10.1016/j.techfore.2007.04.004

Lazzara, G., & Milioto, S. (2010). Dispersions of nanosilica in biocompatible copolymers. *Polymer Degradation & Stability, 95*(4), 610–617. doi:10.1016/j.polymdegradstab.2009.12.007

Lee, S., Seo, Y., & Kim, Y. R. (2010). Effect of hydrated lime on dynamic modulus of asphalt-aggregate mixtures in the state of North Carolina. *KSCE Journal of Civil Engineering, 14*(6), 829–837. doi:10.1007/s12205-010-0944-4

Leydekker, S. (2008). *Nanomaterials in architecture, interior architecture and design*. Birkhaüser Verlag AG, BaselBoston-Berlin.

Li, H., Zhang, M. H., & Ou, J. P. (2006). Abrasion resistance of concrete containing nano-particles for pavement. *Wear*, *260*(11), 1262–1266. doi:10.1016/j.wear.2005.08.006

Li, H., Zhang, M. H., & Ou, J. P. (2007). Flexural fatigue performance of concrete containing nano-particles for pavement. *International Journal of Fatigue*, *29*(7), 1292–1301. doi:10.1016/j.ijfatigue.2006.10.004

Li, W., Buschhorn, S. T., Schulte, K., & Bauhofer, W. (2011). The imaging mechanism, imaging depth, and parameters influencing the visibility of carbon nanotubes in a polymer matrix using an SEM. *Carbon*, *49*(6), 1955–1964. doi:10.1016/j.carbon.2010.12.069

Lim, S. C., Kim, K. S., Jeong, S. Y., Cho, S., Yoo, J. E., & Lee, Y. H. (2005). Nanomanipulator-assisted fabrication and characterization of carbon nanotubes inside scanning electron microscope. *Micron (Oxford, England)*, *36*(5), 471–476. doi:10.1016/j.micron.2005.03.005 PMID:15896968

Lin, S., Reppert, J., Hu, Q., Hudson, J. S., Reid, M. L., Ratnikova, T. A., & Ke, P. C. et al. (2009). Uptake, translocation, and transmission of carbon nanomaterials in rice plants. *Small*, *5*(10), 1128–1132. PMID:19235197

Lovern, S. B., & Klaper, R. (2006). Daphnia magna mortality when exposed to titanium dioxide and fullerene (C60) nanoparticles. *Environmental Toxicology and Chemistry*, *25*(4), 1132–1137. doi:10.1897/05-278R.1 PMID:16629153

Lovern, S. B., Strickler, J. R., & Klaper, R. (2007). Behavioral and physiological changes in Daphnia magna when exposed to nanoparticle suspensions (titanium dioxide, nano-C60, and C60HxC70Hx). *Environmental Science & Technology*, *41*(12), 4465–4470. doi:10.1021/es062146p PMID:17626453

Maher, M., Uzarowski, L., Moore, G., & Aurilio, V. (2006, November). Sustainable Pavements-Making the Case for Longer Design Lives for Flexible Pavements. In *Proceedings of the Fifty-First Annual Conference of the Canadian Technical Asphalt Association (CTAA)*.

Mamalis, A. G., Vogtländer, L. O. G., & Markopoulos, A. (2004). Nanotechnology and nanostructured materials: Trends in carbon nanotubes. *Precision Engineering*, *28*(1), 16–30. doi:10.1016/j.precisioneng.2002.11.002

Maynard, A. D. (2006). *A research strategy for addressing risk. In Nanotechnology*. Woodrow Wilson International Center for Scholars.

Meridian Institute. (2006). *Overview and comparison of conventional water treatment technologies and nano based water treatment technologies. In Global Dialogue on nanotechnology and the poor: opportunities and risks*. Chennai, India: Meridian Institute.

Mojtaba, G., Morteza, M. S., Majid, T., Jalal, K. R., & Reza, T. (2012). Modification of stone matrix asphalt with nano-SiO2. *J. Basic Appl. Sci. Res*, *2*(2), 1338–1344.

Morse, G. (2004). You Don't Have a Nanostrategy? *Harvard Business Review*, *82*(2), 32.

Munoz, J. F., Sanfilippo, J. M., Tejedor, M. I., Anderson, M. A., & Cramer, S. M. (2008). *Preliminary study of the effects of nanoporous films in ITZ properties of concrete*. Poster presented at 87th Annual Transportation Research Board meeting, TRB 2008, Washington, DC.

Niroumanda, H., Zainb, M. F. M., & Alhosseini, S. N. (2013). The Influence of Nano-Clays on Compressive Strength of Earth Bricks as Sustainable Materials. *Procedia - Social and Behavioral Sciences*. doi:10.1016/j.sbspro.2013.08.945

Özen, H. (2011). Rutting evaluation of hydrated lime and SBS modified asphalt mixtures for laboratory and field compacted samples. *Construction & Building Materials*, *25*(2), 756–765. doi:10.1016/j.conbuildmat.2010.07.010

Pacheco-Torgal, F., & Jalali, S. (2011). Nanotechnology: Advantages and drawbacks in the field of construction and building materials. *Construction & Building Materials*, *25*(2), 582–590. doi:10.1016/j.conbuildmat.2010.07.009

Papakonstantinou, C. G. (2005). Protective coatings with nano constituent materials. *2nd International Symposium on Nanotechnology in Construction*, Bilbao, Spain

Rana, A. K., Rana, S. B., Kumari, A., & Kiran, V. (2009). Significance of nanotechnology in construction engineering. *International Journal of Recent Trends in Engineering*, *1*(4), 46–48.

Renwick, L. C., Brown, D., Clouter, A., & Donaldson, K. (2004). Increased inflammation and altered macrophage chemotactic responses caused by two ultrafine particles types. *Occupational and Environmental Medicine*, *61*(5), 442–447. doi:10.1136/oem.2003.008227 PMID:15090666

Roduner, E. (2006). *Nanoscopic Materials Size-Dependent Phenomena* (1st ed.). Stuttgart, Germany: RSC Publishing.

Sahoo, S. K., Parveen, S., & Panda, J. J. (2007). The present and future of nanotechnology in human health care. *Nanomedicine; Nanotechnology, Biology, and Medicine, 3*(1), 20–31. doi:10.1016/j.nano.2006.11.008 PMID:17379166

Saito, R., Dresselhaus, G., & Dresselhaus, M. S. (1998). *Physical Properties of Carbon Nanotubes.* World Scientific.

Salerno, M., Landoni, P., & Verganti, R. (2008). Designing foresight studies for Nanoscience and Nanotechnology (NST) future developments. *Technological Forecasting and Social Change, 75*(8), 1202–1223. doi:10.1016/j.techfore.2007.11.011

Sanchez, F., & Sobolev, K. (2010). Nanotechnology in concrete - A review. *Construction & Building Materials, 24*(11), 2060–2071. doi:10.1016/j.conbuildmat.2010.03.014

Saxl, O. (2001). *Opportunities for Industry in the Application of Nanotechnology.* The Institute of Nanotechnology.

Schneider, T. (2007). Evaluation and control of occupational health risks from nanoparticles. Tema Nord. Nordic Council of Ministers, Copenhagen, 2007:581.

Schulte, P., Geraci, C., Zumwalde, R., Hoover, M., & Kuempel, E. (2008). Occupational risk management of engineered nanoparticles. *Journal of Occupational and Environmental Hygiene, 5*(4), 239–249. doi:10.1080/15459620801907840 PMID:18260001

Scott-Fordsmand, J., Krogh, P., Schaefer, M., & Johansen, A. (2008). The toxicity testing of double-walled nanotubescontaminated food to Eisenia veneta earthworms. *Ecotoxicology and Environmental Safety, 71*(3), 616–619. doi:10.1016/j.ecoenv.2008.04.011 PMID:18514310

Şengül, H., Theis, T., & Ghosh, S. (2008). Towards sustainable nano products: An overview of nanomanufacturing methods. *Journal of Industrial Ecology, 12*(3), 329–359.

Siegel, R. W., Hu, E., & Roco, M. C. (1999, September). Nanostructure science and technology: a worldwide study. *IWGN.*

Sobolev, K., Flores, I., Hermosillo, R., & Torres-Martínez, L. M. (2006). Nanomaterials and Nanotechnology for High-Performance Cement Composites. *ACI Session on Nanotechnology of Concrete: Recent Developments and Future Perspectives,* 91–118.

Steyn, W. J. M. (2008). *Development of auto-luminescent surfacing for concrete pavements.* Paper presented at Transportation Research Board, Washington, DC.

Steyn. (2008). Research and Application of Nanotechnology in Transportation. *27th Southern African Transport Conference.*

Stone, V., (2010). *Engineered nanoparticles: review of health and environmental safety (ENRHES).* Edinburgh, FP7.

Sunter, C. (1996). *The high road: Where are we now.* Cape Town, South Africa: Tafelberg.

Takeda, K., Suzuki, K., Ishihara, A., Kubo-Irie, M., Fujimoto, R., Tabata, M., & Sugamata, M. et al. (2009). Nanoparticles transferred from pregnant mice to their offspring can damage the genital and cranial nerve systems. *Journal of Health Science, 55*(1), 95–102. doi:10.1248/jhs.55.95

Taniguchi, N. (1974). On the basic concept of nanotechnology. *Proceedings of the International Conference on Production Engineering.*

Tegart, G. (2009). Energy and nanotechnologies: Priority areas for Australia's future. *Technological Forecasting and Social Change, 76*(9), 1240–1246. doi:10.1016/j.techfore.2009.06.010

Templeton, R., Ferguson, P., Washburn, K., Scrivens, W., & Chandler, G. (2006). Life-cycle effects of single-walled carbon nanotubes (SWNTs) on an estuarine meiobenthic copepod. *Environmental Science & Technology, 40*(23), 7387–7393. doi:10.1021/es060407p PMID:17180993

Trouiller, B., Reliene, R., Westbrook, A., Solaimani, P., & Schiestl, R. H. (2009). Titanium dioxide nanoparticles induce dna damage and genetic instability in vivo in mice. *Cancer Research, 69*(22), 8784–8789. doi:10.1158/0008-5472.CAN-09-2496 PMID:19887611

Tsuchiya, T., Oguri, I., Yamakoshi, Y., & Miyata, N. (1996). Novel harmful effects of fullerene on mouse embryos in vitro and in vivo. *FEBS Letters, 393*(1), 139–145. doi:10.1016/0014-5793(96)00812-5 PMID:8804443

Uskokovic, V. (2007). Nanotechnologies: What we do not know. *Technology in Society., 29*(1), 43–61. doi:10.1016/j.techsoc.2006.10.005

Wang, M. J., Reznek, S. R., Kutsovsky, Y., & Mahmud, K. (2002). *Elastomeric compounds with improved wet skid resistance and methods to improve wet skid resistance.* US Patent 6469089. Retrieved from www.nanoandme.org/nano-products/paints-and-coatings

Wynand, J. M. (2009). Potential Applications of Nanotechnology in Pavement Engineering. *Journal of Transportation Engineering, 135*(10), 764–772. doi:10.1061/(ASCE)0733-947X(2009)135:10(764)

Xiao, F., Amirkhanian, A. N., & Amirkhanian, S. N. (2011). Influence of Carbon Nanoparticles on the Rheological Characteristics of Short-Term Aged Asphalt Binders. *Journal of Materials in Civil Engineering, 23*(4), 423–431. doi:10.1061/(ASCE)MT.1943-5533.0000184

Yang, J., & Tighe, S. (2013). A review of advances of Nanotechnology in asphalt mixtures. *Procedia: Social and Behavioral Sciences, 96*, 1269–1276. doi:10.1016/j.sbspro.2013.08.144

Yang, Y., Lepech, M. D., Yang, E. H., & Li, V. C. (2009). Autogenous healing of engineered cementitious composites under wet–dry cycles. *Cement and Concrete Research, 39*(5), 382–390. doi:10.1016/j.cemconres.2009.01.013

Yao, H., You, Z., Li, L., Goh, S. W., Lee, C. H., Yap, Y. K., & Shi, X. (2013). Rheological properties and chemical analysis of nanoclay and carbon microfiber modified asphalt with Fourier transform infrared spectroscopy. *Construction & Building Materials, 38*(0), 327–337. doi:10.1016/j.conbuildmat.2012.08.004

You, Z., Mills-Beale, J., Foley, J. M., Roy, S., Odegard, G. M., Dai, Q., & Goh, S. W. (2011). Nanoclay-modified asphalt materials: Preparation and characterization. *Construction & Building Materials, 25*(2), 1072–1078. doi:10.1016/j.conbuildmat.2010.06.070

Yu, J., Zeng, X., Wu, S., Wang, L., & Liu, G. (2007). Preparation and properties of montmorillonite modified asphalts. *Materials Science and Engineering A, 447*(1–2), 233–238. doi:10.1016/j.msea.2006.10.037

Yu, J. Y., Feng, P. C., Zhang, H. L., & Wu, S. P. (2009). Effect of organomontmorillonite on aging properties of asphalt. *Construction & Building Materials, 23*(7), 2636–2640. doi:10.1016/j.conbuildmat.2009.01.007

Zare-Shahabadi, A., Shokuhfar, A., & Ebrahimi-Nejad, S. (2010). Preparation and rheological characterization of asphalt binders reinforced with layered silicate nanoparticles. *Construction & Building Materials, 24*(7), 1239–1244. doi:10.1016/j.conbuildmat.2009.12.013

Zhang, B., Xi, M., Zhang, D., Zhang, H., & Zhang, B. (2009). The effect of styrene-butadiene-rubber/montmorillonite modification on the characteristics and properties of asphalt. *Construction & Building Materials, 23*(10), 3112–3117. doi:10.1016/j.conbuildmat.2009.06.011

Zhang, M. H., & Li, H. (2011). Pore structure and chloride permeability of concrete containing nano-particles for pavement. *Construction & Building Materials, 25*(2), 608–616. doi:10.1016/j.conbuildmat.2010.07.032

Zhu, W., Bartos, P. J. M., & Porro, A. (2004). Application of nanotechnology in construction: Summary of a state-of-the-art report. *Materials and Structures, 37*(9), 649–658. doi:10.1007/BF02483294

Chapter 7
Recent Trends and Advancement in Nanotechnology for Water and Wastewater Treatment:
Nanotechnological Approach for Water Purification

Sushmita Banerjee
University of Allahabad, India

Pavan Kumar Gautam
University of Allahabad, India

Ravindra Kumar Gautam
University of Allahabad, India

Amita Jaiswal
University of Allahabad, India

Mahesh Chandra Chattopadhyaya
University of Allahabad, India

ABSTRACT

Fast growing demand of fresh water due to increasing population and industrialization dictated research interests towards development of techniques that offers highly efficient and affordable methods of wastewater treatment. In recent decades water treatment using nano-technological based expertise have gained significant attention. Varieties of nanoparticles were synthesized and proficiently used in treatment of wide range of organic and inorganic contaminants from waste streams. This chapter encompasses recent development in nano-technological approach

DOI: 10.4018/978-1-5225-0344-6.ch007

Copyright ©2016, IGI Global. Copying or distributing in print or electronic forms without written permission of IGI Global is prohibited.

towards water and wastewater treatment. The authors tried to compile up to-date development, properties, application, and mechanisms of the nanoparticles used for decontamination purpose. This piece of work offer a well organized comprehensive assessment of the technology that delineates opportunities as well as its limitation in water management practices moreover few recommendations for future research are also proposed.

INTRODUCTION

"Water water everywhere, not any drop to drink" a famous quote by Samuel Taylor Coleridge always reminds us about the importance of fresh water. It is well-known that most of earth's water is saline and availability of world's fresh water resource is less than 1% on which whole global population relies. Water therefore, signifies as a precious resource which helps in sustaining life on earth. However, galloping population, increasing industrialization and unsustainable agricultural practices makes the freshwater resources highly vulnerable for pollution. It is therefore necessary to protect and conserve these resources and keep it free from pollution stress. In this regard, reuse and recycling of the wastewater is one of the best sustainable practices of water management. Moreover, wastewater recycling considered as a pre-eminent option for the nations reeling under the burden of fresh water crisis. Everyday millions of tons of wastewater have been generated worldwide as a result of various house hold, agricultural and industrial activities (Mane et al., 2011; Gautam et al., 2013; Banerjee et al., 2014). Hence, wastewater comprises of varieties of organic, inorganic and biological pollutants and these pollutants have deleterious effect on biological system (Raouf et al., 2012). Organic pollutants such as dyes, pesticides, fertilizers, hydrocarbons, detergents, oils and pharmaceuticals were detected frequently from polluted aqua reservoirs (Pedersen et al., 2003). These contaminants have severe impact on hydric environment and its organisms, as these compounds are highly toxic, persistent and almost non-degradable in nature (Puvaneshwari et al., 2006). Among inorganic pollutants, some of the heavy metals ions such as arsenic, mercury, chromium, cadmium, selenium and lead considered as highly toxic contaminants if their concentration in waste streams exceeds the safe permissible limits (Mohan & Pittman, 2007; Boening, 2000; Owlad et al., 2009; Godt et al., 2006; Hamilton, 2004; Papanikolaou et al., 2005). Presence of the anionic species such as nitrate, phosphate, fluoride, chloride, sulfate and oxalate in wastewater contributes to significant threat to human health (Bhatnagar & Sillanpaa, 2011; Huang et al., 2008; Mohapatra et al., 2009; Gopal et al., 2007; Silva et al., 2002; Brady, 2011). Moreover, presence of large number of microbial population such as viruses, bacteria, protozoa, algae and fungi in wastewater results in transmission of several water-borne diseases. Therefore wastewater laden with such sorts of pollutants required to be treated immediately prior to its discharge into receiving streams (Kyzas & Kostoglou, 2014). During the past four decades several water treatment techniques have been reported as well as successfully implemented such as adsorption, coagulation, chemical oxidation, cloud point extraction, froth floatation, ozonation, reverse osmosis, membrane filtration, solvent extraction, ion-exchange, ultrafiltration etc. Among treatment technologies, adsorption considered as the most suitable and proven technology having wide prospective applications in both water and wastewater treatment (Ali & Gupta, 2007). This process is becoming an attractive option because of its simplicity, ease of operation, high efficiency, sludge free operation, insensitivity to toxic substances and outstanding regeneration capacity (Jain et al., 2010; Sharma & Uma, 2010). The term "adsorption" defined as a process of accumulation or separation of substance from an aqueous/gaseous phase onto a solid phase (Bajpai & Rajpoot, 1999). The substance that accumulates at the interface is called adsorbate and the solid on which adsorption occurs is adsorbent (Dabrowski, 2001). Therefore in adsorption process physical and chemical characteristics of the adsorbent material such as surface area, porosity, pore size, surface charge density and zero point charge considered to play noteworthy role.

Ranges of adsorbent were synthesized and tested by researchers for its potential in removal of hazardous organic/inorganic substances from aqueous solution. Among several, activated carbon considered as one of the most popular and versatile adsorbent due to its high surface area and porosity therefore efficiently eliminates almost all kinds of pollutants from wastewater (Pelekani & Snoeyink, 2001; Pendleton & Wu, 2003; Ai et al., 2011). Unfortunately large scale application of this adsorbent was restricted due to its high initial and regeneration costs (Asgher & Bhatti, 2012). This wooed scientific worker to explore new adsorbent material which is comparatively economical and produce more efficient result. In recent years there has been explosive growth of nanoscience and technology. The use of nanotechnology in environmental application via wastewater treatment is well known. The most important aspect related to nano-based technology is the manipulation of matter at the scale of a nanometer. The realm of nanotechnology lies between 1 and 100 nm. The nanoparticles have been reportedly used for the removal of varieties of contaminant from wastewater (Salehi et al., 2010; Salem et al., 2012; Yang et al., 2013; Gautam et al., 2015; Srivastava et al., 2015a). Nano scale adsorbents exhibited remarkable adsorption performance which owed to its enhanced characteristic features such as extremely high specific surface area, small size, availability of good number of active sites for different kind of pollutants, short intraparticle diffusion distance, tunable porosity and easily recyclable without significant loss in adsorption capacity (Qu et al., 2013a). This indicates about the versatility of the techniques which need to be implemented in large scale by employing simple infrastructure. As in present time the practice of water treatment and distribution systems suffered from various setbacks and are unsustainable especially for poor countries (Qu et al., 2013b). Accordingly different approaches required for technology transfer for different countries so that benefits of nanotechnology can be easily accessible to all nations. On the whole nanotechnology proposes a new vista that envisaged fast, highly efficient, and economically feasible method of water treatment.

SYNTHESIS OF NANOPARTICLES

The development of uniform nano scale sized particles has been most demanding because of their technological and fundamental scientific importance (Hyeon, 2003). The synthesis of nanometer particles using nano-enabled technology is the most imperative aspect of wastewater remediation. The size of nano particle plays an important role as during manipulation of matter at molecular level, atom by atom; new structures build up with new molecular organization and novel characteristics property at corresponding scale. Consequently a stable preparation route is desirable that results in high percentage yield, good reproducibility, and produces high-quality nanoparticles with properties tailored for specific applications. In an attempt to engineered material at level of nanometer scale several techniques have been employed. The most fundamental aspect of nano particle synthesis diverges into two strategies: the bottom-up path and the top-down approach (Tiwari et al., 2008). In bottom-up method individual atoms and molecules are employed to construct complex entities with useful properties. Under precise conditions, the atoms, molecules and larger units undergo self assembly process, though directed assembly techniques can also be employed (Dhotel et al., 2013). The major techniques that allow fabrication of nano particle via bottom-up method are pyrolysis, inert gas condensation, solvo-thermal reactions, sol-gel, micellar structured media. For top-down approach nanoparticles can be synthesize from larger structures via attrition or milling or by lasers and vaporization followed by cooling. Moreover, fabrication of nanoparticles using different methods exhibit differences in the controllability of particle size, surface density, morphology, crystal structure and purity. The synthesis techniques mainly involve two different processing routes viz. chemical processing and mechanical milling or physical processing. In chemical processing, chemical precursors and solvents in liquid or gas form are the basic ingredients used in the formation of nanoparticles. However the technique has some disadvantages for instance requirement of expensive chemicals, contamination of particles and generation of significant amount of chemical wastes. Mechanical milling entails breaking a large piece of material into small particles normally by grinding. A very high energy input is required to produce nano scale particles and it is difficult to obtain homogenous sized particles using this method. Physical processing technique usually comprises of sputtering and evaporation of chemicals in an inert atmosphere. The technique is complicated and required high energy input but it has certain advantages over other techniques such as negligible production of chemical wastes, good control of particle size and high quality nanoparticles. Some of the important techniques that were most commonly used for the synthesis of nano scaled particles are briefly described as:

Co-Precipitation Method

This is one of the promising and facile chemical routes that are often used to prepare well-defined and less-agglomerated nano scale particles. The technique involve addition of precipitating agent like NaOH, Na_2CO_3, $NaHCO_3$, $(NH_4)_2CO_3$ into the precursor solution at suitable pH range so that controlled chemical co-precipitation should starts to occur. Nano particles can be obtained by ageing, filtering, washing and drying the resultant suspension. The main advantage of the precipitation process is that a large amount of nanoparticles can be synthesized. The reaction that involve during precipitation process can be described as:

$$M^{n+} + n(OH)^- \rightarrow M(OH)_n$$

where n is the valence of metal ions in the solution.

However, the control of particle size distribution is limited (Mohapatra & Ananad, 2010). Kumar et al. reported similar technique using NaOH as a precipitator for the synthesis of nano scale ZnO and SnO_2 with an average particle size of 25 nm and 10 nm respectively (Kumar et al., 2013). Likewise Sheela et al., (2012) successfully synthesize ZnO nanoparticles with an average size of 26 nm, using NaOH as a precipitating agent and added small amount of cetyl trimethylammonium bromide in order to control the growth of metal hydroxide crystallite during precipitation. Zheng et al., (2012) made an effort for the synthesis of highly crystalline nanostructured microspheres of Zn–Mg–Al layered double hydroxides using urea as a precipitator.

Sol-Gel Method

It is a wet chemical process that involves the sol formation based on hydroxylation and partial condensation of metallic precursor followed by the gel formation that result in three dimensional metal oxide networks. The method is extremely versatile since it allows the formation of a large variety of metal oxides at relatively low temperatures via the processing of metallic alkoxide, acetates, acetylacetones and many inorganic salts as precursors. The preparation condition, precursor's type, pH condition and the ion source decides the structure and composition of nano oxides formed by sol-gel method. The process offers advantages, including high phase purity, good compositional homogeneity, and a high surface activity of the resulting powders. Sharma et al., (2008) effortlessly synthesize alumina nano particles using sol gel method, for this aluminum sulfate was dissolved in double distilled water followed by addition of ammonia solution under continuous stirring. The resultant precipitate was kept for 24 hours at 80^0 C in hot air oven. The rigid white mass then milled to obtain gel powder. The gel powder then calcined at 1100^0C in muffle furnace followed by milling and sieving results in alumina nano particles with an average size of 15-20 nm. Moussavi & Mahmoudi (2009) have prepared MgO nano particles using this method and get particle size of 38-44 nm. Bangi et al., (2012) synthesized nano-structured Zirconia powder using similar technique and provide necessary reactions that may takes place during the process given below:

Hydrolysis:

$$Zr(OC_3H_7)_4 + 4H_2O \rightarrow Zr(OH)_4 + 4C_3H_7OH$$

Condensation:

$$Zr(OH)_4 + Zr(OH)_4 \rightarrow \equiv Zr-O-Zr \equiv + 4H_2O$$
$$Zr(OC_3H_7)_4 + Zr(OH)_4 \rightarrow \equiv Zr-O-Zr \equiv + 4C_3H_7OH$$

Hydrothermal Treatment

It is a kind of heterogeneous reaction where chemical precursor in aqueous phase kept in autoclave that undergoes pressure treatment using hot water that results in precipitation of anhydrous and thus well-crystallized nano particles developed. One of the most beneficial features related to this technique is the formation of agglomerate free nano particles with controlled particle size and distribution. Wang et al., (2012) have hydrothermally fabricated Fe_3O_4 water soluble nanoparticles. The preparation process of the same was one pot synthesis which required homogenous solution of $FeCl_3.6H_2O$ and sodium citrate in distilled water. In the mixture ammonia and polyacrylamide was added along with vigorous agitation and the resultant solution sealed in the Teflon-lined stainless steel autoclave and maintained at 200 ^{0}C for 12 hours. The solid product can be obtained after the reaction completion. The high magnification SEM image confirms that the particles obtained were of nano sized i.e. approximately of 20 nm. Cao et al., (2012) have synthesized flower like α-Fe_2O_3 nanostructures by slightly modifying the process with the use of microwave irradiation.

Liquid Phase Reduction

This is one of the simplest techniques which employed frequently for the synthesis of nano scale particles. The method involves direct precipitation of metal from its aqueous solution by means of chemical reduction of a metal cation. Reducing agents that commonly used were gaseous H_2, solvated ABH_4 (A= alkali metal), hydrazine hydrated ($N_2H_4.H_2O$) and hydrazine dihydrochloride (N_2H_4. 2 HCl). The typical reduction reaction of a transition metal cation given as:

$$M^{n+} + ne^- \rightarrow M^0$$

Numerous researchers (Bokare et al., 2007; Singh et al., 2011; Salman et al., 2014, Han & Yan 2014) reported similar procedure for the synthesis of nanoparticles. Salman et al., (2014) obtained cobalt nanoparticles of spherical shape with a diameter of 400 nm at 298 K using hydrazine as a reducing agent and citric acid as capping agent. However Singh et al., (2011) reported much smaller sized nickel nanoparticles of 3.0 nm using hydrazine hydrate and NaOH as reducing agent and Poly Vinyl Pyrrolidone as a capping agent. Likewise, Kadu & Chikate (2013) skillfully fabricated sodium borohydride based reduced bimetallic (Fe-Ni) nanocomposite having spherical shape of 30-40 nm in size.

Microemulsion Technique (or Reverse Micelle)

The term reverse micelle can be defined as self assembled surfactant in non polar solvents that results in formation of globular aggregates (Fendler, 1987). The structure of reverse micelles comprises of a polar core formed by the hydrophilic heads and non polar shell composed of hydrophobic chains. Water can be readily solubilized in the polar core to form water-in-oil (w/o) droplets, which are usually called reverse micelles at low water content and w/o microemulsions at high water content (Malik et al., 2012). This technique considered as highly desirable for the synthesis of nano sized particles as the process offers easy control over the particle size and yields nanoparticles with a narrow size distribution. It is a typical co-precipitation in water-in-oil microemulsions. The preparation procedure usually involves mixing of two microemulsions one enclosed metal salt and other containing reducing agent. The mixing of two microemulsions generally occurs by collisions of water droplets as a result of Brownian motion, the attractive van der Waals forces and repulsive osmotic and elastic forces between reverse micelles. The effective collision results in coalescence, fusion, and efficient mixing of the reactants. During fusion process the metal salt reduced to zerovalent metal atoms which further undergo collision with different metal atoms and finally fix up into stable metal nuclei (Abedini et al. 2013). The growth of particles takes place within the micelles and the size of the nanoparticles controlled by the droplet size of reverse micelles and various parameters such as type of surfactant, water-surfactant molar ratio and temperature. Ganguli et al. (2008) reported synthesis of cerium oxide, zirconia and zinc oxide nanoparticles using reverse micelles system of CTAB (cetyltrimethyl ammonium bromide), 1-butanol, isooctane, metal precursors include cerium oxalate, zirconium oxalate and zinc oxalate, and aqueous reactants which undergo thermal decomposition to produce various nanosized metal oxides. The size of the metal oxides estimated as 3-4 nm, 4-6 nm for ZrO and CeO respectively. While size of ZnO nanorods measured with a dimension of 120 nm in diameter and 600 nm in length. Chandra et al. (2010) performed water-in-oil microemulsion assisted direct precipitation of amorphous zirconium hydroxide into nano sized zirconium powder. The size and phase of the as-synthesized zirconia particles varied significantly with different water-to-surfactant molar ratio. At low molar ratio tetragonal phase particles were dominant while monoclinic phase appear at high molar ratio. The obtained particles have narrow size distribution in the range of 13-31 nm as observed by HRTEM.

Flame Spray Technique

It is an elegant tool for the fabrication of oxide nanoparticles of any elemental construct which allows the use of inexpensive precursors and solvents for synthesis of nanoparticles with uniform characteristics. Now-a-days the technique has been extensively used for metal oxide production at industrial scale. It is a well investigated high temperature aerosol technique for synthesizing well controlled nanoparticles of almost all metals and transition metals via gas-phase aerosol routes (Teoh et al. 2010). The process is scalable and has several advantages over alternative synthesis methods. It can reproducibly deliver high-purity, high-uniformity product at low cost. The technique involves directly feeding a liquid metal-containing precursor/solvent mixture into a spray flame reactor. Liquid precursor is injected into a turbulent, high velocity hydrogen-oxygen flame. Injected liquid droplets evaporate in hot flame and material nucleates and grows in the flame due to rapid decrease in flame temperature. Liquid Flame Spray generated nanoparticles can be collected as a powder by filters of electrostatic precipitators, or they can be directly deposited onto substrate surface (Pyrgiotakis et al., 2013). Tani et al. (2002) prepared ZnO nanoparticles through flame spray pyrolysis (FSP) of zinc acrylate–methanol–acetic acid solution. Investigation of relation between solution feed rate and particle size was reported and it was found that primary particle size increases from 10 nm to 20 nm as the feed rate increases from 1.0 mL/min to 4.0 mL/min. Sahm et al. (2004) using similar method for the synthesis of tin oxide nanoparticles using Tin (II) 2-ethylhexanoic acid diluted in ethanol as a precursor. The particles produced were highly crystalline having a primary particle and crystallite size of 17 nm. Rudin and Pratsinis (2013) synthesized low-cost iron phosphate nanostructured particles by flame-assisted and flame spray pyrolysis of inexpensive precursors (iron nitrate, phosphate), solvents (ethanol), and support gases (acetylene and methane). It was experientially investigated that at low acetylene flow rates in flame assisted spray pyrolysis and low methane flow rates in flame spray pyrolysis the resulting powders were non homogenous however with the increase of precursor solution flow rates powder production rate also increases and nanoparticles powders were homogeneous.

Solution Combustion Method

It is an effective method for the synthesis of nanoscale materials and most commonly used in the production of ceramic powders for a variety of advanced applications. The synthesis of nano oxide powders usually carried out by employing metal nitrates in an aqueous solution along with a fuel. Metal nitrates used are not only act as metal precursors but also as oxidizing agents. The most suitable and frequently applied fuels are glycine and urea because these are capable to form stable complexes with metal ions so that its solubility increases and further prevent selective precipitation of the metal ions during elimination of water. Solution combustion synthesis is an auto-propagated and exothermic process its technique based on the principle that once reaction set off under heating, a self sustaining exothermic reaction also starts after certain time period and eventually results into a powder as a final product. The exothermic reaction initiates at the ignition temperature and generates a certain amount of heat that matches the temperature of combustion. The main advantage from this synthesis process are it is quick and easy, rapidly produces fine and homogeneous powders and off course saves lot of time as well as energy. However formation of agglomerates is the main problem that often incurred during the process. The process parameters such as choice of fuel and fuel-oxidant ratios decide the properties of the resulting powder which includes crystallite size and structure, purity, specific surface area and particle agglomeration. Bhargavi et al. (2014) synthesize alumina nanoparticles using aluminium nitrate as precursor and glycine as a fuel. The equation for the aluminium combustion synthesis given as:

$$4Al(NO_3)_3 + 10C_2H_5O_2N + 15NH_4NO_3 \rightarrow 2Al_2O_3 + 20CO_2 + 55H_2O + 26N_2$$

Similar procedure was adopted by Toksha et al. (2008) for the synthesis of cobalt ferrite nanoparticles in the range of 11-40 nm using citric acid as fuel. Ceria based nanoparticles reported by Ravishankar et al. (2014) using ceric ammonium nitrate as an oxidizer and ethylenediaminetetraacetic acid (EDTA) as fuel at 450 ºC. The resultant product comprises of cubic phase of CeO_2 particles having crystallite size of 35 nm.

Sonochemical Method

Sonochemical preparation also provides a facile approach for the production of homogenous nano sized particles (Xu et al. 2013). This method is highly efficient in controlling the size as well as agglomeration of the synthesized nanomaterials. Preparation procedure is simple which requires metal precursors and precipitating agent. High intensity ultrasound typically around 20 kHz and 10 to 100 W acoustic power requires to applied in the precipitated metal hydroxide solution by using titanium horn. Ultrasound waves results in formation of metal nuclei along with subsequent control over the nuclei growth. Zhu et al., (2000) reported successful synthesis SnO_2 nanoparticles using ultrasonic irradiation of an aqueous solution of $SnCl_4$ and azodicarbonamide under ambient air. The size of nanoparticles was estimated at approximately 3-5 nm by means of XRD and TEM analysis. Size-controlled sonochemically synthesized zirconia nanoparticles were proficiently documented by Manoharan et al. (2015) using Zirconium (IV) oxychloride octa-hydrate as a metal precursor. The narrow particle size distribution (15–25 nm) of the zirconia nanoparticles was analyzed through FE-SEM statistical analysis and further confirmed by TEM.

Chemical Vapor Synthesis

CVS is one of the commonly used gas-phase aerosol processes for producing high-purity nanoparticles. Another technique analogous to CVS is CVD (Chemical Vapor Deposition) method where the process parameters are adjusted to form solid thin film instead of nanoparticles. In this approach precursors like metal organics, carbonyls, hydrides, chlorides and other volatile compounds in gaseous, liquid or solid state introduced It is done in hot wall reactors and cold-wall reactors, at sub-torr total pressures to above-atmospheric pressures, with and without carrier gases, and at temperatures typically ranging from 200-1600°C that results in nucleation of particles within the vapor phase. The chemical reactions among the gas molecules are stimulated by an input of energy source such as corona discharge ionizer, laser, hot filaments and plasma. The main drawback associated with this technique lies in the availability of appropriate precursor materials. As the precursors need to be volatile at near-room temperatures this trait however absent in several elements in the periodic table, although the use of metal-organic precursors alleviate this situation. Metal chlorides were mostly used as precursors for the formation of oxides because of their comparatively low vaporization temperature and low cost. The typical reaction is as following:

$$SnCl_4(gas) + 2H_2O(gas) \rightarrow SnO_2(solid) + 4HCl$$

The most important process parameters that influences the size and performance of nano materials includes the total pressure, the precursor material, the partial pressure of the precursor, precursor deliverance system, the temperature of the energy source, the carrier gas, and the reactor geometry. The nanoparticles can be collected by means of filters, electrostatic precipitators or scrubbing in a liquid. However, nanoparticles generated by this process were generally produced in the form of aggregates due to their coagulation at the high temperature used (Nakaso et al., 2003). Carbon nanomaterials including carbon nanotubes and carbon onions suitably fabricated by He et al., (2009) via chemical vapor deposition of methane over nano-sized Ni/Al catalyst at moderate temperature of 450–600 °C.

Ball Milling

It is the simplest method of nanoparticles synthesis which involves conventional mechanical grinding techniques to break coarse metal grains into nano sized particles. The repeated ball milling leads to particle collision thus the size of individual grains break down to only a few nanometers and also causes repeated deformation, fracture, and welding of the particles (Li et al.,2006). Gateshki et al., (2005) achieved average grain size of ZrO_2 in the range of 250-400 nm by ball milling. Ohara et al., (2010) synthesize fine $FeTiO_3$ nanoparticles using high-speed ball-milling of steel balls and TiO_2 nanoparticles. The formation of $FeTiO_3$ nanoparticles were as a result of the reaction between TiO_2 nanoparticles, oxygen gas, and Fe atoms around the surface of the steel balls under local high temperatures induced by the collision energy in the ball-milling process. Most of the synthesized pure $FeTiO_3$ had a particle size of 15 nm.

CHARACTERIZATION OF NANO ENGINEERED MATERIALS

In order to ascertain whether the as-synthesized material comprises of nano scale features or not it is therefore requisite to characterize it with certain sophisticated instruments which perhaps facilitates in ascertaining the size as well as various physical and chemical properties. Now-a-days numerous techniques are available for the characterization of nano materials some of the important techniques are discussed briefly in this section.

X-Ray Diffraction Technique (XRD)

The technique helps in determining the size and shape of the crystallite by interpreting diffracted X-Ray beams produces by atoms into many specific directions. It also gives an insight about the orientation of a single crystal or grain. The size estimation of crystallite carried out by using well-known Scherrer's formula (Patterson, 1939) given as:

$$d = \frac{0.9\lambda}{fwhm \cos\theta}$$

where d = crystallite size (nm), λ = wavelength of monochromatic X-ray beam, $fwhm$ = full width at half-maximum for the diffraction peaks under consideration (rad), and θ = diffraction angle (degrees). Sheela et al. (2012) estimated the average particles size of as-synthesized zinc oxide nano particles as 26 nm using Scherrer's formula. Recently our group synthesized lanthanum ferrite nanoparticles, the diffractogram pattern of the same reveals characteristics reflection planes (Figure 1) as per the JCPDS card no 15-148. The crystallite size was calculated using Scherrer's formula as 29 nm.

Scanning Electron Microscopy (SEM)

The technique produces micrographs of a sample by scanning it with a focused beam of electrons. The electrons emitted from the instrument starts interacting with the sample's electron and the interaction results in production of various signals which can be easily detected and interpreted in terms of composition, surface topography as well as size of the samples under investigation. Gong et al., (2012) adeptly make use of SEM technique to characterize the sizes and surface morphologies of the prepared shellac-coated iron oxide nanoparticles (SCMN) for removal of cadmium ions from aqueous solution. The SEM image of the SCMN clearly indicates about the

Figure 1. PXRD pattern of LaFeO3

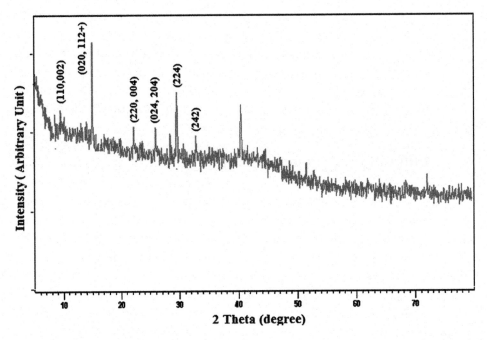

nanosized structure. The sample composition before and after cadmium adsorption was further investigated through EDX (Energy Dispersive X-Ray: an attachment with SEM that provides information related to the elemental composition of the sample). The EDX pattern reveals the presence and weight% of Fe, O and C atom in virgin SCMN while after adsorption the weight and atom % of all elements changes significantly and addition of new peaks of Cd also confirms about the adsorption of Cd^{2+} on SCMN. Figure 2 displays SEM micrograph along with EDX pattern of calcined Zn/Al hydrotalcite particles synthesized by our group. The SEM image (Figure 2a) suggests that the particles were agglomerated and the size appears to be in nano range, meanwhile EDX image (Figure 2b) also confirms about the presence of Zn and Al in hydrotalcite composition.

Transmission Electron Microscopy (TEM)

It is another very useful microscopy technique which requires diffusion of electron beam through the specimen; in due course electrons interact with the sample material and transmitted out through the specimen. The images thus formed during interaction, transmitted through the specimen was captured and magnified followed by focusing onto an imaging device. Sample preparation is however a complex part

Figure 2. a) SEM micrograph of calcined Zn/Al hydrotalcite; b) EDX pattern of Zn/Al hydrotalcite

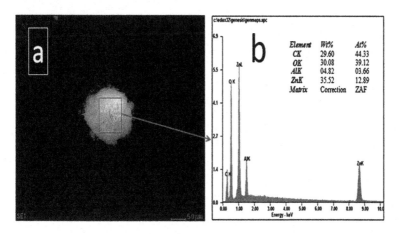

of this instrumentation technique. The technique is highly desirable exclusively for the determination of particle size. Several authors (Huang & Chen, 2009, Gautam et al., 2015, Paul et al., 2015) reported this technique purposefully in order to verify the nano scale size of the as-synthesized material. TEM image of Co/Bi hydrotalcite like compound given in Figure 3, clearly exhibits particle size in nano range. Inset provides SAED (Selected Area Electron Diffraction) pattern of the crystals which can also be obtained through this technique. This pattern indicates about the presence of crystalline feature in the sample and is helpful in determining the crystals reflection planes which may be the characteristics of the synthesized materials. The SAED image in Figure 3, shows eight shining white dots surrounds the large white dot which indicates about the octahedral crystal structure of the synthesized Co/Bi hydrotalcite, octahedral arrangement is the characteristic feature of hydrotalcite like compounds.

The above three techniques namely XRD, SEM and TEM were the most widely used methods for deciphering molecular structure, composition as well as size of the particles. Beside these three, other techniques such as FTIR, BET surface area analyzer, AFM, VSM etc were also extensively used to determine qualitative characters of the synthesized nano materials. The appealing features related to these aforementioned techniques were concisely explained below.

Fourier Transform Infra Red Spectroscopy (FTIR)

It is an absorption spectroscopy which aims at measuring beam absorbed by sample when a beam containing many frequencies of light shining upon the sample. The

Figure 3. TEM image of Co/Bi hydrotalcite compound along with SAED pattern (inset)

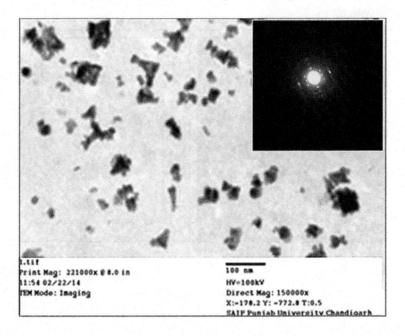

FTIR profile usually plotted as transmittance % versus wave number. The technique is helpful in determining the chemical composition or in particular functional groups present in the synthesized nanoparticles. For instance Huang & Chen, (2009) have synthesized poly acrylic acid coated amino-functionalized Fe_3O_4 nanoparticles. The confirmation of presence amino group in amino-functionalized magnetic nanoparticles was further ensured through FTIR analysis. It has been observed that after amino-functionalization, that the characteristic peaks of PAA at 1708cm^{-1} (C=O stretch) 1448 and 1409cm^{-1} (C-O stretch) were disappeared and appearance of two new peaks at 1331cm^{-1} (N-H bend) and 1044cm^{-1} (C-N stretch) confirms addition of amino groups from diethylenetriamine. FTIR analysis also helps during investigation of adsorptive removal of pollutants from water or wastewater. The adsorbents used for the scavenging of pollutants were analyzed through this technique to check the performance of adsorbents for contaminants removal. FTIR spectra of pure magnetite nanoparticles synthesized by Gautam et al. (2015) for Ni^{2+} removal from aqueous solutions is represented in Figure 4. The spectrum exhibits strong peaks at 539 cm^{-1} are due to the Fe-O functional group that substantiates the as-prepared nanoparticles was magnetite. The Figure 4b represents change in spectra of magnetite after Ni^{2+} loading. After Ni^{2+} adsorption it is clearly evident that the several peak's intensity alters especially peak at 539 cm^{-1} as a result of Fe-O interaction which ascertain Ni^{2+} adsorption.

Figure 4. a) FTIR spectra for pure magnetite particles; b) FTIR spectra of Ni2+ loaded magnetite particles

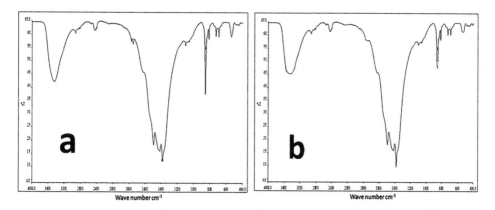

Brunauer-Emmett-Teller (BET) N$_2$ Adsorption Isotherm

The most widely used technique for estimating surface area is the so-called BET method. The determination of surface area and porosity are essentially required to assess the potential of synthesized materials as these characteristics are capable of affecting the quality and utility of many materials. For this reason it is quite imperative to determine and control them accurately. The BET method involves the determination of the amount of the adsorbate or adsorptive gas usually N$_2$ to cover the external and the accessible internal pore surfaces of a synthesized materials at 77 K with a complete monolayer of adsorbate. This monolayer capacity can be calculated from the adsorption isotherm by means of the BET equation. The technique also used for the prediction of distribution of porosity through computational method given by Barrett, Joyner and Halenda (BJH). The method allows the computation of pore sizes from equilibrium gas pressures. Fei et al., 2011; Tian et al., 2013; Banerjee et al., 2015; Srivastava et al., 2015b & Li et al., 2015 reported the surface area and pore size distribution of engineered nanoparticles through BET and BJH method. This method is tremendous in interpreting the qualitative aspect of adsorbent material used for wastewater treatment. Surface area of the adsorbent material plays significant role in sorption process, higher the surface area greater will be the capacity of adsorbent for uptaking pollutants from liquid phase (Fei et al., 2011). As for instance Fei et al., 2011 synthesized 3D self-assembled iron hydroxide and oxide hierarchical nanostructures for Congo red removal and the same were analyzed for BET surface area measurements. The specific surface area of 239 m^2/g was obtained for 3D self assembled iron hydroxide nanostructure while iron oxide hierarchical nanostructures possess specific surface area of 69 m^2/g. During application of these two nanostructures in Congo red removal it has been found that material possesses large surface area exhibits high adsorption capacity of about 239 mg/g while other one shows sorption capacity of 66 mg/g.

Atomic Force Microscopy (AFM)

This technique is particularly devoted to gain information about the surface morphology of the solid samples under investigation. The AFM is one of the leading tools for imaging and measuring matter at the nanoscale level. The information is gathered by "feeling" the surface using a mechanical sensor. The technique is highly beneficial in acquiring three dimensional surface profiles and the samples do not required any special treatment before analysis unlikely SEM and TEM. Du et al., (2008) suitably interpreted the morphology of as-synthesized chitosan nanoparticles using atomic force microscopy. The image obtained through the technique clearly reveals that the prepared chitosan particles were spherical in shape and sizes were in nano range with an average diameter of 69.33 nm. The AFM image of the magnetic nanoparticles (Gautam et al. 2015) given in Figure 5. The image suggests that the synthesized material bears highly undulated, rough and jagged surface features. The sizes of the particles were obtained in the range of 5 to 40 nm.

Figure 5. (a) AFM assisted interpretation of surface morphology of synthesized magnetic nanoparticles; (b) size distribution of magnetic nanoparticles

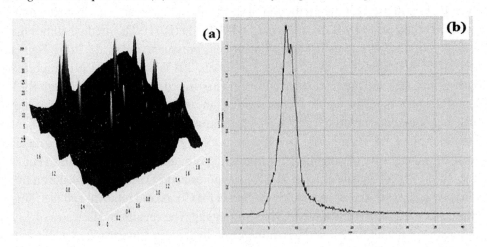

Vibrating Sample Magnetometry (VSM)

This instrumentation technique is exclusively dedicated for the measurement of magnetic property of the engineered materials. The technique is extremely helpful in ascertaining the magnetic behavior as well as quantification of the magnetic property of the synthesized nano materials. The nano materials that possess magnetic property termed as "magnetic nanoparticles". The Vibrating Sample Magnetometer works on the principle of Faraday's law of Induction which states that change in magnetic field will produce an electric field. The measurement of this as-produced electric field helps in determining the changing magnetic field. The results were obtained in the form of magnetization curve as shown in Figure 6. The synthesis of magnetic nanoparticles for wastewater treatment is an imperative approach as the method provides an easiest way of magnetic separation of adsorbent loaded with pollutants from liquid phase. Therefore synthesis of magnetic nanoscale adsorbent offers a new vista in remediation of water pollutants through adsorption technology. So far lot many works have been reported in the related area (Wu et al., 2004; Liu et al., 2008; Feng et al., 2010; Kaur et al., 2014; Ghasemi & Sillanpaa, 2015). The main theme upon which the success of this technique relies is the quantum of magnetic property incorporated into the synthesized magnetic nano materials. Greater the magnetic property exhibited by the synthesized magnetic adsorbent more easily it will be separated from the waste streams after treatment. VSM technique therefore offers a promising role in deciphering the quantum of magnetic property possessed by the as prepared magnetic adsorbent. The Figure 6 represents the magnetization profile of magnetite nanoparticles (Gautam et al., 2015). The sample is kept in the external magnetic and magnetic field applied in range of -50,000 to +50,000 Oe. The value of magnetic saturation obtained through VSM curve estimated as 84 emu/g indicates superparamagnetic behavior of the magnetite nanoparticles.

STATE-OF-ART-APPLICATIONS

In material science shape, size and morphology of the solid nano materials plays crucial role as these factors determine the utility of these nano structured materials for application in different fields. It is therefore important to classify the nano materials in order to enhance the scope related to its use in various other areas. The classification of nano structured materials was firstly proposed by H. Gleiter in 1995 and he was the first to introduce the term nanocrystal and nanostructured material (Glezer, 2011). Gleiter's classification was based on crystalline forms and chemical composition of nanostructured material. His classification may be divided into three

Figure 6. Magnetic saturation curve of magnetite nanoparticles

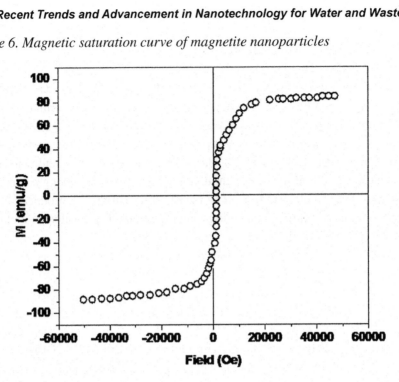

basic categories. The first category comprises of materials having low dimensions or dimensionality which may be in the form of isolated, substrate supported or embedded nanosized particles, thin wires or thin films. The second category consists of materials in which nano sized microstructure in form of thin film were synthesize on the surface region of a bulk material. The third category includes bulk solids with a nano sized microstructure. These solid varies on a length scale of a few nanometers throughout the bulk. Shortly the classification faces disapproval due to non inclusion of zero and one dimensional nanostructures in the schemes. Skorokhod (2001) introduced dimensionality as the basis of classification of nanostructures. According to Pokropivny & Skorokhod, (2007) nanostructures are different from nanostructured materials as the former can be characterize by a form and dimensionality while later better describe on the basis of composition in addition to dimensionality. Therefore they categorized nanostructures into four different types on the basis of number of dimensions which includes 0D, 1D, 2D and 3D.

- **Zero Dimensional Nanomaterials (0D):** Materials wherein all the dimensions are measured within the nanoscale and no dimensions are larger than 100 nm. These zero dimensional structures were also called as nanoparticles. Metallic nanoparticles such as silver, copper, zinc nanoparticles and semiconductor like quantum dots are the perfect example of this kind of nanoparticles. The 0D nanoparticles were enormously used for water treatment in either batch mode or through column setup.

- **One Dimensional Nanomaterial (1D):** Materials wherein two dimensions confined to nanoscale while one of the dimensions is outside the nanoscale. These materials are having length of several micrometers but with a diameter of only a few nanometers. The structures formed were of needle like shape. One dimensional material mainly includes nanorods, nanotubes and nanowires of metals and oxides. The 1D nanomaterials were also used as an potential adsorbents for waste water treatment. Carbon nanorods are the most popular 1D nanomaterials used in water pollution studies.

- **Two Dimensional Nanomaterials (2D):** Materials wherein one dimension at the nanoscale and the other two of the dimensions not confined to nanoscale range. These materials mostly possess plate like shapes for instance nanofilms, nanolayers, and nanocoatings. These materials were mostly used in fabrication of membranes for water treatment purposes.

- **Three Dimensional Nanomaterials (3D):** These materials are characterized by having all dimensions at the macroscale sized. These are usually bulk nanomaterials composed of a multiple arrangement of nano sized crystals in different orientations. Therefore 3D nanomaterials contain dispersions of nanoparticles, bundles of nanowires and nanotubes. Among 3D materials graphene is the most admired material which exhibits exceptional potential for adsorption of variety of organics and inorganics from wastestreams.

On the basis of chemical composition, the most widely use nanomaterials for water treatment can be classified into four types:

Carbonaceous nano materials; Metal based nano materials; Composites and Dendrimers

- **Carbonaceous Nano Materials:** These nanomaterials are composed of carbon atoms as its main component. The unique characteristic property such as high surface area to volume ratio, high porosity with controllable pore size, high thermal stability, manipulatable surface chemistry makes these nanoparticles a proficient adsorbent. Sorption studies employing carbonaceous nanoparticles shows fast scavenging efficiency due to rapid equilibrium rates and high sorption capacity. For instance Teng et al., (2012) synthesize electrospun mesoporous carbon nanofibres and employed the same for the adsorption of basic dye methylthionyl chloride molecules. The high adsorption potential of 567 mg/g was observed and adsorption equilibrium was attained within 20 min. The rapid equilibrium shown by carbonaceous nanoparticles may be due to several interaction mechanisms such as hydrophobic interaction, electrostatic interaction, polarizability of π electrons, π- π electron-donar-acceptor interaction and hydrogen boding (Pan & Xing, et al., 2008). Yang et al., (2008) synthesized multiwalled carbon nanotubes with an average surface area of 174 m^2/g, mesopore and micropore volume of 0.597 cm^3/g and 0.068 cm^3/g, respectively. They measured the sorption potential of aniline and phenol at 25 ^{0}C with maximum sorption capacity of 2.06 and 1.81 mg/g respectively. Pillay et al., have evaluated and compare the performance of functionalized multi-walled carbon nanotubes and unfunctionalized multiwalled carbon nanotubes for Cr (VI) sorption at ppb level. The surface area analysis exhibited higher surface of 130 m^2/g for functionalized multiwalled carbon nanotubes while unfunctionalised have surface area of 30 m^2/g. The functionalized adsorbents have relatively low sorption capacity for Cr (IV) as the adsorbent contains electron rich atom in their functional groups which repels negatively charged dichromate ions. The unfunctionalised adsorbent possesses high adsorption capability of about 98%.

- **Metal Based Nanoparticles:** The high running costs of carbonaceous nanoparticles have limited direct, widespread use and large scale application of carbon based adsorbents for water treatment. This let the scientific community to explore certain alternates which are equally effective like that of carbonaceous materials in terms of high surface area and porosity. In recent decade use of metal based nanoparticles for waste water treatment have gain significant interest. These materials showed an excellent ability to remove various water pollutants. Metal nanoparticles possess certain fascinating properties, such as large surface area, adsorption capacity, large number of active sites, excellent catalytic potential, simple operation, easy and large scale production. Among various metal nanoparticles zerovalent iron is one of the most investigated adsorbent materials. Boparai et al., (2011) have evaluated the potential of nano zerovalent iron particles for sorption of Cd^{2+}. It has been found that within 15 minutes Cd^{2+} sorption on zero valent iron reach equilibrium condition with maximum sorption capacity of 769 mg/g at 24 ^{0}C. The attainment of equilibrium within short period of time is highly desirable in water treatment process. It has been also investigated by several workers that efficiency of adsorbent enhances by surface modification with various functional groups. The surface functionalized adsorbents have selective adsorption affinity to targeted pollutants. Afkhami et al., (2011) successfully synthesized alumina nanoparticles grafted with 2,4-Dinitrophenylhydrazine (DNPH) for removal of formaldehyde from water samples. The grafting of DNPH greatly improves the accessibility of the sorption sites to formaldehyde thus sorbent becomes selective towards the adsorbate.

- **Composite Nanomaterials:** Nanocomposites are one of the most important classes of nanomaterials which comprises of multiphase solid materials such as metals, polymers, ceramics where at least one of the constituent phase has one, two or three dimensions of less than 100 nm. Typically nanocomposites classified into three main categories inorganic-inorganic nanocomposites, organic-organic nanocomposites and organic-inorganic nanocomposites. Recently, studies of composite nanoparticles have been attracting much attention due to its unusual features and multifunctional properties. A significant number of studies on composite nanomaterials have focused to exploit its exclusive properties to develop high capacity and selective adsorbents for metal ions and anions. Do et al., (2011) have investigated the potential of Activated carbon/Fe_3O_4 nanoparticle composite for removal of methyl orange. The synthesized nanocomposite posses combine feature of activated carbon and Fe_3O_4 nanoparticles. These materials in its independent form have both advantage and disadvantage for instance activated carbon considered to have high surface area and porosity but its application are limited due to difficulties occur during its separation and regeneration similarly Fe_3O_4 nanoparticles can be easily separated magnetically from liquid phase but its specific surface area is comparatively low. Therefore in order to enjoy the benefit of both materials the composite of the material have been synthesized. The composites have high surface area of 1110 m^2/g and saturation magnetization of 7.0 emu/g thus indicated superparamagnetic behavior. The nanoparticles composite exhibits tremendous sorption capacity of 303 mg/g for methyl orange. Paul et al., (2015) also synthesized Graphene/Fe_3O_4 nanocomposite, the adsorbent shows enhancing effect in removal of Arsenic. Paul et al., investigated that arsenic removal is largely affected in presence of humic acid this negative effect can be control using graphene/Fe_3O_4 nanoparticles. As in the removal process surface area of the composite doesn't play any role. The major role contributed by humic acid as its coating modified the mechanism through π-π interaction of aromatic fraction of amine groups of humic acid with graphene. This makes the graphene surface more positive thus enhancing adsorption of arsenic.
- **Dendrimers:** These are unique class of polymers, have been regarded as significant promising molecules in many fields because of their identical groups, highly branched molecular structure, and controllable size and shape (Eskandarian et al., 2013). These are hyper-branched macromolecules with a carefully tailored architecture that can be effortlessly functionalized, thus leads to modification of their physicochemical or biological properties (Kim & Zimmerman, 1998). Dendrimers are characterized by special features that render them promising candidates for a lot of applications. Their

interior shells can be hydrophobic for sorption of organic compounds while the exterior branches can be tailored (e.g., hydroxyl- or amine-terminated) for adsorption of heavy metals (Qu et al., 2013). Several researchers reported efficient use of dendrimer in remediation of water pollutants. Eskandarian et al., 2013 efficiently remove direct blue 86 and direct red 23 using multiwalled carbon nanotubes modified with poly(propylene imine) dendrimer. The synthesized adsorbent of 0.1 g/L capable to decolorize over 95% of direct dyes within 60 minutes of contact time at 25 ^{0}C. The maximum sorption capacities for DB 86 and DR 23 evaluated as 1000 and 666 mg/g respectively. Shen et al., (2015) reported dendrimer based prepration of mesoporous alumina nanofibres by electrospinning method for removal of methyl orange. Dendrimer polyamidoamine was used in modification of mesoporous alumina which provides the alumina a fibrous morphology. Sorption process reaches equilibrium condition at about 70-80 min with sorption capacity of 351 mg/g. The adsorbent exhibits high reuse performance without loss of significant sorption capacity.

NANOTECHNOLOGICAL PRINCIPAL APPLIED FOR WATER TREATMENT

Adsorption process is the principal phenomenon which causes separation of pollutant species from liquid bulk phase and its accumulation over solid surface offered by nanoparticles. The adsorption process is a surface phenomenon therefore surface area of the adsorbent has high influence on the removal of pollutants. The smaller the size of adsorbent greater will be its surface area in this perspective nanoparticles furnished high surface area (> 1000 m^2/g) with remarkable removal capacity. However, besides surface area of the adsorbent other process variables such as pH, temperature, ionic strength, pollutant concentration, adsorbent dose plays significant role in sorption process. The removal efficiency can be enhances to several fold through optimizing these parameters. The effects of these parameters on removal of pollutants were briefly described in the following sections.

Effect of pH

It is well known that the pH of the aqueous solution is an important controlling parameter in the adsorption process (Öztürk & Malkoc, 2014). The adsorptive process was affected by the change in solution pH because of the protonation and deprotonation of active functional groups of the adsorbent and adsorbate in aqueous solution at a certain pH. It has been observed that adsorption behavior of pollutant species in aqueous solution always alters with the change in pH values and at certain specific pH range maximum sorption takes place. This observation can be explained on the basis of pH_{ZPC} of the adsorbent. pH_{ZPC} is the condition when the net surface charge of the adsorbent becomes electrically neutral or zero. Therefore at pH > pH_{ZPC} adsorbent surface becomes positively charge which is favorable for sorption of anionic species, while at pH < pH_{ZPC} adsorbent surface bear negative charges thus favors sorption of cationic species. Similar observation also reported by Zhang et al., (2013) for removal of arsenic (V) using nanostructured iron (III) and copper (II) binary oxide. The maximum removal of arsenic oxyanions observed at acidic medium. The pH_{ZPC} of the binary oxide reported as 7.9 therefore at pH lower than 7.9 the adsorbent becomes highly protonated thus negatively charged arsenic ions attracted electrostatically. While at pH > 7 the removal capacity of binary oxide decreases significantly due dominance of coulumbic repulsion between arsenic anions and negatively charged surface of binary oxide. Gautam et al., (2015) also reported the removal mechanism of Ni^{2+} on the basis of pH_{ZPC}. They reported higher pH favors sorption of Ni^{2+} on Fe_3O_4 nanoparticles as the pH_{ZPC} of the adsorbent evaluated as 6.2 and therefore the adsorbent surface becomes negatively charged above this value which helps in binding of positively charge Ni^{2+} ions electrostatically.

Effect of Temperature

Temperature of the aqueous solution plays considerable role in sorption process of water pollutants. The temperature has two major effects on the sorption process. Increasing the temperature is known to augments the rate of diffusion of the adsorbed molecules across the external boundary layer and the internal pores of the adsorbent particles, due to the decrease in the viscosity of the solution. Moreover changing temperature will change the equilibrium capacity of the adsorbent material for a particular adsorbate owing to formation or weakening of hydrogen bonds or van der Waals interaction between adsorbate and adsorbent species. When sorption process increases with the rise of temperature then the removal process considered as endothermic in nature conversely when rise in temperature leads to decline in adsorption efficiency then the removal process termed as exothermic in nature. Zhao et al., (2010) evaluated the efficiency of $Fe_3O_4@Al(OH)_3$ magnetic nanoparticles for fluoride removal. The sorption of fluoride ions found to be highly dependent on temperature. The fluoride sorption capacity increases steadily from 33 mg/g to 37 mg/g with the rise of temperature from 298 K to 323 K which indicates endothermic nature of sorption process. Al-Johani & Salam (2011), tested sorption percentage of multiwalled carbon nanotubes for aniline removal at different temperatures. The percentage removal of found to decrease from 82% to 38% with the rise of solution temperature from 280 K to 330 K. The observation indicates exothermic nature of sorption process.

Effect of Ionic Strength

Wastewater characteristically contains a complex mixture of both organic and inorganic pollutants whereas natural water resources usually comprises of various anions such as chloride, nitrate, sulfate, bicarbonate, oxalate, fluoride, silicate and phosphate. Therefore presence of these ions severely impairs the treatment process of natural or waste water by competing with the active sites present on the adsorbent. It is therefore required to examine the performance of adsorbent in the presence of interfering ions. Ge et al., (2012), observed the effect of background electrolytes such as Na^+, K^+ and Mg^{2+} on sorption of Cu^{2+} ions over polymer-modified magnetic nanoparticles. The presence of Na^+ and K^+ ions not exhibit pronounced effect on sorption process but divalent Mg^{2+} ions severely retards the sorption process because of the stronger complexation ability of carbonyl groups with Mg^{2+} ions. Zhang et al., (2013), evaluated the performance of nanostructured Fe-Cu binary oxide for removal of arsenic (V) from natural water in presence of coexisting anions. The presence of sulfate and bicarbonate ions slightly reduced the sorption of As (V) but phosphate ions have pronounced effect on As (V) sorption. This mainly attributed to high charge density of phosphate ions results in strong competition between arsenic and phosphate ions for the surface active sites of the sorbent.

Effect of Pollutant Concentration and Contact Time

The sorption capacity of adsorbent for pollutants from liquid phase invariably depends on pollutant concentration and its duration of contact with the adsorbate. It has been found that during initial time period of sorption process the uptake capacity of the adsorbent is rapid but as the sorption proceeds for longer duration the adsorbent became exhausted because of the gradual occupancy of the active sites, and the adsorption was replaced by the transportation of adsorbate from the external sites to the internal sites of the adsorbent particles. Thus, the uptake rate starts decline progressively (Roy et al., 2013). Dhillon & Kumar (2015) observed comparable trend for adsorption of fluoride ions on nanoporous Fe–Ce–Ni adsorbent. It has been reported that more than 90% of fluoride ion was removed within 20 minutes of contact time and complete saturation of the sorption process takes place up to a time period of 30 minutes.

Effect of Adsorbent Dose

The investigation of the effect of adsorbent dose is important from the economic viewpoint, which requires the minimum amount of adsorbent for effective removal of pollutants from the aqueous phase. It has been found that percentage removal of pollutants increases with the increase of adsorbent dose. This is mainly attributed to increase of sorption sites for interaction with pollutant species at higher adsorbent dosages. However, after fixed adsorbent dosage the percentage removal of pollutant species attain equilibrium condition still in some cases removal percent found to be declines at higher sorbent dosage. This happens mainly due to overlapping of surface active sites that leads to decrease of surface area or due to the significant unsaturation of adsorption sites at the constant pollutant concentration. Alipour et al., (2014) have examined the effect of adsorbent dosage of oyster shell supported zero valent iron nanoparticles on removal of humic acid from aqueous solutions. The nanoadsorbent dose ranges from 0.5 g/L to 5g/L were selected for removal studies. The sorption capacity decreases from 0.6 mg/g to 0.05 mg/g with the increase of sorbent dose from 0.5 g/L to 5.0g/L. The reason behind this observation elucidate on the basis of decrease of surface area with the increase of adsorbent dosage.

PREDICTION OF INTERACTION BEHAVIOR BETWEEN NANOPARTICLES AND POLLUTANT SPECIES THROUGH EQUILIBRIUM ISOTHERM AND SORPTION KINETICS STUDY

Adsorption Isotherm

The study of adsorption isotherm is important as it gives an idea about the interaction behavior between adsorbent and adsorbate which further helps in optimization of the adsorption mechanism process, determination of the surface properties as well as sorption capacities of the adsorbents and effective designing of an operating adsorption system for industrial effluents. Therefore beside exploring novel adsorbents with high efficacy for sorption system meanwhile it is also imperative to ascertain the most suitable adsorption equilibrium isotherm. Adsorption equilibrium may be defined as the ratio between the amount of pollutants adsorbed on the adsorbent and remaining in the liquid phase established dynamic balance at solid liquid interface with the pollutants present in liquid phase contacted with the adsorbent for certain period of time. The plot representing amount of adsorbate adsorbed on the solid adsorbent versus residual concentration of adsorbate remaining in liquid phase at constant temperature and pH were investigated and the resultant slope and intercept values extracted from the plots were further analyzed through mathematical expression of the isotherm model. The most fitted plots with high correlation values obtained against the typical isotherm model were finally used for the prediction of sorption isotherm data. Treatment of sorption data with appropriate isotherm models facilitate predication of adsorption parameters in more précised manner and also offer quantitative evaluation of adsorbent performance at different environmental conditions (Foo & Hameed, 2010). Till date varieties of equilibrium isotherm models such as Freundlich (1906), Hill (1910), Langmuir (1916), Bruanuer-Emmett-Teller (1938), Tempkin (1940), Dubinin-Radushkevich (1947), Sips (1948), Redlich-Peterson (1959), Toth (1981), were proposed and effectively used for the treatment of sorption isotherm data. Out of several the four most widely used isotherm model have been discussed below in brief:

Freundlich Isotherm Model

The model assumes a heterogeneous surface with a nonuniform distribution of the heat of adsorption over the surface. The Freundlich isotherm in its nonlinear form can be expressed by equation given below:

$$q_e = K_F C_e^{1/n}$$

The above equation can be rearranged to a linearized form given as:

$$\log q_e = \log K_F + (1/n)\log C_e$$

The isotherm data can be plotted as $\log q_e$ versus $\log C_e$. Where q_e (mg/g) is the amount of adsorbate adsorbed on solid adsorbent and C_e is the residual concentration of adsorbate remains in the solution. The parameter K_F (mg/g(L/mg)$^{1/n}$) and $1/n$ can be evaluated from the intercept and slope of the plot respectively. K_F is roughly an indicator of the sorption capacity as well as the strength of adsorptive bond and $1/n$ is the adsorption intensity. $1/n$ values are typical indicator of beneficial adsorption if the value of $1/n > 1$ then adsorption is unfavorable for favorable adsorption $1/n < 1$.

Langmuir Isotherm Model

The model based on certain assumptions that the adsorption process takes place at specific homogeneous sites within the adsorbent surface and once adsorbate molecule occupies a site, no further adsorption can takes place at that site. The adsorbent has a finite capacity for the adsorbate and all sites are identical and energetically equivalent. Maximum adsorption corresponds to a saturated monolayer of adsorbate molecules on the adsorbent surface. The expression for Langmuir isotherm in nonlinearized form given as:

$$\frac{q_e}{q_m} = \frac{(C_e K_L)}{(1 + C_e K_L)}$$

The linearized Langmuir isotherm profile can be plotted using following expression:

$$\frac{C_e}{q_e} = \left(\frac{1}{K_L Q_0}\right) + \left(\frac{C_e}{q_m}\right)$$

where q_e is the equilibrium concentration of adsorbate on the adsorbent (mg/g), C_e is the equilibrium concentration of adsorbate remains in liquid phase, q_m is the monolayer capacity of the adsorbent (mg/g) and K_L is the Langmuir adsorption constant. The plot of C_e/q_e against C_e was used to evaluate the Langmuir parameters q_m (mg/g) and K_L from slope and intercept respectively.

Dubinin-Radushkevich Isotherm Model

The model assumes that the adsorption of subcritical vapor onto microporous sorbent occurs through filling of pores. The model helps in predicting the adsorption mechanism through Gaussian energy distribution onto a heterogeneous surface of the adsorbent (Foo & Hameed 2010). The model also helps in interpreting the interaction behavior of adsorbent with the adsorbate. The nonlinear expression of D-R isotherm given as:

$$q_e = q_m \exp\left(-\beta \varepsilon^2\right)$$

The linear form of the above model can be expressed by:

$$\ln\left(q_e\right) = \ln\left(q_m\right) - \beta \varepsilon^2$$

where q_e is the amount of adsorbate adsorbed per unit dosage of the adsorbent (mg/g), q_m is the theoretical monolayer saturation capacity (mg/g), β is the activity coefficient (mol²kJ⁻²) related to the mean sorption energy and ε is the Polanyi potential described as:

$$\varepsilon = RT \ln\left[1 + \left(\frac{1}{C_e}\right)\right]$$

where R is the gas constant in kJ/mol/K, T is the temperature in Kelvin, and C_e is as defined above. The D-R parameters β and q_m can be obtained from the slope and intercept of the plot $\ln q_e$ versus ε^2.

Tempkin Isotherm Model

The model assumes that heat of adsorption of all molecules in the layer would decrease linearly rather than logarithmic with coverage due to adsorbate-adsorbent interaction. Moreover adsorption is characterized by a uniform distribution of binding energies. The Tempkin isotherm in nonlinear form can be given as:

$$q_e = \frac{RT}{b \ln(K_T C_e)}$$

The linear form of the Tempkin isotherm can be expressed as follows:

$$q_e = B \ln A + B \ln C_e$$

where B is the Tempkin constant related to the heat of adsorption, A is the equilibrium binding constant corresponding to the maximum binding energy (L/mg) and R is the gas constant (8.314 J/mol/K). The constants A and B can be determined from the intercept and slope of the linear plot of data q_e versus $\ln C_e$.

Adsorption Kinetics

The kinetic studies help in understanding the rates of chemical processes and also facilitates in determining the exact mechanism of the sorption process. The investigation of chemical kinetics involves examination of various process variables that influences the rate of chemical reaction. The study gives an idea about the probable mechanism related to adsorbate-adsorbent interaction. The interaction mechanism can be examined by fitting the sorption kinetics data into mathematical models. The results obtained from the model helps in predicting reaction rates and the dependent factors which further helps in developing suitable adsorbent materials for industrial applications. The three prominent kinetics models namely Lagergren's first order (1898), and pseudo second order (1999), kinetics and intra-particle diffusion model (1963), has been extensively use for the processing of kinetics data and to understand the dynamics of the adsorption process in terms of the order of the rate constant. The data were plotted for the determination of most fitted model, the model with linearized plots and high values of correlation coefficient were considered to be the most suitable model for predicting sorption kinetics data.

Lagergren's First Order Kinetics

The rate expression for the first order kinetics given as:

$$\ln(q_e - q_t) = \ln q_e - k_1 t$$

where q_e and q_t are the amount of adsorbate adsorbed at equilibrium and at time t (min), respectively, and k_1 is the rate constant of adsorption (min^{-1}). The parameters for the first order kinetics can be calculated from the slope and intercept of the plot of $\ln(q_e - q_t)$ versus t.

Pseudo-Second Order Kinetics

The rate expression of the model represented as:

$$\frac{t}{q_t} = \frac{1}{k_2 q_e^2} + \frac{t}{q_e}$$

where k_2 (g/mg•min^{-1}) is the rate constant of second-order adsorption. The linear plot of t/q_t against t helps in determining k_2 values from the intercept of the plot. The sorption kinetics data if found consistent with pseudo-second order model then chemisorption process considered to play dominant role in adsorbate-adsorbent interaction (Zhang et al., 2013).

Intra-particle Diffusion Model

The model is used for investigating the adsorption mechanism for the designing of treatment plants. The effect of intra-particle diffusion resistance on adsorption can be determined by the following relationship:

$$q_t = k_{id}t^{1/2} + C$$

The model assumes that if the plot of q_t versus $t^{1/2}$ produces linear line while passing through the origin then pore diffusion controls the sorption process but if the plot results in straight line with non zero intercept then film diffusion is the rate controlling step of sorption process.

LIMITATIONS OF NANOPARTICLES

The nanoparticles played multifarious role in several branch of science and technology. The rising number of works on the toxic effects of nanoparticles suggested that apart from the benefits, nanotechnologies produce uncertainties and risks. Regardless of its huge utility and wide acceptability, nanoparticles found to inflict toxicological impact on human beings as well as on environment (Dhawan & Sharma, 2010; Videa et al., 2011). However, still there is a lack of information regarding the probable vulnerability and risks associated with the use of nanomaterials for human health and environment. The large scale application of nanomaterials in various industries, pathological clinics, hospitals and research centers leads to direct entry of nanoparticles in the environment however their fate and behavior are largely unknown (Farre´et al., 2011). It has been anticipated that nanoparticles possess enormous potential to cross the cell membranes of plants and animals and after entering into the cell the particles are capable enough to interact actively with several important biomolecules such as proteins and nucleic acids. This interaction results in disruption of various vital activities such as malfunctioning of enzymatic action, altering gene transcription and in some cases interfere protein synthesis. The large surface area of the nanoparticles makes them highly reactive and thus provided easy and rapid transportation of contaminants carried by them on their surfaces. Therefore assessment of the toxicity of the as-synthesized nanoparticles is essentially required. Various works on toxicity assessment of nano materials have been documented by several researchers. Lin & Xing (2007) carried out detailed investigation on phytotoxicity of nanoparticles. For this study they have selected five nanoparticles specifically multi-walled carbon nanotubes, aluminium, alumina, zinc and zinc oxide for determining its effect on seed germination and root growth of six higher plant species (radish, rape, rye grass, lettuce, corn and cucumber). The study reveals that among five nanoparticles nano scale zinc and zinc oxide nanoparticles drastically affected root elongation. Fifty percent inhibitory concentrations of nano zinc and nano zinc oxide were estimated to be 50 mg/L for radish and nearly 20 mg/L for rape and rye grass. These results clearly indicated that careless disposal of these engineered nanoparticles could adversely effects the growth and development of several plant species and if we extrapolate the consequence then the nanoparticles perhaps directly challenges our food security in near future. However, Tripathi et al., (2011) reported an interesting study revealing synergistic behavior of water soluble carbon nanotubes. The group performed their study on the gram plant by investigating the growth rate of the plants treated with water soluble muti-walled carbon nanotubes. The nanoparticles treated plants shows an increased growth rate in every part including the roots, shoots and branching. Moreover, the treated plant also exhibits better water absorption and retention. The work thus represents the positive aspect of nanoparticles. Zhao et al., (2014) presented an exemplary review on the toxicity of graphene which is one of the highly investigated carbonaceous materials. Their investigation instigate that graphene nanoparticles have undesirable impacts on aquatic organisms such as bacteria, algae, plants, invertebrates, and fish. They reported damage of cell membrane as the probable mechanism that causes toxicity in aquatic organisms. However, they conclude that behavior of graphene nanoparticles affected by solution chemistry therefore future research in this area is required for addressing toxicity aspect more acutely.

Nanotechnology is a major, innovative, scientific and economic growth area, but the increasing production and use of nanomaterials warrants more study and extensive research regarding the thorough assessment of the possible impacts on human health and the environment.

FUTURE OUTLOOK OF NANOPARTICLES AND CONCLUSION

The future perspective of nanoparticles deemed to be very intense and commendable only if proper safety measures were taken into consideration right from its synthesis, utility, marketing and ultimately to its disposal. Till date numerous research papers have been published based on nanoecotoxicological aspect but the exact quantity of release and in which form do the nanoparticles release is remain uncertain and unaddressed. Therefore future research must strive on the behavioral aspect of nanoparticles in changing environment such as change in pH, ionic strength. Moreover exploration of the chemical and physical attribute of the nanoparticles in different environment such as in soil, water and sewage must be a wishful investigation. Beside development of environmental sustainable, eco-friendly nanoparticles which can be easily degrade through microbial action is also an imperative approach which guarantees safe use and disposal of nanoparticles without affecting human and environmental health (Ali, 2012). Furthermore lifecycle assessment of the synthesize nanoparticles into the environment is highly crucial for the establishment and the implementation of effective, protective regulatory policy (Farre et al., 2011). The management of exhausted nanoparticles is also a matter of concern which needs urgent attention however, considerable efforts being made for the regeneration of nanoparticles. The study of nanoparticles for water treatment were largely investigated using batch sorption process therefore, is a great need to develop and optimized treatment process by using column operations at commercial levels.

The nanotechnology based water treatment therefore offers tremendous scope in removal of varieties of pollutants from aqueous solutions even present in low concentration. However, lot more research in the toxicological aspect of the nanoparticles is still required. The exhausted nanoparticles create disposal problem so there is need to develop some eco-friendly method of waste management. On the whole nanotechnology serves a lot in maintaining the long-term quality, availability, and viability of water.

ACKNOWLEDGMENT

SB and RKG express their thanks to Council of Scientific and Industrial Research and University Grants Commission, New Delhi for awarding Senior Research Fellowship.

REFERENCES

Abdel-Raouf, N., Al-Homaidan, A. A., & Ibraheem, I. B. M. (2012). Microalgae and wastewater treatment. *Saudi Journal of Biological Sciences*, *19*(3), 257–275. doi:10.1016/j.sjbs.2012.04.005 PMID:24936135

Abedini, A., Daud, A. R., Abdul Hamid, M., Kamil Othman, N., & Saion, E. (2013). A review on radiation-induced nucleation and growth of colloidal metallic nanoparticles. *Nanoscale Research Letters*, *8*(1), 474. doi:10.1186/1556-276X-8-474 PMID:24225302

Afkhami, A., Bagheri, H., & Madrakian, T. (2011). Alumina nanoparticles grafted with functional groups as a new adsorbent in efficient removal of formaldehyde from water samples. *Desalination*, *281*, 151–158. doi:10.1016/j.desal.2011.07.052

Ai, L., Li, M., & Long, L. (2011). Adsorption of Methylene Blue from Aqueous Solution with Activated Carbon/Cobalt Ferrite/Alginate Composite Beads: Kinetics, Isotherms, and Thermodynamics. *Journal of Chemical & Engineering Data*, *56*(8), 3475–3483. doi:10.1021/je200536h

Al-Johani, H., & Salam, M. A. (2011). Kinetics and thermodynamic study of aniline adsorption by multi-walled carbon nanotubes from aqueous solution. *Journal of Colloid and Interface Science*, *360*(2), 760–767. doi:10.1016/j.jcis.2011.04.097 PMID:21620412

Ali, I. (2012). New Generation Adsorbents for Water Treatment. *Chemical Reviews*, *112*(10), 5073–5091. doi:10.1021/cr300133d PMID:22731247

Ali, I., & Gupta, V. K. (2007). Advances in water treatment by adsorption technology. *Nature Protocols*, *1*(6), 2661–2667. doi:10.1038/nprot.2006.370 PMID:17406522

Alipour, V., Nasseri, S., Nodehi, R. N., Mahvi, A. H., & Rashidi, A. (2014). Preparation and application of oyster shell supported zero valent nano scale iron for removal of natural organic matter from aqueous solutions. *Journal of Environmental Health Science & Engineering*, *12*(146), 1–8. PMID:25648623

Asgher, M., & Bhatti, H. N. (2012). Evaluation of thermodynamics and effect of chemical treatments on sorption potential of Citrus waste biomass for removal of anionic dyes from aqueous solutions. *Ecological Engineering*, *38*(1), 79–85. doi:10.1016/j.ecoleng.2011.10.004

Bajpai, A. K., & Rajpoot, M. (1999). Adsorption Techniques – A Review. *Journal of Scientific and Industrial Research*, *58*(11), 844–860.

Banerjee, S., Gautam, R. K., Jaiswal, A., & Chattopadhyaya, M. C. (2014). Adsorption characteristics of a metal-organic framework material for the removal of Acid Orange 10 from aqueous solutions. *Journal of the Indian Chemical Society*, *91*(8), 1491–1499.

Banerjee, S., Gautam, R. K., Jaiswal, A., Chattopadhyaya, M. C., & Sharma, Y. C. (2015). Rapid scavenging of methylene blue dye from a liquid phase by adsorption on alumina nanoparticles. *RSC Advances*, *5*(19), 14425–14440. doi:10.1039/C4RA12235F

Bangi, U. K. H., Park, C. S., Baek, S., & Park, H. (2013). Sol–gel synthesis of high surface area nanostructured zirconia powder by surface chemical modification. *Powder Technology*, *239*, 314–318. doi:10.1016/j.powtec.2013.02.014

Bhargavi, R. J., Maheshwari, U., & Gupta, S. (2015). Synthesis and use of alumina nanoparticles as an adsorbent for the removal of Zn(II) and CBG dye from wastewater. *International Journal of Industrial Chemistry*, *6*(1), 31–41. doi:10.1007/s40090-014-0029-1

Bhatnagar, A., & Sillanpää, M. (2011). A review of emerging adsorbents for nitrate removal from water. *Chemical Engineering Journal, 168*(2), 493–504. doi:10.1016/j.cej.2011.01.103

Boening, D. W. (2000). Ecological effects, transport, and fate of mercury: A general review. *Chemosphere, 40*(12), 1335–1351. doi:10.1016/S0045-6535(99)00283-0 PMID:10789973

Bokare, A. D., Chikate, R. C., Rode, C. V., & Paknikar, K. M. (2007). Effect of Surface Chemistry of Fe-Ni Nanoparticles on Mechanistic Pathways of Azo Dye Degradation. *Environmental Science & Technology, 41*(21), 7437–7443. doi:10.1021/es071107q PMID:18044523

Boparai, H. K., Joseph, M., & O'Carrol, D. M. (2011). Kinetics and thermodynamics of cadmium ion removal by adsorption onto nano zerovalent iron particles. *Journal of Hazardous Materials, 186*(1), 458–465. doi:10.1016/j.jhazmat.2010.11.029 PMID:21130566

Bruanuer, S., Emmett, P. H., & Teller, E. (1938). Adsorption of gases in multimolecular layers. *Journal of the American Chemical Society, 60*(2), 309–316. doi:10.1021/ja01269a023

Cao, C. Y., Qu, J., Yan, W. S., Zhu, J. F., Wu, Z. Y., & Song, W. G. (2012). Low-Cost Synthesis of Flowerlike α-Fe2O3 Nanostructures for Heavy Metal Ion Removal: Adsorption Property and Mechanism. *Langmuir, 28*(9), 4573–4579. doi:10.1021/la300097y PMID:22316432

Chandra, N., Singh, D. K., Sharma, M., Upadhyay, R. K., Amritphale, S. S., & Sanghi, S. K. (2010). Synthesis and characterization of nano-sized zirconia powder synthesized by single emulsion-assisted direct precipitation. *Journal of Colloid and Interface Science, 342*(2), 327–332. doi:10.1016/j.jcis.2009.10.065 PMID:19942226

Dabrowski, A. (2001). Adsorption--from theory to practice. *Advances in Colloid and Interface Science, 93*(1-3), 135–224. doi:10.1016/S0001-8686(00)00082-8 PMID:11591108

Dhawan, A., & Sharma, V. (2010). Toxicity assessment of nanomaterials: Methods and challenges. *Analytical and Bioanalytical Chemistry, 398*(2), 589–605. doi:10.1007/s00216-010-3996-x PMID:20652549

Dhillon, A., & Kumar, D. (2015). Development of a nanoporous adsorbent for the removal of health-hazardous fluoride ions from aqueous systems. *Journal of Materials Chemistry A, 3*(8), 4215–4228. doi:10.1039/C4TA06147K

Do, M. H., Phan, N. H., Nguyen, T. D., Pham, T. T. S., Nguyen, V. K., Vu, T. T. T., & Nguyen, T. K. P. (2011). Activated carbon/Fe_3O_4 nanoparticle composite: Fabrication, methyl orange removal and regeneration by hydrogen peroxide. *Chemosphere, 85*(8), 1269–1276. doi:10.1016/j.chemosphere.2011.07.023 PMID:21840037

Du, W. L., Xu, Z. R., Han, X. Y., Xu, Y. L., & Miao, Z. G. (2008). Preparation, characterization and adsorption properties of chitosan nanoparticles for eosin Y as a model anionic dye. *Journal of Hazardous Materials, 153*(1-2), 152–156. doi:10.1016/j.jhazmat.2007.08.040 PMID:17890000

Dubinin, M. M., & Radushkevich, L. V. (1960). The equation of the characteristic curve of the activated charcoal. In Proceedings Academy of Sciences USSR (vol. 55, pp. 331–337).

Eskandarian, L., Arami, M., & Pajootan, E. (2014). Evaluation of Adsorption Characteristics of Multiwalled Carbon Nanotubes Modified by a Poly (propylene imine) Dendrimer in Single and Multiple Dye Solutions: Isotherms, Kinetics, and Thermodynamics. *Journal of Chemical & Engineering Data, 59*(2), 444–454. doi:10.1021/je400913z

Farre, M., Sanchis, J., & Barcelo, D. (2011). Analysis and assessment of the occurrence, the fate and the behavior of nanomaterials in the environment. *Trends in Analytical Chemistry, 30*(2), 517–527. doi:10.1016/j.trac.2010.11.014

Fei, J., Cui, Y., Zhao, J., Gao, L., Yang, Y., & Li, J. (2011). Large-scale preparation of 3D self-assembled iron hydroxide and oxide hierarchical nanostructures and their applications for water treatment. *Journal of Materials Chemistry, 21*(32), 11742–11746. doi:10.1039/c1jm11950h

Fendler, J. H. (1987). Atomic and molecular clusters in membrane mimetic chemistry. *Chemical Reviews, 87*(5), 877–899. doi:10.1021/cr00081a002

Feng, Y., Gong, J. L., Zeng, G. M., Niu, Q. Y., Zhang, H. Y., Niu, C. G., & Yan, M. et al. (2010). Adsorption of Cd (II) and Zn (II) from aqueous solutions using magnetic hydroxyapatite nanoparticles as adsorbents. *Chemical Engineering Journal, 162*(2), 487–494. doi:10.1016/j.cej.2010.05.049

Foo, K. Y., & Hameed, B. H. (2010). Insights into the modeling of adsorption isotherm systems. *Chemical Engineering Journal, 156*(1), 2–10. doi:10.1016/j.cej.2009.09.013

Freundlich, H.M.F. (1906). Over the adsorption in solution. *Journal of Physical Chemistry, 57*(A), 385–471.

Ganguli, A. K., Vaidya, S., & Ahmad, T. (2008). Synthesis of nanocrystalline materials through reverse micelles:A versatile methodology for synthesis of complex metal oxides. *Bulletin of Materials Science, 31*(3), 415–419. doi:10.1007/s12034-008-0065-6

Gateshki, M., Petkov, V., Williams, G., Pradhan, S. K., & Ren, Y. (2005). Atomic-scale structure of nanocrystalline ZrO_2 prepared by high-energy ball milling. *Physical Review B: Condensed Matter and Materials Physics, 71*(22), 224107–224109. doi:10.1103/PhysRevB.71.224107

Gautam, R. K., Gautam, P. K., Banerjee, S., Soni, S., Singh, S. K., & Chattopadhyaya, M. C. (2015). Removal of Ni (II) by magnetic nanoparticles. *Journal of Molecular Liquids, 204*(4), 60–69. doi:10.1016/j.molliq.2015.01.038

Gautam, R. K., Mudhoo, A., & Chattopadhyaya, M. C. (2013). Kinetic, equilibrium, thermodynamic studies and spectroscopic analysis of Alizarin Red S removal by mustard husk. *Journal of Environmental Chemical Engineering, 1*(4), 1283–1291. doi:10.1016/j.jece.2013.09.021

Ge, F., Li, M. M., Ye, H., & Zhao, B. X. (2012). Effective removal of heavy metal ions Cd^{2+}, Zn^{2+}, Pb^{2+}, Cu^{2+} from aqueous solution by polymer-modified magnetic nanoparticles. *Journal of Hazardous Materials, 211– 212*(4), 366– 372.

Ghasemi, E., & Sillanpaa, M. (2015). Magnetic hydroxyapatite nanoparticles: An efficient adsorbent for the separation and removal of nitrate and nitrite ions from environmental samples. *Journal of Separation Science, 38*(1), 164–169. doi:10.1002/jssc.201400928 PMID:25376506

Glezer, A. M. (2011). Structural Classification of Nanomaterials. *Russian Metallurgy (Metally), 2011*(4), 263–269. doi:10.1134/S0036029511040057

Godt, J., Scheidig, F., Siestrup, C. G., Esche, V., Brandenburg, P., Reich, A., & Groneberg, D. A. (2006). The toxicity of cadmium and resulting hazards for human health. *Journal of Occupational Medicine and Toxicology (London, England), 1*(22), 1–6. PMID:16961932

Gong, J., Chen, L., Zeng, G., Long, F., Deng, J., Niu, Q., & He, X. (2012). Shellac-coated iron oxide nanoparticles for removal of cadmium(II) ions from aqueous solution. *Journal of Environmental Sciences (China), 24*(7), 1165–1173. doi:10.1016/S1001-0742(11)60934-0 PMID:23513435

Gopal, K., Tripathy, S. S., Bersillon, J. L., & Dubey, S. P. (2007). Chlorination by-products, their toxicodynamics and removal from drinking water. *Journal of Hazardous Materials, 140*(1-2), 1–6. doi:10.1016/j.jhazmat.2006.10.063 PMID:17129670

Hamilton, S. J. (2004). Review of selenium toxicity in the aquatic food chain. *The Science of the Total Environment*, *326*(1-3), 1–31. doi:10.1016/j.scitotenv.2004.01.019 PMID:15142762

Han, Y., & Yan, W. (2014). Bimetallic nickel-iron nanoparticles for groundwater decontamination: Effect of groundwater constituents on surface deactivation. *Water Research*, *66*, 149–159. doi:10.1016/j.watres.2014.08.001 PMID:25201338

He, C. N., Zhao, N. Q., Shi, C. S., & Song, S. Z. (2009). Fabrication of carbon nanomaterials by chemical vapor deposition. *Journal of Alloys and Compounds*, *484*(1-2), 6–11. doi:10.1016/j.jallcom.2009.04.088

Ho, Y. S., & McKay, G. (1999). Pseudo-second order model for sorption processes. *Process Biochemistry*, *34*(5), 451–465. doi:10.1016/S0032-9592(98)00112-5

Horsfall, M., & Spiff, A. I. (2005). Equilibrium sorption study of Al^{3+}, Co^{2+} and Ag^{2+} in aqueous solutions by fluted pumpkin (Telfairia occidentalis HOOK) waste biomass. *Acta Chimica Slovenica*, *52*(2), 174–181.

Huang, S. H., & Chen, D. H. (2009). Rapid removal of heavy metal cations and anions from aqueous solutions by an amino-functionalized magnetic nano-adsorbent. *Journal of Hazardous Materials*, *163*(1), 174–179. doi:10.1016/j.jhazmat.2008.06.075 PMID:18657903

Huang, W., Wang, S., Zhu, Z., Li, L., Yao, X., Rudolph, V., & Haghseresht, F. (2008). Phosphate removal from wastewater using red mud. *Journal of Hazardous Materials*, *158*(1), 35–42. doi:10.1016/j.jhazmat.2008.01.061 PMID:18314264

Hyeon, T. (2003). Chemical synthesis of magnetic nanoparticles. *Chemical Communications (Cambridge)*, (8): 927–934. doi:10.1039/b207789b PMID:12744306

Jain, S., & Jayaram, R. V. (2010). Removal of basic dyes from aqueous solution by low-cost adsorbent: Wood apple shell (Feronia acidissima). *Desalination*, *250*(3), 921–927. doi:10.1016/j.desal.2009.04.005

Kadu, B. S., & Chikate, R. C. (2013). Improved adsorptive mineralization capacity of Fe–Ni sandwiched montmorillonite nanocomposites towards magenta dye. *Chemical Engineering Journal*, *228*, 308–317. doi:10.1016/j.cej.2013.04.103

Kaur, R., Hasan, A., Iqbal, N., Alam, S., Saini, M. K., & Raza, S. K. (2014). Synthesis and surface engineering of magnetic nanoparticles for environmental cleanup and pesticide residue analysis: A review. *Journal of Separation Science*, *37*(14), 1805–1825. doi:10.1002/jssc.201400256 PMID:24777942

Kim, Y., & Zimmerman, S. C. (1998). Applications of dendrimers in bio-organic chemistry. *Current Opinion in Chemical Biology*, 2(6), 733–742. doi:10.1016/S1367-5931(98)80111-7 PMID:9914193

Kyzas, G. Z., & Kostoglou, M. (2014). Green Adsorbents for Wastewaters: A Critical Review. *Materials (Basel)*, 7(1), 333–364. doi:10.3390/ma7010333

Lagergren, S. (1898). About the theory of so-called adsorption of soluble substances, *Kungliga Svenska Vetenskapsakademiens. Handlingar*, 24(4), 1–39.

Langmuir, I. (1916). The constitution and fundamental properties of solids and liquids. *Journal of the American Chemical Society*, 38(11), 2221–2295. doi:10.1021/ja02268a002

Li, L., Fan, M., Brown, R. C., & Leeuwen, J. V. (2006). Synthesis, Properties, and Environmental Applications of Nanoscale Iron-Based Materials: A Review. *Critical Reviews in Environmental Science and Technology*, 36(5), 405–431. doi:10.1080/10643380600620387

Li, Z. J., Wang, L., Yuan, L. Y., Xiao, C. L., Mei, L., Zheng, L. R., & Shi, W. Q. et al. (2015). Efficient removal of uranium from aqueous solution by zero-valent iron nanoparticle and its graphene composite. *Journal of Hazardous Materials*, 290, 26–33. doi:10.1016/j.jhazmat.2015.02.028 PMID:25734531

Lin, D., & Xing, B. (2007). Phytotoxicity of nanoparticles: Inhibition of seed germination and root growth. *Environmental Pollution*, 150(2), 243–250. doi:10.1016/j.envpol.2007.01.016 PMID:17374428

Liu, J. F., Zhao, Z. S., & Jiang, G. B. (2008). Coating Fe_3O_4 Magnetic Nanoparticles with Humic Acid for High Efficient Removal of Heavy Metals in Water. *Environmental Science & Technology*, 42(18), 6949–6954. doi:10.1021/es800924c PMID:18853814

Malik, M. A., Wani, M. Y., & Hashim, M. A. (2014). Microemulsion method: A novel route to synthesize organic and inorganic nanomaterials: 1st Nano Update. *Arabian Journal of Chemistry*, 5(4), 397–417. doi:10.1016/j.arabjc.2010.09.027

Mane, V. S., & Vijay Babu, P. V. (2011). Studies on the adsorption of Brilliant Green dye from aqueous solution onto low-cost NaOH treated saw dust. *Desalination*, 273(2-3), 321–329. doi:10.1016/j.desal.2011.01.049

Manoharan, D., Loganathan, A., Kurapati, V., & Nesamony, V. J. (2015). Unique sharp photoluminescence of size-controlled sonochemically synthesized zirconia nanoparticles. *Ultrasonics Sonochemistry, 23*, 174–184. doi:10.1016/j.ultsonch.2014.10.004 PMID:25453213

Mohan, D., & Pittman, C. U. Jr. (2007). Arsenic removal from water/wastewater using adsorbents—A critical review. *Journal of Hazardous Materials, 142*(1-2), 1–53. doi:10.1016/j.jhazmat.2007.01.006 PMID:17324507

Mohapatra, M., & Anand, S. (2010). Synthesis and applications of nano-structured iron oxides/hydroxides – a review. *International Journal of Engineering Science and Technology, 2*(8), 127–146.

Mohapatra, M., Anand, S., Mishra, B. K., Giles, D. E., & Singh, P. (2009). Review of fluoride removal from drinking water. *Journal of Environmental Management, 91*(1), 67–77. doi:10.1016/j.jenvman.2009.08.015 PMID:19775804

Moussavi, G., & Mahmoudi, M. (2009). Removal of azo and anthraquinone reactive dyes from industrial wastewaters using MgO nanoparticles. *Journal of Hazardous Materials, 168*(2-3), 806–812. doi:10.1016/j.jhazmat.2009.02.097 PMID:19303210

Nakaso, K., Han, B., Ahn, K. H., Choi, M., & Okuyama, K. (2003). Synthesis of non-agglomerated nanoparticles by an electrospray assisted chemical vapor deposition (ES-CVD) method. *Aerosol Science, 34*(7), 869–881. doi:10.1016/S0021-8502(03)00053-3

Ohara, S., Sato, K., Tan, Z., Shimoda, H., Ueda, M., & Fukui, T. (2010). Novel mechanochemical synthesis of fine $FeTiO_3$ nanoparticles by a high-speed ball-milling process. *Journal of Alloys and Compounds, 504*(1), L17–L19. doi:10.1016/j.jallcom.2010.05.090

Owlad, M., Aroua, M. K., Daud, W. A. W., & Baroutian, S. (2009). Removal of Hexavalent Chromium-Contaminated Water and Wastewater: A Review. *Water, Air, and Soil Pollution, 200*(1-4), 59–77. doi:10.1007/s11270-008-9893-7

Öztürk, A., & Malkoc, E. (2014). Adsorptive potential of cationic Basic Yellow 2 (BY2) dye onto natural untreated clay (NUC) from aqueous phase: Mass transfer analysis, kinetic and equilibrium profile. *Applied Surface Science, 299*, 105–115. doi:10.1016/j.apsusc.2014.01.193

Papanikolaou, N. C., Hatzidaki, E. G., Belivanis, S., Tzanakakis, G. N., & Tsatsakis, A. M. (2005). Lead toxicity update. A brief review. *Medical Science Monitor, 11*(10), 329–336. PMID:16192916

Paul, B., Parashar, V., & Mishra, A. (2015). Graphene in the Fe_3O_4 nano-composite switching the negative influence of humic acid coating into an enhancing effect in the removal of arsenic from water. *Environmental Science and Water Research Technology*, *1*(1), 77–83. doi:10.1039/C4EW00034J

Pedersen, J. A., Yeager, M. A., & Suffet, I. H. (2003). Xenobiotic Organic Compounds in Runoff from Fields Irrigated with Treated Wastewater. *Journal of Agricultural and Food Chemistry*, *51*(5), 1360–1372. doi:10.1021/jf025953q PMID:12590482

Pelekani, C., & Snoeyink, V. L. (2001). A kinetic and equilibrium study of competitive adsorption between atrazine and congo red dye on activated carbon: The importance of pore size distribution. *Carbon*, *39*(1), 25–37. doi:10.1016/S0008-6223(00)00078-6

Pendleton, P., & Wu, S. H. (2003). Kinetics of dodecanoic acid adsorption from caustic solution by activated carbon. *Journal of Colloid and Interface Science*, *266*(2), 245–250. doi:10.1016/S0021-9797(03)00575-7 PMID:14527446

Pillay, K., Cukrowsk, E. M., & Coville, N. J. (2009). Multi-walled carbon nanotubes as adsorbents for the removal of parts per billion levels of hexavalent chromium from aqueous solution. *Journal of Hazardous Materials*, *166*(2-3), 1067–1075. doi:10.1016/j.jhazmat.2008.12.011 PMID:19157694

Pokropivny, V. V., & Skorokhod, V. V. (2007). Classification of nanostructures by dimensionality and concept of surface forms engineering in nanomaterial science. *Materials Science and Engineering C*, *27*(5), 990–993. doi:10.1016/j.msec.2006.09.023

Puvaneshwari, N., Muthukrishnan, J., & Gunasekaran, P. (2006). Toxicity assessment and microbial degradation of azo dyes. *Indian Journal of Experimental Biology*, *44*(8), 618–626. PMID:16924831

Pyrgiotakis, G., Blattmann, C. O., Sotiris, P., & Philip, D. (2013). Nanoparticle-nanoparticle interactions in biological media by Atomic Force Microscopy. *Langmuir*, *29*(36), 11385–11395. doi:10.1021/la4019585 PMID:23978039

Qu, X., Alvarez, J. J. P., & Li, Q. (2013a). Applications of nanotechnology in water and wastewater treatment. *Water Research*, *47*(12), 3931–3946. doi:10.1016/j.watres.2012.09.058 PMID:23571110

Qu, X. L., Brame, J., Li, Q., & Alvarez, J. J. P. (2013b). Nanotechnology for a safe and sustainable water supply: Enabling integrated water treatment and reuse. *Accounts of Chemical Research*, *46*(3), 834–843. doi:10.1021/ar300029v PMID:22738389

Ravishankar, T. N., Ramakrishnappa, T., Nagaraju, G., & Rajanaika, H. (2014). Synthesis and Characterization of CeO_2 Nanoparticles via Solution Combustion Method for Photocatalytic and Antibacterial Activity Studies. *Chemistry Open*, *4*(2), 146–154. PMID:25969812

Raziyeh, S., Arami, M., Mahmoodi, N. M., Bahrami, H., & Khorramfar, S. (2010). Novel biocompatible composite (Chitosan–zinc oxide nanoparticle): Preparation,characterization and dye adsorption properties. *Colloids and Surfaces. B, Biointerfaces*, *80*(1), 86–93. doi:10.1016/j.colsurfb.2010.05.039 PMID:20566273

Redlich, O., & Peterson, D. L. (1959). A useful adsorption isotherm. *Journal of Physical Chemistry*, *63*(6), 1024–1026. doi:10.1021/j150576a611

Roy, A., Adhikari, B., & Majumder, S. B. (2013). Equilibrium, Kinetic, and Thermodynamic Studies of Azo Dye Adsorption from Aqueous Solution by Chemically Modified Lignocellulosic Jute Fiber. *Industrial & Engineering Chemistry Research*, *52*(19), 6502–6512. doi:10.1021/ie400236s

Rudin, T., & Pratsinis, S. E. (2012). Homogeneous Iron Phosphate Nanoparticles by Combustion of Sprays. *Industrial & Engineering Chemistry Research*, *51*(23), 7891–7900. doi:10.1021/ie202736s PMID:23407874

Sahm, T., Mädler, L., Gurlo, A., Barsan, N., Pratsinis, S. E., & Weimar, U. (2004). Flame spray synthesis of tin dioxide nanoparticles for gas sensing. *Sensors and Actuators. B, Chemical*, *98*(2-3), 148–153. doi:10.1016/j.snb.2003.10.003

Salam, M. A., Gabal, M. A., & Obaid, A. Y. (2012). Preparation and characterization of magnetic multi-walled carbon nanotubes/ferrite nanocomposite and its application for the removal of aniline from aqueous solution. *Synthetic Metals*, *161*(23), 2651–2658. doi:10.1016/j.synthmet.2011.09.038

Salman, S. A., Usami, T., Kuroda, K., & Okido, M. (2014). Synthesis and characterization of cobalt nanoparticles using hydrazine and citric acid. Journal of Nanotechnology. Article ID 525193, 6 pages.

Sharma, Y. C., Srivastava, V., Upadhyay, S. N., & Weng, C. H. (2008). Alumina Nanoparticles for the Removal of Ni(II) from Aqueous Solutions. *Industrial & Engineering Chemistry Research*, *47*(21), 8095–8100. doi:10.1021/ie800831v

Sharma, Y. C., & Uma. (2010). Optimization of Parameters for Adsorption of Methylene Blue on a Low-Cost Activated Carbon. *J. Chem. Eng. Data, 55*(11), 435–439.

Sheela, T., Nayaka, Y. A., Viswanatha, R., Basavanna, S., & Venkatesha, T. G. (2012). Kinetics and thermodynamics studies on the adsorption of Zn (II), Cd (II) and Hg (II) from aqueous solution using zinc oxide nanoparticles. *Powder Technology, 217,* 163–170. doi:10.1016/j.powtec.2011.10.023

Shen, J., Li, Z., Wu, Y., Zhang, B., & Li, F. (2015). Dendrimer-based preparation of mesoporous alumina nanofibers by electrospinning and their application in dye adsorption. *Chemical Engineering Journal, 264,* 48–55. doi:10.1016/j.cej.2014.11.069

Silva, A. J., Varesche, M. B., Foresti, E., & Zaiat, M. (2002). Sulphate removal from industrial wastewater using a packed-bed anaerobic reactor. *Process Biochemistry, 37*(9), 927–935. doi:10.1016/S0032-9592(01)00297-7

Singh, M., Kumar, M., Štěpánek, F., Ulbrich, P., Svoboda, P., Santava, E., & Singla, M. L. (2011). Liquid-Phase Synthesis of Nickel Nanoparticles stabilized by PVP and study of their structural and magnetic properties. *Advanced Material Letters, 2*(6), 409–414. doi:10.5185/amlett.2011.4257

Sips, R. (1948). Combined form of Langmuir and Freundlich equations. *The Journal of Chemical Physics, 16*(5), 490–495. doi:10.1063/1.1746922

Skorokhod, V., Ragulya, A., & Uvarova, I. (2001). *Physico-chemical Kinetics in Nanostructured Systems.* Kyiv: Academperiodica.

Srivastava, V., Sharma, Y. C., & Sillanpää, M. (2015 a). Green synthesis of magnesium oxide nanoflower and its application for the removal of divalent metallic species from synthetic wastewater. *Ceramics International, 44*(5), 6702–6709. doi:10.1016/j.ceramint.2015.01.112

Srivastava, V., Sharma, Y. C., & Sillanpää, M. (2015 b). Application of nano-magnesso ferrite (n-MgFe2O4) for the removal Co^{2+} ions from synthetic wastewater: Kinetic, equilibrium and thermodynamic studies. *Applied Surface Science, 338,* 42–54. doi:10.1016/j.apsusc.2015.02.072

Tani, T., Lutz, M., & Pratsinis, S. E. (2002). Homogeneous ZnO nanoparticles by flame spray pyrolysis. *Journal of Nanoparticle Research, 4*(4), 337–343. doi:10.1023/A:1021153419671

Tempkin, M. I., & Pyzhev, V. (1940). Kinetics of ammonia synthesis on promoted iron catalyst. *Acta Physicochimica U.R.S.S., 12,* 327–356.

Teng, M., Qiao, J., Li, F., & Bera, P. K. (2012). Electrospun mesoporous carbon nanofibers produced from phenolic resin and their use in the adsorption of large dye molecules. *Carbon, 50*(8), 2877–2886. doi:10.1016/j.carbon.2012.02.056

Teoh, W. Y., Amal, R., & Lutz, M. (2010). Flame spray pyrolysis: An enabling technology for nanoparticles design and fabrication. *Nanoscale*, *2*(8), 1324–1347. doi:10.1039/c0nr00017e PMID:20820719

Tian, P., Han, X. Y., Ning, G., Fang, H., Ye, J., Gong, W., & Lin, Y. (2013). Synthesis of Porous Hierarchical MgO and Its Superb Adsorption Properties. *Applied Materials and Interfaces*, *5*(23), 12411–12418. doi:10.1021/am403352y PMID:24224803

Tiwari, D. K., Behari, J., & Sen, P. (2008). Application of Nanoparticles in Waste Water Treatment. *World Applied Sciences Journal*, *3*(3), 417–433.

Toksha, B. G., Shirsath, S. E., Patange, S. M., & Jadhav, K. M. (2008). Structural investigations and magnetic properties of cobalt ferrite nanoparticles prepared by sol–gel auto combustion method. *Solid State Communications*, *147*(11), 479–483. doi:10.1016/j.ssc.2008.06.040

Toth, J. (1981). A uniform interpretation of gas/solid adsorption. *Journal of Colloid and Interface Science*, *79*(1), 85–95. doi:10.1016/0021-9797(81)90050-3

Tripathi, S., Sonkar, S. K., & Sarkar, S. (2011). Growth stimulation of gram (Cicer arietinum) plant by water soluble carbon nanotubes. *Nanoscale*, *3*(3), 1176–1181. doi:10.1039/c0nr00722f PMID:21253651

Videa, J. R. P., Zhao, L., Morenoc, M. L. L., de la Rosa, G., Hong, J., & Torresdey, J. L. G. (2011). Nanomaterials and the environment: A review for the biennium 2008–2010. *Journal of Hazardous Materials*, *186*(1), 1–15. PMID:21134718

Weber, W. J. Jr, & Morris, J. C. (1963). Kinetics of adsorption on carbon from solution. *Journal of Sanitation Engineering Division American Society of Civil Engineering*, *89*(1), 31–60.

Wu, R., Qu, J., He, H., & Yu, Y. (2004). Removal of azo-dye Acid Red B (ARB) by adsorption and catalytic combustion using magnetic $CuFe_2O_4$ powder. *Applied Catalysis B: Environmental*, *48*(1), 49–56. doi:10.1016/j.apcatb.2003.09.006

Xu, H., Zeiger, B. W., & Suslick, K. S. (2013). Sonochemical synthesis of nanomaterials. *Chemical Society Reviews*, *42*(7), 2555–2567. doi:10.1039/C2CS35282F PMID:23165883

Yang, J., Zeng, Q., Peng, L., Lei, M., Song, H., Tie, B., & Gu, J. (2013). La-EDTA coated Fe_3O_4 nanomaterial: Preparation and application in removal of phosphate from water. *Journal of Environmental Sciences (China)*, *25*(2), 413–418. doi:10.1016/S1001-0742(12)60014-X PMID:23596964

Yang, K., Wu, W., Jing, Q., & Zhu, L. (2008). Aqueous Adsorption of Aniline, Phenol, and their Substitutes by Multi-Walled Carbon Nanotubes. *Environmental Science & Technology*, *42*(21), 7931–7936. doi:10.1021/es801463v PMID:19031883

Yogesh, K. K., Muralidhara, H. B., Nayaka, Y. A., Balasubramanyam, J., & Hanumanthappa, H. (2013). Low-cost synthesis of metal oxide nanoparticles and their application in adsorption of commercial dye and heavy metal ion in aqueous solution. *Powder Technology*, *246*, 125–136.

Zhang, G., Ren, Z., Zhang, X., & Chen, J. (2013). Nanostructured iron(III)-copper(II) binary oxide: A novel adsorbent for enhanced arsenic removal from aqueous solutions. *Water Research*, *47*(12), 4022–4031. doi:10.1016/j.watres.2012.11.059 PMID:23571113

Zhao, J., Wang, Z., White, J. C., & Xing, B. (2014). Graphene in the Aquatic Environment: Adsorption, Dispersion, Toxicity and Transformation. *Environmental Science & Technology*, *48*(17), 9995–10009. doi:10.1021/es5022679 PMID:25122195

Zhao, X., Wang, J., Wu, F., Wang, T., Cai, Y., Shi, Y., & Jiang, G. (2010). Removal of fluoride from aqueous media by Fe_3O_4@$Al(OH)_3$ magnetic nanoparticles. *Journal of Hazardous Materials*, *173*(1-3), 102–109. doi:10.1016/j.jhazmat.2009.08.054 PMID:19747775

Zheng, Y., Li, N., & Zhang, W.-D. (2012). Preparation of nanostructured microspheres of Zn–Mg–Al layered double hydroxides with high adsorption property. *Colloids and Surfaces. A, Physicochemical and Engineering Aspects*, *415*, 195–201. doi:10.1016/j.colsurfa.2012.10.014

Zhu, J., Lu, Z., Aruna, S. T., Aurbach, D., & Gedanken, A. (2000). Sonochemical Synthesis of SnO2 Nanoparticles and Their Preliminary Study as Li Insertion Electrodes. *Chemistry of Materials*, *12*(9), 2557–2566. doi:10.1021/cm9906831

Chapter 8
Risks and Preventive Measures of Nanotechnology

Waqas Anwar
Mirpur University of Science and Technology, Pakistan

Anwar Khitab
Mirpur University of Science and Technology, Pakistan

ABSTRACT

Application of Nanotechnology in Civil Engineering is a rapidly growing field. It has brought improvements in construction materials as well as practices. Moreover, further developments are foreseeable in this field based on the positive outcomes of the current research works. Utilization of nanoparticles in Civil Engineering has been proved advantageous from several aspects of strength, durability and sustainability. Unfortunately, there are not only benefits associated with the application of nanotechnology. According to various studies, nanoparticles are supposed to damage human organs through physical contact and inhalation. Considering the environmental impacts, atmospheric transport, as well as transport in saturated and unsaturated regions in the subsurface are possible. Nowadays, nanoparticles are progressively produced and they could easily be released in air, water, and eventually contaminate the soil which is harmful for the environment and its habitats. The following chapter would address these issues as well as preventive measures in order to improve benefit-risk ratio.

DOI: 10.4018/978-1-5225-0344-6.ch008

Copyright ©2016, IGI Global. Copying or distributing in print or electronic forms without written permission of IGI Global is prohibited.

INTRODUCTION

Whether Nanotechnology is good or bad for the environment is totally based on the nature of its use and considerations made during its application. Several projects which involve the use of nanotechnology for betterment of the environment have been successfully established. However, the use of nanotechnology in any field requires great care and any sort of negligence is likely to bring negative effects for the environment and its habitats. For example, nano particles have the potential for the treatment of water and waste water to a great degree. CNTs, nano sized magnetite, CeO_2 and TiO_2 have been considered as prime nanoparticles to remove pollutants from water (Deliyanni et al., 2003; Mayo et al., 2007; Nawrocki et al., 2010). However, any negligence may leave undesirable quantities of these nanoparticles in water; thus instead of doing well, it may cause harmful effects on the environment and health of consumers.

EXPOSURE TO ENVIRONMENT AND HABITATS

Generally, nanomaterials become threat for the environment and its habitats when they are discharged in undesirable quantities into the wrong destinations. Nanomaterials may be released from point or non-point sources. Point sources include industries, storage units etc. and non-point sources include storm water runoff or wet deposition from the atmosphere. Exposure to nanomaterials may occur unintentionally in the environment or through the use of nanotechnology based products in our daily lives. Human exposure to these nanoparticles is more likely to happen during the manufacturing process. However, inhalation of nanomaterials released to the atmosphere and use of drinking water or food having accumulated nano particles is also possible. Moreover, absorption by soil and then transportation in saturated and unsaturated regions in the subsurface is also possible. This is very likely to affect the ground water table which then needs proper treatment before it is used for drinking and irrigation purposes (Wiesner et al., 2006). Furthermore from soil, nanoparticles may easily become the element of the vegetations; thus becoming a serious health threat for all consumers including the tiny insects.

Risks of Nanotechnology

As size of nano-particles is very small, they can easily stay in atmosphere and can cause air born diseases and several harmful environmental effects (Maynard et al,

Risks and Preventive Measures of Nanotechnology

2011; Oberdörster et al., 2005). Size of nano-particles can be as small as biological molecules such as proteins. They can easily be absorbed and may reach the inner bio molecules in the body (European Comission). Availability of limited knowledge about this technology and its impacts on the habitats is a major reason to consider risks seriously. It is an emerging field at the moment and not all nanomaterials have been studied in detail regarding their harmful effects. So, there is need to adopt wide precautions and consider all available research findings very critically during the whole life cycle of the nano-based products. Basically, for any material its exposure and effects throughout the life cycle are important to be considered. Figure 1 integrates the life cycle stages of nano particles with their pathways, transportation, exposure and effects in a simplified way.

Figure 1. Nanotechnology life-cycle stages integrated with pathways, transformation, and exposure

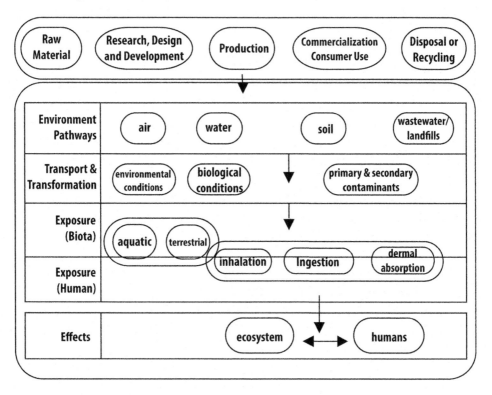

Types of Risks

Risks can be divided into two main categories; known risks and potential risks.

- **Known Risks:** When the relation between the cause and its impact is established, the risks are 'known risks'. In case of known risks, the significance of danger is well known and prevention is easy to make.
- **Potential Risks:** When a relation between the cause and the impact is not established, the risks are categorized as 'potential risks'. In potential risks, the significance and the certainty of dangers is not known.

Risks from nanotechnology fall in both categories. Though the potential risks are suspicious however as a matter of public health and environmental safety, precautionary measurements are mandatory to be taken for both types of risks.

Risk Assessment of ENPs

Cumulative Risk Assessment (CRA) is a process in which scientific and regulatory principles are applied in a systematic way to identify and quantify the risks. It is considered as the most relevant system and is based on four major steps: Hazard identification, Dose response assessment, Exposure assessment, and, lastly, risk characterization (Hansen, 2009; IPCS, 2009). Its main result is a statement of the probability which depicts whether humans or other environmental receptors will be harmed or not when exposed to a pollutant and what would be the intensity of this harm. The CRA methodology is internationally recognized and employed by many well reputed organizations, such as the World Health Organization (WHO) and the Organization for Economic Co-operation and Sustainability 2009, as well as by several European and U.S. agencies (Nielsen et al., 2008). It is considered as a very valuable tool for the regulation of chemicals including nanoparticles. CRA is also a fundamental ingredient of the new European Union (EU) chemical regulation policy, known as Registration, Evaluation and Authorization of Chemicals (REACH).

- **Hazard Identification:** Hazard identification (HI) is defined as the recognition of the undesirable effects, which a matter has an intrinsic capacity to cause (Hoshino et al., 2004; Lovric et al., 2005). In 2007, 428 studies reporting on toxicity of ENPs were identified (Green et al., 2005). In these studies, adverse health effects of 965 tested ENPs of different chemical compositions were observed (Green et al., 2005). It is very important to note that the vast majority of the reviewed studies demonstrate some degree of hazardous effects on the tested organisms. Toxicity has been reported for many ENPs, but for most of them further investigation and confirmation are needed before hazard can be identified. For proper hazard identification data characterization and standardized tests are mandatory.

- **Dose-Response Assessment:** Dose-response assessment (DRA) is defined as an approximation of the relationship between dose, or level of exposure to a substance and the occurrence as well as severity of an effect which occurs (Hoshino et al., 2004; Lovric et al., 2005). The dose is generally measured in units of mass units (i.e., μg, mg, g), however it has been concluded that toxicity of some engineered nanoparticles is not mass dependent and is influenced by several physico-chemical properties including surface area, morphology and chemical composition. Some studies done in the past used the concept of mass units while others considered the surface area or other characteristics for the assessment purposes.
- **Exposure Assessment:** Exposure assessment (EA) is defined as an estimation of the concentrations/doses to which the specific human populations (i.e., workers, consumers and people exposed indirectly via the environment) or environmental compartments (aquatic environment, terrestrial environment and air) may be exposed (Hoshino et al., 2004; Lovric et al., 2005). Figure 2 links each stage of a nanoparticle life with the specific exposure it may cause to the human beings and the environment.

Figure 2. Life-cycle stages of a nanoparticle integrated with specific vulnerable exposure

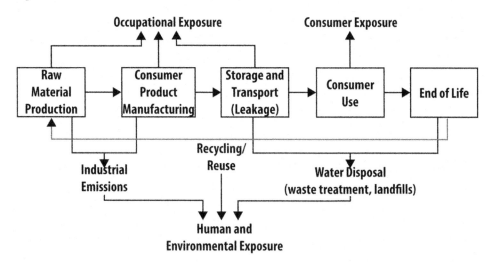

In another study, ENP-containing products were divided into several categories (including appliances, food and beverages, health and fitness, home and garden and goods for children). The researchers found out that the expected consumer exposure is more for products which fall in the categories of appliances, home & garden and health & fitness.

- **Risk Characterization:** The last step of Risk assessment procedure is the Risk Characterization which is defined as the estimation of the occurrence and severity of the harmful effects expected to occur in human population or the environment based on the 'actual exposure' to the substance. The phase may also include risk estimation. In this last phase, all the information which is collected during the first three steps of risk assessment is combined, weighted and the then the risk is quantified. The quantitative rick characterization compares the predicted environmental concentration (PEC) of a chemical agent with its predicted no-effect concentration (PNEC). The PNEC is the concentration, below which the exposure to the substance is not probable to cause any adverse effects, while the PEC is the prognosticated concentration of a chemical in the environment. The ratio of PEC to PNEC is called risk quotient (RQ). If the RQ is lower than 1, no further testing or risk reduction measures are needed (ECJRC, 2003). If it is greater than 1, further testing or remedial measures are mandatory to reduce the RQ (ECJRC, 2003; Nielsen et al., 2007).

Risk assessment of nanoparticles is so important that scientists have developed a separate field by the name of 'Nano (eco)-toxicology'. It is a new field of research which deals with the engineering of nanodevices and nanostructures to completely understand their impacts in living organisms (Oberdorster et al., 2005). The reason behind the development of this separate scientific discipline is the request of a number of scientists with the purpose of generating data and knowledge about nanotechnology effects on the environment and its habitats (Oberdorster et al., 2005; Pierce, 2004).This field is still in the development phase, however it can have sufficient contribution towards the identification and assessment of risks.

Nanotechnology Risks to the Human Health

Nanotechnology if used properly can have positive effects on the health of humans. Tiny nanotechnology based sensors in the future may help to detect cancer in its early stages (Rachel, 2003). European commission has also talked about nano scale atom- robots which may have the potential to cure several diseases (Hazards Magazine). Furthermore, gold nanoparticles have been found to help in early diagnosis of heart attacks (Eurekalert, 2015). Nano-technological transport capsules make it possible to release medications specifically to the target organ(s) only (Siegrist et al., 2007). Just like any other product, ENP based products if taken in toxic quantities or applied to those who do not need it would definitely lead to the occurrence of disastrous effects. Currently a considerable number of people are exposed to the nanoparticles. In 2004, a British report (Occupational Hygiene Review, 2004) following estimates for the United Kingdom were established:

1. 500 workers are directly exposed during nanoparticle production, and 102,000 are possibly exposed during their handling in the industries that use them
2. Within 15 years, approximately 660, 000 workers will be exposed to nanoparticles either by industrial production and/or use.

Not all nanoparticles are equally dangerous to the human health if taken in undesirable quantities either by direct or indirect contact. Some can have acute effect on human health while others can be fatal to the human beings and other living organisms. Moreover, upper intake limits also vary from one nanoparticle to the other. Nanoparticles can affect different parts of the human beings from skin to the brain. Moreover their presence on many consumables is very difficult to find and using any products having excess quantities of nano particles can affect several organs within the human body. A potential route of inhaled nanoparticles within the body is the olfactory nerve; nanoparticles may cross the mucous membrane inside the nose and then reach the brain through the olfactory nerve.

Studies have shown that inhaling nanoparticles can even affect the central nervous system(Maynard et al, 2011; Oberdörster et al., 2005). Moreover, their extremely small size is very tangible to affect the skin and eyes of people exposed to them. (Goddard et al., 2004) It has also been found that increased breathing rate due to exercise further promotes the deposition of these particles in lungs (Jaques et al., 2000). Deposition of nano particles may occur in the cardiovascular system, liver, brain, testis, spleen, stomach, and kidneys. This in turn may lead to apoptosis of cells, inflammation and changes in the immune responses (Maynard et al, 2011; Oberdörster et al., 2005; Fubini, 2010).

However, the potential hazard is always dependent on the type and concentration of the particles, an individual is exposed. Some important nano particles along with their potential hazards are briefly discussed below,

- **TiO_2:** TiO2 nanoparticles are widely used for a variety of applications. Many procedures have been reported for producing TiO2 nanoparticles; most commonly used procedures involve the synthesis by hydrolysis and calcinations (Pottier et al., 2001). However, these procedures might induce toxic amounts of TiO2 in the workers dealing with them. It has been found that, micrometer-sized particles of TiO2 can easily get through the human outermost layer of the epidermis and even into some hair follicles (Lademann et al., 1998). Moreover, The International Agency for Research on Cancer (IARC) has classified Titanium dioxide dust as an IARC Group 2B carcinogen, meaning it is possibly carcinogenic to humans (4). In addition to cancer, titanium dioxide toxicity is also associated with the DNA damage (Fubini et al., 2010).

- **CNTs:** CNTs are widely used in civil engineering applications specifically for ground improvement, water cleaning and air purification. However, their use in civil engineering projects is associated with some risks. To assess the risks of CNTs, several studies on animals in laboratories have been carried out. Moreover, advanced intensive research is also in process. Based on the data available, some of the conclusions regarding adverse effects of CNTs toxicity have been made; a few of them are discussed here. One study showed that carbon nanotubes that look like asbestos can cause asbestos related diseases in humans that inhale them and asbestos related diseases may even include mesothelioma and lung cancer. With nanotubes being used in new technologies like solar panels, construction materials, batteries, medical devices and plastics, the probability that they will enter into our environment is becoming more certain. Furthermore, according to (Lam et al., 2004), it has been found that, single walled carbon nanotubes can cause interstitial inflammation and lesions in rats and topical application of raw single walled carbon nanotubes to nude mice has been shown to cause dermal irritation (Murray et al., 2007). Moreover, multi walled CNTs have also shown some toxicity in rats and have lead to significant inflammation as well as damage to the tissues (Carrero-Sanchez et al., 2006; Poland et al., 2008). Most studies on the toxicological effects of C60 fullerenes suggest that these materials tend to induce oxidative stress in living organisms (Lai et al., 2000; Oberdörster, 2004; Zhu et al., 2006) which is a condition associated with the oxidative damage in a cell, tissue, or any organ, caused by the reactive oxygen species. Though use of CNTs is revolutionary in various disciples, the risks found in lab studies have made it clear to adopt necessary precautions during their application.
- **Zn ENPs:** ZnO ENPs are widely used in pigments, photo catalysts, semiconductors, plastics, ceramics, lubricants, paints and coatings (Chaúque et al., 2014). The harmful effects of zinc ENPs specifically on humans have also been studied. After experiencing a specific dose for a certain time period, the individuals started feeling sore throat, chest tightness, headache, chills and fever (Gordon et al., 1992). Moreover, one study on mice concluded that environmental exposure to Zn ENPs causes lung inflammatory response (Sayes et al., 2007). Wang et al. (2006) also found that Zn ENPs can cause severe

symptoms of lethargy, vomiting, anorexia, and diarrhea, reduction in weight and even death in mice. All these tests on animals have unveiled the potential risks associated with Zinc engineered nano particles and as these are used in lubricants, paints and coatings which due to wreathing and passage of time may peel off from the applied surfaces; chances are that their toxic amount may enter into the environment and affect the human beings.

- **Alumina Nanoparticles:** Alumina nanoparticles are more inflammatory and they also penetrate into brain very easily adopting a number of routes which may include the blood and olfactory nerve. Olfactory neural tracts connect directly to the area of the brain that is most effected by Alzheimer's disease (Blaylock, R. L., 2013).

Nanotechnology Risks to the Environment

Nanotechnology has the potential to improve the inventory storage as well as enhance the ability to grow at higher yields and more variety of crops. This can be a good positive contribution to the environment. Moreover, nanotechnology based improvements in the energy technology can reduce the dependence on fossil fuels by making the photovoltaic energy production competitive with other energy sources. This may improve the renewable energy systems including biomass (Roco et al., 2005).

But as explained earlier, nanotechnology if misused can have serious impacts on the environment and these harmful impacts on the environment are definitely linked with the habitats. It is more like a cycle, nanotechnology may directly affect the human beings or it may first disturb the environment thus ultimately affecting the habitats. Sometimes, nanotechnology is used for the betterment of environment and carelessness and lack of knowledge leads to more harm than good. Water is the most important matter on earth not merely for human beings but for all creatures. One of the fundamental Humanitarian aims is the provision of clean and affordable water to the community. However, it is still a major challenge for the 21st century (Mayo et al., 2007). In recent decades, water treatment using nano-technological based expertise has gained significant attention of researchers all around the world. However, it has been proved that there are abundant challenges faced by water/wastewater treatment nanotechnologies including misuse of nanoparticles, technical hurdles, high cost, and potential environmental and human risk (Qu et al., 2013). The use of engineered nanoparticles and nanomaterials for water treatment and groundwater remediation, as explained earlier in the book, has raised concerns for human exposure. These concerns are based on the fact that nanoparticles will be highly mobile in porous media because of their small size; thus implying a greater potential for exposure as they are dispersed over greater distances and their effective persistence in the environment increases (Guzman et al., 2006). So, it can be said that nanoparticles can enter into the environment from a number of sources but the most likely doorways for nanomaterials into the environment are sewage water and wastes. The production of the raw materials, the manufacture of products with nanomaterials, as well as at the end of the products' lifecycle generates all these wastes. Some specific nanoparticles and their harmful effects on the environment are discussed below,

- **CNTs:** As far as environmental negative effects of CNTs are concerned, certain studies have been done for their evaluation. However, the results are controversial; some studies were even unable to determine any negative effect however others did. One of the important parameter to find the environmental hazards of nanotechnology is to study aqueous creatures in detail. The data available has shown considerable negative effects on the aqueous creatures including the fishes and amphibian larvae when exposed to the nanoparticles.

Actually carbon nano tubes usually do not settle under the action of gravity in water thus they are regarded as surface functionalized and their extremely fine distribution in water remains stable. Such behavior promotes the accumulation of several heavy metals, which then influences their transport in the water bodies and in biological systems (Schierz et al., 2009).

- **Nano-TiO_2:** Titanium dioxide is one of the most commonly investigated nanoparticles and its impacts on environment are well concluded. Research has come up with the establishment of a number of standardized nano TiO_2 tests for fishes, crustaceans and even algae. According to (Battin et al., 2009), microorganisms are very sensitive to nano TiO_2. Furthermore, Nano-$TiO2$ shows photo catalytic behavior under ultra violet radiations which causes the development of reactive oxygen species. These species have the potential to damage the cell membrane of several microorganisms. Furthermore, many studies have also been conducted for the simulation of actual condition in natural running waters on the laboratory scale, which have showed that $TiO2$ nanoparticles and small concentrations of larger naturally developed agglomerates can both damage the cell membranes of living organisms. Damage to these living organisms is definitely a big threat to the environment.

The $TiO2$ nanoparticles can also attach themselves to the chitinous exoskeleton of the animals leading to obstruct molting, which is necessary for the growth in juveniles. This phenomenon may kill such animals thus having serious negative effects on the environmental balance in long term. Regarding this particular molting obstruction effect in juveniles, dose of the nano-TiO2 was kept at 0.24mg/liter in one study and a comparison was also established between dosages of nano and larger form of the TiO2 particles; nano- TiO2 proved to be twice harmful at this dose as compared to the larger forms (Dabrunz et al., 2011).

- **Nanosilver:** The use of nanosilver technology has more recently been expanded from medical field, textiles to personal electronic devices and appliances such as humidifiers, air cleaners, room sanitizers and water purification units. Silver nanoparticles have antimicrobial properties and may also have some role to play in preventing or managing infections (Risk Science Center). However, their uncontrolled discharge in the environment is harmful. The primary pathway for nanosilver in the environment is the waste-water washed out of textiles, cosmetic units and several other industries. Furthermore, nanosilver in waste water has been identified by an international group of researchers from different scientific disciplines as one of 15 areas of concern that can threaten biological diversity (Sutherland et al., 2009) and it is well known that biodiversity is very important for maintaining balance of the ecosystem, provision of biological sources and for social benefits (recreation, cultural value, research). Once bio diversity is disturbed, the whole environment starts to suffer.

Microorganisms including bacteria, fungi and algae are all vital part of the environment. They have their own roles and importance in our world. For example, bacteria is the most diverse group of microorganisms; although some are parasitic for the habitats, most of them are either neutral or have beneficial relationship with the humans, animals and plants. Similarly, algae is a source of oxygen for aquaculture and natural food for the cultured animals. It is therefore very important to not disturb their presence up to a certain degree. As the world is developing day by day, these microorganisms are getting very close to lose their balance in the environment. As far as nanotechnology is concerned, it has also shown some negative effects on these microorganisms.

Silver ions from silver compounds or those which develop from nanosilver particles through contact with water are very toxic to the several microorganisms including bacteria, fungi and algae. In addition to these, there are many useful organisms in soil, so when sewage sludge having nanosilver pollutants is spread on fields, it negatively affects these microorganisms. For fishes and crustaceans, even low concentrations of nanosilver are enough to cause considerable damages however for mammals, the material is toxic only at high concentrations. According to (Yin et al., 2011), very limited research work is available on harmful impacts on nanosilver on plants. However the available data at the moment concludes the growth impairment of grass seedlings due to cell damage when exposed to nanosilver in abundant quantities.

- **Fe ENPS:** Zero valent iron ENPs undergo chemical transformation when used in environmental remediation techniques (Zhang, 2003). After transformation these are oxidized to FeO in the reaction path. The oxidized form may be more harmful than their corresponding free metals. Same behavior is also observed when some other metal ENPs are converted to oxides in air or water e.g. Cu, Si. (US EPA, 2007).
- **Alumina Nanoparticles:** It has been found that the root growth of five plant varieties (corn, beans, cucumber, carrots and cabbage) is affected by a brief exposure to alumina nanoparticles (Yang et al., 2005).

Social Risks of Nanotechnology

- **Terrorism:** There are some social concerns about nanotechnology that it may also allow us to create more powerful weapons, both lethal and non-lethal. Some organizations are concerned that the implications of nanotechnology in weaponry should remain in ethical limits. They urge scientists and politicians to examine carefully all the possibilities of nanotechnology before planning to use nanotechnology in a large number of weapons. It is also important to note how uncomplicated is the application of nanotechnology in weaponry. Any easily doable or portable use of nanotechnology in weapons can raise some concerns for illegal use of this technology. Furthermore according to Altmann J. (2002), new options for nuclear artillery might also include nanotechnology materials extraction and processing, weapons production, and perhaps new types of nuclear weapons.
- **Privacy:** As nanotechnology based products reduce in size, spy devices too can become invisible to the naked eye and even more mobile. This would make it easier to invade one's privacy. These devices can even be planted into human bodies and as technology is thriving in the recent era, it won't be outlandish to state that mind controlling may be developed to affect one's thoughts by manipulating the brain processes.

However, social risks are never too difficult to cope with; all it needs is to keep the right information among the right people to avoid any misuse of the technology. As far as other risks are concerned, there is need to adopt a number of specific remedial measures which are discussed in the 'Remedial Measures' section.

Fire and Explosion Risk

Combustible nanomaterial possesses higher risk of fire explosion as compared to the same material in coarser size. Rate of combustion also increases leading to the possibility of relatively inert materials becoming more reactive in the nanometer size range. Futhermore, combustion leads to the dispersion of materials in air and here again naoparticles offer more safety risk when dispersed in air as compared to the non-nanomaterials having same compositions. For instance, nanoscale Al/MoO3 thermites ignite more than 300 times faster than corresponding micrometer-scale material (Granier et al., 2004)

REMEDIAL MEASURES

Advancement in Analysis Techniques

Conventionally, it is a good approach to measure the toxicity of nano-particles in terms of their mass. However, it may also be reasonable to measure the doses in terms of the number of nanoparticles as well as their surface area. Conclusions about the toxicity potential of nano-particles on human beings are concluded from the tests carried out on animals. However, it is possible that certain nano-particles affect only specific form of living organisms and these animal test based results may lead to erroneous conclusions. Moreover, sometimes tests are carried out on healthy animals. Studies have shown that some toxins may not be harmful for healthy organisms but may have ruinous effects on unhealthy or already diseased organisms (European Comission). There is need to bring new advancements particularly related to the nanoparticles toxicity tests to detect their possible harmful effects on human beings. Long term studies are also necessary to determine delayed impacts of engineered nano particles and to help determine prospective adaptive mechanisms. Additionally, more studies on bioaccumulation of ENPs in the food chain and their interaction with other pollutants are also mandatory.

As far as instruments for the exposure assessment are concerned, there is need to bring new advancements because nanotechnologies are diverse and the exposure to nanoparticles also vary widely, so multiple sensors operating under different conditions are required. Three zones stand out as productive ground for this particular novel research: monitors for airborne exposure, detectors for waterborne nanomaterials, and smart sensors that can measure both exposure and potential hazards.

Continuous Research and Education

Careful examination of the existing data and planning of the new research in this field is required. Research should also examine how social and economic forces affect allocation of benefits and risks, both across social classes and across societies of the world. There is need to introduce several worker transition programs and postdoctoral trainings in the physical sciences, but more important is to inaugurate trainings that combine specialties in the social sciences and humanities with knowledge of nanoscience and nano-engineering. For the completion of these projects, there is need to develop infrasturtucre of nanoscience laboratories, shared social science information systems and simulated virtual laboratories. Moreover, several new statistical softwares should be developed for creating linkage between nanoparticles and their harmful effects to human health and the environment.

It is also important how early a risk is identified and strategy is established; it may save additional costs which usually occur for risk remediation once the technology has already in use (Dunphy et al., 2006). For this purpose, different well funded pilot projects in different countries must be introduced.

Rules and Legislations

Most of the government regulatory frameworks which are used today were generated almost 40 years ago when nanotechnology was not in application. Therefore, these frameworks lack the coverage of many unique properties which are sole feature of nanoparticles.

In order to mitigate the risks associated with the use of nano building materials, specific guidelines for nanotechnology should be developed at national level to regulate the use as well as disposal of nanoparticles (Lee et al., 2010). Another suitable approach would be to only allow professionals to work with nanotechnology (Reynolds, 2003). Moreover, it has been found that there is no obligatory registration for any nanoparticle and most of the companies producing the nanomaterials are not willing to disclose their production volumes; this aspect should also be considered critically.

In addition to all this, a federal registry should be established for all companies and organizations manufacturing, importing and supplying products containing nanomaterials. It must register the organization's name and its products containing nanomaterials and all this information should be made publicly available.

Public Awareness and Feedback

Engineers, corporate management and societal policy-makers may not know very much about the best implementation of the possible technological developments, nor about what the indirect or second-order societal effects might occur; for example at the intersection of nanomaterials and nano systems with biological and ecological systems (Roco et al., 2005). So, feedback from well-informed public and international partners has become essential for progress in this field. More interactions between scientists, engineers, economists, health professionals and the public are needed to identify and reach the robust balance between benefits and limiting factors of nanotechnology.

There is need to inform general public about this technology. In recent years, several studies examined public opinion of nanotechnology in Europe and United States. A U.S. survey showed that more than 80% of the respondents indicated that they had heard "little" or "nothing" about nanotechnology (Cobb et al., 2004). In addition to this, results of a 2001 EC Euro-barometer showed that over 65% of the respondents did not understand the topic of nanotechnologies (European Commission, 2001). Recent studies suggest that the public's awareness of nanotechnology is low, and that knowledge about nanotechnology is very limited (Lee et al., 2005). One way to cope with lack of knowledge is to employ social trust when assessing the risks of a new technology (Lam et al., 2004).

However, there are still some people who are aware of this technology and the products based on it. These people should be involved in feedbacks through various survey programs. The consumption of nanotechnology based products by a community must be recorded for study purposes. The record must include the type, duration and quantity of products people use. Any harmful effects people experienced due to nanotechnology based product must also be recorded and analyzed. These measurements would further unclear the risks and uncertainties associated with this technology.

Workers Protection

Like many other emerging technologies, nanotechnology also poses several risks to the workers and their protection remains a key issue. People who work in the molecular nanotechnology field should develop and utilize professional guidelines that are grounded in reliable technology. Persons dealing with the nanotechnology should use PPE and go through regular check ups to avoid any serious harmful impacts.

Workers continuous training is also an important aspect which could minimize the risks of the technology. Furthermore, there is need to develop specific procedures for the installment of various engineering controls including exhaust ventilations and good working practices at those locations where exposure to nanomaterials may occur. Examples of good working practices include cleaning of work areas using HEPA vacuum pickup and wet wiping methods, preventing the consumption of food/beverages in workplaces, providing hand-washing facilities, and providing facilities for showering and changing clothes. Exclusion of pregnant or nursing women from the jobs and availability of enhanced medical surveillance at sites is also mandatory. These are very simple steps but can reduce risks to a considerable degree.

Suitable steps should be taken to lower the risks of worker exposure through the implementation of a risk management program (Schulte et al., 2008). Risk management programs for nanomaterials should be seen as a fundamental part of an overall occupational safety and health program in any organization producing or using nanomaterials or nano-enabled products.

Engineering Control

Engineering control techniques such as isolation of the generation source from the workers to minimize the interaction and application of efficient exhaust ventilation systems for capturing airborne nanomaterials can play an important role in worker's health and safety. Current knowledge indicates that a well-designed exhaust ventilation system with a HEPA filter should effectively remove nanoparticles (Hinds, 1999). Furthermore continuous monitoring of nanoparticles in air using appropriate devices is also very important to avoid any unnecessary effects on workers. (OSHA)

Development of Models and Robust Systems

There is requirement to develop models for predicting the potential impact of engineered nano materials on the human health and the environment. These models would help scientists to assess the safety of many complex multi component and multi functional nano materials. Thinking in terms of life cycles leads to a comprehensive approach to managing risks and benefits. Developing robust ways of evaluating the potential impacts both good and bad of a nano product from its initial manufacture, through its use, to the final disposal will lead to new methodologies that would be widely applicable.

Maintenance of Workplaces

Buildings, machines and all the equipment requires regular maintenance to keep the environment reliable and safe. Maintenance can be proactive for the prevention of machine & structure failures or reactive to repair equipment or building modules. Maintenance can be done in a variety of ways including servicing, inspecting, repairing, adjusting and replacing parts. For example, opening closed production units, replacing filters, removing paints affected by nanoparticles in air, grinding and sanding etc. (OSHA) According to a research based on the questionnaire jointly conducted by the Japan National Institute of Occupational Safety and Health (JNIOSH) and the National Institute of Advanced Industrial Science and Technology (AIST) from September 2007 through to February 2008, it was observed that more than 50% of the occupational health supervisors in workplaces gave a notion that nanoparticles may be leaked outside of the workplaces during the production process and workers may get exposed to the nanoparticles (Report of review panel meetings, 2008).

Waste Monitoring

As nanotechnology is rapidly emerging so the waste in recent era may have considerable amount of nanomaterials in it which can easily become the part of the environment. For example, LEDs contain nano-scale coatings of the semiconductor materials arsenic, phosphorus, gallium and their compounds. Therefore they belong to the waste category requiring special treatment or monitoring. In particular the semiconductor material gallium arsenide is very problematic and could create environmental damage in a normal landfill (Steinfeldt et al., 2004). Therefore, it is very important to monitor the waste. Solid wastes should be packed in preserved containers for handling until incineration or other processing is carried out. Gas emissions and liquid effluents from nanoparticle industries must also be treated and final releases must be monitored (MOE & SD, 2006).

Recycling

An approach to reduce the undesirable quantities of nano-waste in the environment is to recycle it. However sufficient precautions and considerations should be made during recycling. At present knowledge regarding the recycling of nanoparticles may not be sufficient however the scheme seems to have very high potential in it and ongoing research has also shown many positive results.

Manufacturers of nano based materials should make plans for the efficient recovering and recycling of nanomaterials into the product life cycle. They should emphasis the production of those products which allow easy separation and re-use of nanoparticles. This recovery can be done using a variety of ways, two important procedures are briefly explained below:

- **Nanoparticle Recovery using a Micro-Emulsion:** In 2010, researchers from Bristol University published work on the separation of cadmium and zinc nanoparticles using a special solvent. The solvent is a stable micro emulsion of oil in water but when heated it breaks down into two layers and all of the nanoparticles in the solution end up in one of the layers, which is simply separated.
- **Nanoparticle Recovery by Cloud Point Extraction:** Another process was reported by a research team of Pakistan in 2011. The researchers used a technique known as cloud point extraction (CPE) to separate gold and palladium nanoparticles from an aqueous solution. The cloud point of an emulsion is the point when the two phases are on the verge between mixing fully and forming two layers, causing clouding of the solution. In their method, the nanoparticle solution was heated to this cloud point and later on centrifuged for the efficient separation of the layers. This lead to the recovery of nanoparticles from the solution.

REFERENCES

Aitken, R. J., Creely, K. S., & Tran, C. L. (2004). Nanoparticles: an occupational hygiene review, Research report 274. *Institute of Occupational Medicine.* Retrieved May 04, 2015 from http://www.hse.gov.uk/research/rrpdf/rr274.pdf

Altmann, J., & Gubrud, M. A. (2002). *Risks from military uses of nanotechnology–the need for Technology Assessment and Preventive Control. In Nanotechnology–Revolutionary Opportunities and Societal Implications* (pp. 144–148). Luxembourg: European Communities.

Battin, T. J., Kammer, F. V., Weilhartner, A., Ottofuelling, S., & Hofmann, T. (2009). Nanostructured TiO2: Transport behavior and effects on aquatic microbial communities under environmental conditions. *Environmental Science & Technology, 43*(21), 8098–8104. doi:10.1021/es9017046 PMID:19924929

Benn, T. M., & Westerhoff, P. (2008). Nanoparticle silver released into water from commercially available sock fabrics. *Environmental Science & Technology*, *42*(11), 4133–4139. doi:10.1021/es7032718 PMID:18589977

Blaylock, R. L. (n.d.). Impacts of Chemtrails on Human Health. *Nanoaluminum: Neurodegenerative and Neurodevelopmental Effects.*

Carrero-Sanchez, J., Elias, A., Mancilla, R., Arrellin, G., Terrones, H., Laclette, J., & Terrones, M. (2007). Biocompatibility and toxicological studies of carbon nanotubes doped with nitrogen. *Nano Letters*, 1609–1616. PMID:16895344

Chaúque, E. F. C., Zvimba, J. N., Ngila, J. C., & Musee, N. (2014). Stability studies of commercial ZnO engineered nanoparticles in domestic wastewater. *Physics and Chemistry of the Earth Parts A/B/C*, *67*, 140–144. doi:10.1016/j.pce.2013.09.011

Cobb, M. D., & Macoubrie, J. (2004). Public perceptions about nanotechnology: Risks, benefits and trust. *Journal of Nanoparticle Research*, *6*(4), 395–405. doi:10.1007/s11051-004-3394-4

Dabrunz, A., Duester, L., Prasse, C., Seitz, F., Rosenfeldt, R., Schilde, C., & Schulz, R. et al. (2011). Biological surface coating and molting inhibition as mechanisms of TiO2 nanoparticle toxicity in Daphnia magna. *PLoS ONE*, *6*(5), e20112. doi:10.1371/journal.pone.0020112 PMID:21647422

Dangers Come in Small Particles. (n.d.). *Hazards Magazine*. Retrieved 13 September, 2014 from http://www.hazards.org/nanotech/safety.htm

Deliyanni, E. A., Bakoyannakis, D. N., Zouboulis, A. I., & Matis, K. A. (2003). Sorption of As(V) ions by akaganeite-type nanocrystals. *Chemosphere*, *50*(1), 155–163. doi:10.1016/S0045-6535(02)00351-X PMID:12656241

Dunphy Guzman, K. A., Taylor, M. R., & Banfield, J. F. (2006). Environmental risks of nanotechnology: *National nanotechnology initiative funding, 2000-2004*. *Environmental Science & Technology*, *40*(5), 1401–1407. PMID:16568748

European Commission Technical Guidance Document (TGD) on Risk Assessment. (2003). *European Commission Joint Research Center (ECJRC)*. Retrieved from http://ecb.jrc.ec.europa.eu/tgd/

Fubini, B., Ghiazza, M., & Fenoglio, I. (2010). Physico-chemical features of engineered nanoparticles relevant to their toxicity. *Nanotoxicology*, *4*(4), 347–363. doi:10.3109/17435390.2010.509519 PMID:20858045

Goddard, W. A. III, Brenner, D. W., Lyshevski, S. E., & Iafrate, G. J. (2004). Properties of High-Volume Fly Ash Concrete Incorporating Nano-SiO2. *Cement and Concrete Research, 34*(6), 1043–1049. doi:10.1016/j.cemconres.2003.11.013

Gold nanoparticles show promise for early detection of heart attacks. (n.d.). New York University Polytechnic School of Engineering. Retrieved June 20, 2015 from, http://www.eurekalert.org/pub_releases/2015-01/nyup-gns011515.php

Gordon, T., Chen, L., Fine, J., Schlesinger, R., Su, W., Kimmel, T., & Amdur, M. (1992). Pulmonary effects of inhaled zinc-oxide in human-subjects, guinea-pigs, rats, and rabbits. *American Industrial Hygiene Association Journal, 53*(8), 503–509. doi:10.1080/15298669291360030 PMID:1509990

Granier, J. J., & Pantoya, M. L. (2004). Laser ignition of nanocomposite thermites. *Combustion and Flame, 138*(4), 373–382. doi:10.1016/j.combustflame.2004.05.006

Green, M., & Howman, E. (2005). Semiconductor quantum dots and free radical induced DNA nicking. *Chemical Communications (Cambridge), 1*(1), 121–123. doi:10.1039/b413175d PMID:15614393

Hansen, S. F. (2009). *Regulation and risk assessment of nanomaterials: too little, too late.* Department of Environmental Engineering, Technical University of Denmark. Retrieved August 14, 2015 from http://www2.er.dtu.dk/publications/fulltext/2009/ENV2009-069.pdf

Health and Food Safety, Scientific Committees. (n.d.). European Comission. Retrieved August 20, 2015, from http://ec.europa.eu/health/scientific_committees/?opinions_layman/en/nanotechnologies/1-2/

Hinds, W. C. (2012). *Aerosol technology: properties, behavior, and measurement of airborne particles.* John Wiley & Sons.

Hoshino, A., Hanaki, K., Suzuki, K., & Yamamoto, K. (2004). Applications of t-lymphoma labeled with fluorescent quantum dots to cell tracing markers in mouse body. *Biochemical and Biophysical Research Communications, 314*(1), 46–53. doi:10.1016/j.bbrc.2003.11.185 PMID:14715244

International Programme on Chemical Safety. (2004). *IPCS Risk Assessment Terminology.* World Health Organization. Retrieved from http://www.who.int/ipcs/methods/ harmonization/areas/ipcsterminologyparts1and2.pdf

Jaques, P. A., & Kim, C. S. (2000). Measurement of total lung deposition of inhaled ultrafine particles in healthy men and women. *Inhalation Toxicology, 12*(8), 715–731. doi:10.1080/08958370050085156 PMID:10880153

Karn, B., Roco, M., Masciangioli, T., & Savage, N. (2003, May). *Nanotechnology and the Environment*. National Nanotechnology Initiative Workshop.

Kuempel, E. D. (2006). *Titanium dioxide 93*. International Agency for Research on Cancer.

Lademann, J., Weigmann, H. J., Rickmeyer, C., Barthelmes, H., Schaefer, H., Mueller, G., & Sterry, W. (1998). Penetration of titanium dioxide microparticles in a sunscreen formulation into the horny layer and the follicular orifice. *Skin Pharmacology and Applied Skin Physiology*, *12*(5), 247–256. doi:10.1159/000066249 PMID:10461093

Lai, H., Chen, W., & Chiang, L. (2000). Free radical scavenging activity of fullerenol on the ischemia-reperfusion intestine in dogs. *World Journal of Surgery*, *24*(4), 450–454. doi:10.1007/s002689910071 PMID:10706918

Lam, C. W., James, J. T., McCluskey, R., & Hunter, R. L. (2004). Pulmonary toxicity of single-wall carbon nanotubes in mice 7 and 90 days after intratracheal instillation. *Toxicological Sciences*, *77*(1), 126–134. doi:10.1093/toxsci/kfg243 PMID:14514958

Lee, C.-J., Scheufele, D. A., & Lewenstein, B. V. (2005). Public attitudes toward emerging technologies. *Science Communication*, *7*(2), 240–267. doi:10.1177/1075547005281474

Lee, J., Mahendra, S., & Alvarez, P. J. J. (2010). Nanomaterials in the construction industry: A review of their applications and environmental health and safety considerations. *ACS Nano*, *4*(7), 3580–3590. doi:10.1021/nn100866w PMID:20695513

Lovric, J., Bazzi, H., Cuie, Y., Fortin, G., Winnik, F., & Maysinger, D. (2005). Differences in subcellular distribution and toxicity of green and red emitting CdTe quantum dots. *Journal of Molecular Medicine (Berlin, Germany)*, *83*(5), 377–385. doi:10.1007/s00109-004-0629-x PMID:15688234

Maynard, A. D., Warheit, D. B., & Philbert, M. A. (2010). The new toxicology of sophisticated materials: Nanotoxicology and beyond. *Toxicological Sciences*. PMID:21177774

Mayo, J. T., Yavuz, C., Yean, S., Cong, L., Shipley, H., Yu, W., & Colvin, V. L. et al. (2007). The effect of nanocrystalline magnetite size on arsenic removal. *Science and Technology of Advanced Materials*, *8*(1-2), 71–75. doi:10.1016/j.stam.2006.10.005

Murray, A. R., Kisin, E., Kommineni, C., Kagan, V. E., Castranova, V., & Shvedova, A. A. (2007). Single-walled carbon nanotubes induce oxidative stress and inflammation in skin. *The Toxicologist*, *96*, A1406.

Nanomaterials in Maintenance Work: Occupational Risks and Prevention. (n.d.). *European Agency for safety and health at work*. Retrieved July 11, 2015 from https://osha.europa.eu/en/publications/e-facts/e-fact-74-nanomaterials-in-maintenance-work-occupational-risks-and-prevention

Nanotechnologies, Nanoparticles: What Hazards – What Risks? (2006). Ministry of Ecology and Sustainable Development (MOE & SD). Retrieved from http://www.developpement-durable.gouv.fr/IMG/pdf/CPP_NanotechnologiesNanoparticles.pdf

Nawrocki, J., & Kasprzyk-Hordern, B. (2010). The efficiency and mechanisms of catalytic ozonation. *Applied Catalysis B: Environmental*, *99*(1-2), 27–42. doi:10.1016/j.apcatb.2010.06.033

Nielsen, E., Ostergaard, G., & Larsen, J. (2007). *Toxicological Risk Assessment of Chemicals: A Practical Guide* (pp. 2–3). New York, NY: Informa Healthcare.

Nielsen, E., Ostergaard, G., & Larsen, J. C. (2008). *Toxicological risk assessment of chemicals: A practical guide*. CRC Press. doi:10.1201/9781420006940

Oberdörster, E. (2004). Manufactured nanomaterials (fullerenes, C60) induce oxidative stress in juvenile largemouth bass. *Environmental Health Perspectives*, *112*(10), 1058–1062. doi:10.1289/ehp.7021 PMID:15238277

Oberdörster, E. (2004). *Toxicity of C60 Fullerenes to Two Aquatic Species: Daphnia and Largemouth Bass*. Anaheim, CA: American Chemical Society.

Oberdörster, G., Oberdörster, E., & Oberdörster, J. (2005). Nanotoxicology: An emerging discipline evolving from studies of ultrafine particles. *Environmental Health Perspectives*, *113*(7), 823–839. doi:10.1289/ehp.7339 PMID:16002369

Oberdorster, G., Oberdorster, E., & Oberdorster, J. (2005). Nanotoxicology: An emerging discipline evolving from studies of ultrafine particles. *Environmental Health Perspectives*, *113*(7), 823–839. doi:10.1289/ehp.7339 PMID:16002369

Pan, B., & Xing, B. (2008). Adsorption mechanisms of organic chemicals on carbon nanotubes. *Environmental Science & Technology*, *42*(24), 9005–9013. doi:10.1021/es801777n PMID:19174865

Pierce, J. (2004). Safe as sunshine? *The Engineer*. Retrieved August 23, 2015 from http://www2.er.dtu.dk/publications/fulltext/2007/MR2007-239.pdf

Poland, C., Duffin, R., Kinloch, I., Maynard, A., Wallace, W., Seaton, A., & Donaldson, K. et al. (2008). Carbon nanotubes introduced into the abdominal cavity of mice show asbestos-like pathogenicity in a pilot study. *Nature Nanotechnology*, *3*(7), 423–428. doi:10.1038/nnano.2008.111 PMID:18654567

Pottier, A., Chanéac, C., Tronc, E., Mazerolles, L., & Jolivet, J. P. (2001). Synthesis of brookite TiO2 nanoparticlesby thermolysis of TiCl4 in strongly acidic aqueous media. *Journal of Materials Chemistry, 11*(4), 1116–1121. doi:10.1039/b100435m

Qu, X., Alvarez, P. J., & Li, Q. (2013). Applications of nanotechnology in water and wastewater treatment. *Water Research, 47*(12), 3931–3946. doi:10.1016/j.watres.2012.09.058 PMID:23571110

Rachel's Environment and Health News. (2003). Retrieved from http://www.rachel.org/files/rachel/Rachels_Environment_Health_News_2362.pdf

Reynolds, G. H. (2003). Nanotechnology and regulatory policy: Three futures. *Harvard Journal of Law & Technology, 17*, 179.

Roco, M. C., & Bainbridge, W. S. (2005). Societal implications of nanoscience and nanotechnology: Maximizing human benefit. *Journal of Nanoparticle Research, 7*(1), 1–13. doi:10.1007/s11051-004-2336-5

Sayes, C., Marchione, A., Reed, K., & Warheit, D. (2007). Comparative pulmonary toxicity assessments of C60 water suspensions in rats: Few differences in fullerene toxicity in vivo in contrast to in vitro profiles. *Nano Letters, 7*(8), 2399–2406. doi:10.1021/nl0710710 PMID:17630811

Schierz, A., & Zänker, H. (2009). Aqueous suspensions of carbon nanotubes: Surface oxidation, colloidal stability and uranium sorption. *Environmental Pollution, 157,* 1088-1094.

Siegrist, M., & Cvetkovich, G. (2000). Perception of hazards: The role of social trust and knowledge. *Risk Analysis, 20*(5), 713–719. doi:10.1111/0272-4332.205064 PMID:11110217

Siegrist, M., Keller, C., Kastenholz, H., Frey, S., & Wiek, A. (2007). Laypeople's and experts' perception of nanotechnology hazards. *Risk Analysis, 27*(1), 59–69. doi:10.1111/j.1539-6924.2006.00859.x PMID:17362400

Steinfeldt, M., Von Gleich, A., Petschow, U., Haum, R., Chudoba, T., & Haubold, S. (2004). Nachhaltigkeitseffekte durch Herstellung und Anwendung nanotechnischer Produkte. Institut für ökologische Wirtschaftsforschung GmbH, Berlin, Germany.

Sutherland, W. J., Bardsley, S., Bennun, L., Clout, M., Côté, I. M., Depledge, M. H., & Gibbons, D. W. et al. (2011). Horizon scan of global conservation issues for 2011. *Trends in Ecology & Evolution, 26*(1), 10–16. doi:10.1016/j.tree.2010.11.002 PMID:21126797

Trouiller, B., Reliene, R., Westbrook, A., Solaimani, P., & Schiestl, R. H. (2009). Titanium dioxide nanoparticles induce DNA damage and genetic instability in vivo in mice. *Cancer Research, 69*(22), 8784–8789. doi:10.1158/0008-5472.CAN-09-2496 PMID:19887611

U.S. Environmental Protection Agency Nanotechnology White Paper. (2007). Retrieved May 17, 2015 from http://www.epa.gov/osa/pdfs/nanotech/epa-nanotechnology-whitepaper-0207.pdf

Wang, B., Feng, W., Wang, T., Jia, G., Wang, M., Shi, J., & Chai, Z. et al. (2006). Acute toxicity of nano- and micro-scale zinc powder in healthy adult mice. *Toxicology Letters, 161*(2), 115–123. doi:10.1016/j.toxlet.2005.08.007 PMID:16165331

Wiesner, M. R., Lowry, G. V., Alvarez, P., Dionysiou, D., & Biswas, P. (2006). Assessing the risks of manufactured nanomaterials. *Environmental Science & Technology, 40*(14), 4336–4345. doi:10.1021/es062726m PMID:16903268

Yang, L., & Watts, D. J. (2005). Particle surface characteristics may play an important role in phytotoxicity of alumina nanoparticles. *Toxicology Letters, 158*(2), 122–132. doi:10.1016/j.toxlet.2005.03.003 PMID:16039401

Yin, L., Cheng, Y., Espinasse, B., Colman, B. P., Auffan, M., Wiesner, M., & Bernhardt, E. S. et al. (2011). More than the ions: The effects of silver nanoparticles on Lolium multiflorum. *Environmental Science & Technology, 45*(6), 2360–2367. doi:10.1021/es103995x PMID:21341685

Zhang, W. (2003). Nanoscale iron particles for environmental remediation: An overview. *Journal of Nanoparticle Research, 5*(3/4), 323–332. doi:10.1023/A:1025520116015

Zhu, S., Oberdorster, E., & Haasch, M. (2006). Toxicity of an engineered nanoparticle (Fullerene, C60) in two aquatic species, daphnia and fathead minnow. *Marine Environmental Research, 62*, S5–S9. doi:10.1016/j.marenvres.2006.04.059 PMID:16709433

Compilation of References

Abdel Hameed, R. S., Abu-Nawwasb, A. H., & Shehataa, H. A. (2013). Nano-composite as corrosion inhibitors for steel alloys in different corrosive media. *Advances in Applied Science*, *4*, 126–129.

Abdel-Raouf, N., Al-Homaidan, A. A., & Ibraheem, I. B. M. (2012). Microalgae and wastewater treatment. *Saudi Journal of Biological Sciences*, *19*(3), 257–275. doi:10.1016/j.sjbs.2012.04.005 PMID:24936135

Abdullah, M. E., Zamhari, K. A., Buhari, R., Kamaruddin, N. H. M., Nayan, N., Hainin, M. R., ... & Yusoff, N. I. M. (2015). A review on the exploration of nanomaterials application in pavement engineering. *Jurnal Teknologi, 73*(4).

Abdullahi, M. (2006). Properties of Some Natural Fine Aggregates in Minna, Nigeria and Environs. *Leonardo. Journal of Science*, (8), 1–6.

Abedini, A., Daud, A. R., Abdul Hamid, M., Kamil Othman, N., & Saion, E. (2013). A review on radiation-induced nucleation and growth of colloidal metallic nanoparticles. *Nanoscale Research Letters*, *8*(1), 474. doi:10.1186/1556-276X-8-474 PMID:24225302

Adityamadhav. (2011). Retrieved August 12, 2015, from http://commons.wikimedia.org/wiki/File%3AIgneousrock_(Tuff_type)_at_Rushikonda%2C_Visakhapatnam.jpg

Afkhami, A., Bagheri, H., & Madrakian, T. (2011). Alumina nanoparticles grafted with functional groups as a new adsorbent in efficient removal of formaldehyde from water samples. *Desalination*, *281*, 151–158. doi:10.1016/j.desal.2011.07.052

Ai, L., Li, M., & Long, L. (2011). Adsorption of Methylene Blue from Aqueous Solution with Activated Carbon/Cobalt Ferrite/Alginate Composite Beads: Kinetics, Isotherms, and Thermodynamics. *Journal of Chemical & Engineering Data*, *56*(8), 3475–3483. doi:10.1021/je200536h

Aitken, R. J., Creely, K. S., & Tran, C. L. (2004). Nanoparticles: an occupational hygiene review, Research report 274. *Institute of Occupational Medicine*. Retrieved May 04, 2015 from http://www.hse.gov.uk/research/rrpdf/rr274.pdf

Akkaya, Y., Shah, S. P., & Ghandehari, M. (2003). Influence of fiber dispersion on the performance of microfiber reinforced cement composites. *ACI Special Publications*, *216*, 1–18.

Ali, I. (2012). New Generation Adsorbents for Water Treatment. *Chemical Reviews*, *112*(10), 5073–5091. doi:10.1021/cr300133d PMID:22731247

Ali, I., & Gupta, V. K. (2007). Advances in water treatment by adsorption technology. *Nature Protocols*, *1*(6), 2661–2667. doi:10.1038/nprot.2006.370 PMID:17406522

Alipour, V., Nasseri, S., Nodehi, R. N., Mahvi, A. H., & Rashidi, A. (2014). Preparation and application of oyster shell supported zero valent nano scale iron for removal of natural organic matter from aqueous solutions. *Journal of Environmental Health Science & Engineering*, *12*(146), 1–8. PMID:25648623

Al-Johani, H., & Salam, M. A. (2011). Kinetics and thermodynamic study of aniline adsorption by multi-walled carbon nanotubes from aqueous solution. *Journal of Colloid and Interface Science*, *360*(2), 760–767. doi:10.1016/j.jcis.2011.04.097 PMID:21620412

Alkhateb, H., Al-Ostaz, A., Cheng, A. H. D., & Li, X. (2013). Materials Genome for Graphene-Cement Nanocomposites. *Journal of Nanomechanics and Micromechanics*, *3*(3), 67–77. doi:10.1061/(ASCE)NM.2153-5477.0000055

Al-Mosawe, A., Al-Mahaidi, R., & Zhao, X. L. (2015). Effect of CFRP properties, on the bond characteristics between steel and CFRP laminate under quasi-static loading. *Construction & Building Materials*, *98*, 489–501. doi:10.1016/j.conbuildmat.2015.08.130

Al-Rub, R. K. A., Ashour, A. I., & Tyson, B. M. (2012). On the aspect ratio effect of multi-walled carbon nanotube reinforcements on the mechanical properties of cementitious nanocomposites. *Construction & Building Materials*, *35*, 647–655.

Al-Rub, R. K. A., Tyson, B. M., Yazdanbakhsh, A., & Grasley, Z. (2012). Mechanical Properties of Nanocomposite Cement Incorporating Surface-Treated and Untreated Carbon Nanotubes and Carbon Nanofibers. *Journal of Nanomechanics and Micromechanics*, *2*(1), 1–6.

Altmann, J., & Gubrud, M. A. (2002). *Risks from military uses of nanotechnology–the need for Technology Assessment and Preventive Control. In Nanotechnology–Revolutionary Opportunities and Societal Implications* (pp. 144–148). Luxembourg: European Communities.

Aly, M., Hashmi, M. S. J., Olabi, A. G., Messeiry, M., & Hussain, A. I. (2011). Effect of nano clay particles on mechanical, thermal and physical behaviours of waste-glass cement mortars. *Materials Science and Engineering A*, *528*(27), 7991–7998.

Amirkhanian, A. N., Xiao, F., & Amirkhanian, S. N. (2011). Characterization of unaged asphalt binder modified with carbon nano particles. *International Journal of Pavement Research and Technology*, *4*(5), 281.

Anderson, J., & Shiers, D. (2009). *Green guide to specification*. John Wiley & Sons.

Anwar Khitab. (2012). *Materials of Construction (Classical and Novel)*. Allied Book Company Brick-History. Retrieved July 13, 2015 from http://science.jrank.org/pages/1022/Brick-History.html

Compilation of References

Arafa, M. D., DeFazio, C., & Balaguru, B. (2005). Nano-composite coatings for transportation infrastructures: Demonstration projects.*2nd International Symposium on Nanotechnology in Construction*, Bilbao, Spain.

Aragão, F. T. S., Lee, J., Kim, Y. R., & Karki, P. (2010). Material-specific effects of hydrated lime on the properties and performance behavior of asphalt mixtures and asphaltic pavements. *Construction & Building Materials*, *24*(4), 538–544. doi:10.1016/j.conbuildmat.2009.10.005

Arefi, M. R., Javeri, M. R., & Mollaahmadi, E. (2011). To study the effect of adding Al_2O_3 nanoparticles on the mechanical properties and microstructure of cement mortar. *Life Science Journal*, *8*(4), 613–617.

ARI News. (2005). *Nanotechnology in Construction – One of the Top Ten Answers to World's Biggest Problems.* Retrieved July 7, 2015, from www.aggregateresearch.com/article.asp?id=6279

Arkema. (2011). *Analytical Techniques of Structural Characterization* [Online]. Retrieved on October 2011, from http://www.arkema-inc.com/

Arulraj, G. P., & Carmichael, M. J. (2011). Effect of nano-fly ash on strength of concrete. *International Journal of Civil & Structural Engineering*, *2*(2), 475–482.

Asgher, M., & Bhatti, H. N. (2012). Evaluation of thermodynamics and effect of chemical treatments on sorption potential of Citrus waste biomass for removal of anionic dyes from aqueous solutions. *Ecological Engineering*, *38*(1), 79–85. doi:10.1016/j.ecoleng.2011.10.004

Ashby, M. F., Ferreira, P. J., & Schodek, D. L. (2009). *Nanomaterials, Nanotechnologies and Design. An Introduction for Engineers and Architects*. Boston: Elsevier.

Australian Microscopy and Microanalysis Research Facility (AMMRF). (n.d.). Retrieved from http://www.ammrf.org.au/myscope/sem/practice/principles/layout.php

Azzam, E. M. S., Sayyah, S. M., & Taha, A. S. (2013). Fabrication and characterization of nanoclay composites using synthesized polymeric thiol surfactants assembled on gold nanoparticles. *Egyptian Journal of Petroleum*, *22*(4), 493–499. doi:10.1016/j.ejpe.2013.11.011

Baeza, F. J., Galao, O., Zornoza, E., & Garcés, P. (2013). Multifunctional Cement Composites Strain and Damage Sensors Applied on Reinforced Concrete (RC) Structural Elements. *Materials (Basel)*, *6*(3), 841–855.

Bai, C. (2000). *Scanning tunneling microscopy and its application* (Vol. 32). New York: Springer-Verlag Berlin Heidelberg.

Bajpai, A. K., & Rajpoot, M. (1999). Adsorption Techniques – A Review. *Journal of Scientific and Industrial Research*, *58*(11), 844–860.

Balaguru, P. N. (2005). *Nanotechnology and concrete: Background, opportunities and challenges*. Paper presented at the International Conference on Applications of Nanotechnology in Concrete Design. doi:10.1680/aonicd.34082.0012

Balaguru, P., & Chong, K. (2006). Nanotechnology and concrete: Research opportunities. *Proceedings of ACI Session on Nanotechnology of Concrete: Recent Developments and Future Perspectives.*

Ballari, M. M., Hunger, M., Hüsken, G., & Brouwers, H. (2010). Modelling and experimental study of the NOx photocatalytic degradation employing concrete pavement with titanium dioxide. *Catalysis Today, 151*(1), 71–76.

Banerjee, S., Gautam, R. K., Jaiswal, A., & Chattopadhyaya, M. C. (2014). Adsorption characteristics of a metal-organic framework material for the removal of Acid Orange 10 from aqueous solutions. *Journal of the Indian Chemical Society, 91*(8), 1491–1499.

Banerjee, S., Gautam, R. K., Jaiswal, A., Chattopadhyaya, M. C., & Sharma, Y. C. (2015). Rapid scavenging of methylene blue dye from a liquid phase by adsorption on alumina nanoparticles. *RSC Advances, 5*(19), 14425–14440. doi:10.1039/C4RA12235F

Bangi, U. K. H., Park, C. S., Baek, S., & Park, H. (2013). Sol–gel synthesis of high surface area nanostructured zirconia powder by surface chemical modification. *Powder Technology, 239*, 314–318. doi:10.1016/j.powtec.2013.02.014

Barbhuiya, S., Mukherjee, S., & Nikraz, H. (2014). Effects of nano-Al_2O_3 on early-age microstructural properties of cement paste. *Construction & Building Materials, 52*, 189–193.

Barnett, S. J., Lataste, J. F., Parry, T., Millard, S. G., & Soutsos, M. N. (2010). Assessment of fibre orientation in ultra high performance fibre reinforced concrete and its effect on flexural strength. *Materials and Structures/Materiaux et Constructions, 43*(7), 1009-1023.

Barone, V., Hod, O., & Scuseria, G. E. (2006). Electronic structure and stability of semiconducting graphene nanoribbons. *Nano Letters, 6*(12), 2748–2754. doi:10.1021/nl0617033 PMID:17163699

Bartos, P. J. (2009). Nanotechnology in construction: a roadmap for development. In Z. Bittnar, P. J. M. Bartos, J. Nemecek, V. Smilauer, & J. Zeman (Eds.), *Nanotechnology in Construction 3* (pp. 15–26). Berlin, Germany: Springer.

Bartos, P. J. M. (2006). Nanotechnology in construction: A roadmap for development. *Proceedings of ACI Session on Nanotechnology of Concrete: Recent Developments and Future Perspectives.*

Battin, T. J., Kammer, F. V., Weilhartner, A., Ottofuelling, S., & Hofmann, T. (2009). Nanostructured TiO2: Transport behavior and effects on aquatic microbial communities under environmental conditions. *Environmental Science & Technology, 43*(21), 8098–8104. doi:10.1021/es9017046 PMID:19924929

Bear, J. (2013). *Dynamics of fluids in porous media*. Courier Corporation.

Becher, P. F. (1991). Microstructural design of toughened ceramics. *Journal of the American Ceramic Society, 74*(2), 255–269. doi:10.1111/j.1151-2916.1991.tb06872.x

Beetz, T., & Jacobsen, C. (2003). Soft X-ray radiation-damage studies in PMMA using a cryo-STXM. *Synchrotron Radiation, 10*(3), 280–283. doi:10.1107/S0909049503003261 PMID:12714762

Compilation of References

Behfarnia, K., & Salemi, N. (2013). The effects of nano-silica and nano-alumina on frost resistance of normal concrete. *Construction & Building Materials*, *48*, 580–584.

Belytschko, T., Xiao, S. P., Schatz, G. C., & Ruoff, R. S. (2002). Atomistic simulations of nanotube fracture. *Physical Review B: Condensed Matter and Materials Physics*, *65*(23), 2354301–2354308. doi:10.1103/PhysRevB.65.235430

Benn, T. M., & Westerhoff, P. (2008). Nanoparticle silver released into water from commercially available sock fabrics. *Environmental Science & Technology*, *42*(11), 4133–4139. doi:10.1021/es7032718 PMID:18589977

Bensted, J., & Barnes, P. (2002). *Structure and Performance of Cements*. Spon Press.

Berber, S., Kwon, Y.-K., & Tomanek, D. (2000). Unusually high thermal conductivity of carbon nanotubes. *Physical Review Letters*, *84*(20), 4613–4616. doi:10.1103/PhysRevLett.84.4613 PMID:10990753

Berkovich, E. (1950). Three-faceted diamond pyramid for studying microhardness by indentation. *Zavodskaya Laboratoria*, *13*(3), 345–347.

Berra, M., Carassiti, F., Mangialardi, T., Paolini, A. E., & Sebastiani, M. (2012). Effects of nano-silica addition on workability and compressive strength of Portland cement pastes. *Construction & Building Materials*, *35*, 666–675.

Bhargavi, R. J., Maheshwari, U., & Gupta, S. (2015). Synthesis and use of alumina nanoparticles as an adsorbent for the removal of Zn(II) and CBG dye from wastewater. *International Journal of Industrial Chemistry*, *6*(1), 31–41. doi:10.1007/s40090-014-0029-1

Bhatnagar, A., & Sillanpää, M. (2011). A review of emerging adsorbents for nitrate removal from water. *Chemical Engineering Journal*, *168*(2), 493–504. doi:10.1016/j.cej.2011.01.103

Bhuvaneshwari, B., Sasmal, S., & Iyer, N. R. (2011, July). Nanoscience to nanotechnology for civil engineering: proof of concepts. In *Proceedings of the 4th WSEAS International Conference on Recent Researches in Geography, Geology, Energy, Environment and Biomedicine (GEMESED'11)* (pp. 230-235).

Biedermann, L. B., Bolen, M. L., Capano, M. A., Zemlyanov, D., & Reifenberger, R. G. (2009). Insights into few-layer epitaxial graphene growth on 4 H-SiC (000 1⁻) substrates from STM studies. *Physical Review B: Condensed Matter and Materials Physics*, *79*(12), 125411. doi:10.1103/PhysRevB.79.125411

Bilba, K., Arsene, M. A., & Ouensanga, A. (2003). Sugar cane bagasse fibre reinforced cement composites. Part I. Influence of the botanical components of bagasse on the setting of bagasse/cement composite. *Cement and Concrete Composites*, *25*(1), 91–96. doi:10.1016/S0958-9465(02)00003-3

Bio-imaging Unit. (n.d.). Newcastle Biomedicine, Newcastle University. Retrieved from http://www.ncl.ac.uk/bioimaging/techniques/confocal/

Birgisson, B., Taylor, P., Armaghani, J., & Shah, S. P. (2010). American road map for research for nanotechnology-based concrete materials. *Transportation Research Record*, *2142*, 130–137. doi:10.3141/2142-20

Blaylock, R. L. (n.d.). Impacts of Chemtrails on Human Health. *Nanoaluminum: Neurodegenerative and Neurodevelopmental Effects*.

Boening, D. W. (2000). Ecological effects, transport, and fate of mercury: A general review. *Chemosphere*, *40*(12), 1335–1351. doi:10.1016/S0045-6535(99)00283-0 PMID:10789973

Bokare, A. D., Chikate, R. C., Rode, C. V., & Paknikar, K. M. (2007). Effect of Surface Chemistry of Fe-Ni Nanoparticles on Mechanistic Pathways of Azo Dye Degradation. *Environmental Science & Technology*, *41*(21), 7437–7443. doi:10.1021/es071107q PMID:18044523

Boparai, H. K., Joseph, M., & O'Carrol, D. M. (2011). Kinetics and thermodynamics of cadmium ion removal by adsorption onto nano zerovalent iron particles. *Journal of Hazardous Materials*, *186*(1), 458–465. doi:10.1016/j.jhazmat.2010.11.029 PMID:21130566

Borja, R. I. (2011). *Multiscale and Multiphysics Processes in Geomechanics: Results of the Workshop on Multiscale and Multiphysics Processes in Geomechanics, Stanford, June 23-25, 2010*. Springer.

Borm, P. J., Robbins, D., Haubold, S., Kuhlbusch, T., Fissan, H., Donaldson, K., & Krutmann, J. et al. (2006). The potential risks of nanomaterials: A review carried out for ECETOC. *Particle and Fibre Toxicology*, *3*(1), 1. doi:10.1186/1743-8977-3-11 PMID:16907977

Bouhicha, M., Aouissi, F., & Kenai, S. (2005). Performance of composite soil reinforced with barley straw. *Cement and Concrete Composites*, *27*(5), 617–621. doi:10.1016/j.cemconcomp.2004.09.013

Brakenhoff, G. J., Voort, H. T. M., Spronsen, E. A., & Nanninga, N. (1989). Three-dimensional imaging in fluorescence by confocal scanning microscopy. *Journal of Microscopy*, *153*(2), 151–159. doi:10.1111/j.1365-2818.1989.tb00555.x PMID:2651673

Brandt, A. M. (2008). Fibre reinforced cement-based (FRC) composites after over 40 years of development in building and civil engineering. *Composite Structures*, *86*(1-3), 3–9. doi:10.1016/j.compstruct.2008.03.006

Broekhuizen, P. V., Broekhuizen, F. V., Cornelissen, R., & Reijnders, L. Use of nanomaterials in the European construction industry and some occupational health aspects there of. *Journal of Nanoparticle Research*. doi:10.1007/s11051-010-0195-9

Brown, L., Sanchez, F., Kosson, D., & Arnold, J. (2013). Performance of carbon nanofiber-cement composites subjected to accelerated decalcification. In *Proceedings of EPJ Web of Conferences*. London, UK: EDP Sciences.

Bruanuer, S., Emmett, P. H., & Teller, E. (1938). Adsorption of gases in multimolecular layers. *Journal of the American Chemical Society*, *60*(2), 309–316. doi:10.1021/ja01269a023

Compilation of References

Bushby, A. J., P'ng, K. M., Young, R. D., Pinali, C., Knupp, C., & Quantock, A. J. (2011). Imaging three-dimensional tissue architectures by focused ion beam scanning electron microscopy. *Nature Protocols*, *6*(6), 845–858. doi:10.1038/nprot.2011.332 PMID:21637203

Cai, W., Gao, T., Hong, H., & Sun, J. (2008). Applications of gold nanoparticles in cancer nanotechnology. *Nanotechnology, Science and Applications*, *1*, 17. PMID:24198458

Cai, W., Hsu, A. R., Li, Z. B., & Chen, X. (2007). Are quantum dots ready for in vivo imaging in human subjects? *Nanoscale Research Letters*, *2*(6), 265–281. doi:10.1007/s11671-007-9061-9 PMID:21394238

Cai, W., Shin, D. W., Chen, K., Gheysens, O., Cao, Q., Wang, S. X., & Chen, X. et al. (2006). Peptide-labeled near-infrared quantum dots for imaging tumor vasculature in living subjects. *Nano Letters*, *6*(4), 669–676. doi:10.1021/nl052405t PMID:16608262

Camiletti, J., Soliman, A. M., & Nehdi, M. L. (2013). Effects of nano- and micro-limestone addition on early-age properties of ultra-high-performance concrete. *Materials and Structures*, *46*(6), 881–898.

Campillo, I., Dolado, J. S., & Porro, A. (2004). High-performance nanostructured materials for construction. In Nanotechnology in Construction. Royal Society of Chemistry.

Campillo, I., Guerrero, A., Dolado, J. S., Porro, A., Ibáñez, J. A., & Goñi, S. (2007). Improvement of initial mechanical strength by nanoalumina in belite cements. *Materials Letters*, *61*(8), 1889–1892.

Cao, C. Y., Qu, J., Yan, W. S., Zhu, J. F., Wu, Z. Y., & Song, W. G. (2012). Low-Cost Synthesis of Flowerlike α-Fe2O3 Nanostructures for Heavy Metal Ion Removal: Adsorption Property and Mechanism. *Langmuir*, *28*(9), 4573–4579. doi:10.1021/la300097y PMID:22316432

Cao, G. (2006). *Nanostructures & Nanomaterials: Synthesis, Properties & Applications* (1st ed.). London: Imperial College Press.

Cárdenas, C., Tobón, J. I., García, C., & Vila, J. (2012). Functionalized building materials: Photocatalytic abatement of NOx by cement pastes blended with TiO_2 nanoparticles. *Construction & Building Materials*, *36*, 820–825.

Carlsson, K., Danielsson, P. E., Liljeborg, A., Majlöf, L., Lenz, R., & Åslund, N. (1985). Three-dimensional microscopy using a confocal laser scanning microscope. *Optics Letters*, *10*(2), 53–55. doi:10.1364/OL.10.000053 PMID:19724343

Carrero-Sanchez, J., Elias, A., Mancilla, R., Arrellin, G., Terrones, H., Laclette, J., & Terrones, M. (2007). Biocompatibility and toxicological studies of carbon nanotubes doped with nitrogen. *Nano Letters*, 1609–1616. PMID:16895344

Cassar, L. (2005). Nanotechnology and photocatalysis in cementitious materials.*2nd International Symposium on Nanotechnology in Construction*, Bilbao, Spain.

Chandra, N., Singh, D. K., Sharma, M., Upadhyay, R. K., Amritphale, S. S., & Sanghi, S. K. (2010). Synthesis and characterization of nano-sized zirconia powder synthesized by single emulsion-assisted direct precipitation. *Journal of Colloid and Interface Science, 342*(2), 327–332. doi:10.1016/j.jcis.2009.10.065 PMID:19942226

Chang, T., Shih, J., Yang, K., & Hsiao, T. (2007). Material properties of Portland cement paste with nano-montmorillonite. *Journal of Materials Science, 42*(17), 7478–7487.

Chaunsali, P., & Peethamparan, S. (2013). Influence of the composition of cement kiln dust on its interaction with fly ash and slag. *Cement and Concrete Research, 54*, 106–113. doi:10.1016/j.cemconres.2013.09.001

Chaúque, E. F. C., Zvimba, J. N., Ngila, J. C., & Musee, N. (2014). Stability studies of commercial ZnO engineered nanoparticles in domestic wastewater. *Physics and Chemistry of the Earth Parts A/B/C, 67*, 140–144. doi:10.1016/j.pce.2013.09.011

Chen, L., Mindess, S., & Morgan, D. R. (1995). Comparative toughness testing of fiber reinforced concrete. *Testing of Fiber Reinforced Concrete, 47*-75.

Chen, S. J., & Zhang, X. N. (2012, October). Mechanics and Pavement Properties Research of Nanomaterial Modified Asphalt. In *Advanced Engineering Forum* (Vol. 5, pp. 259-264). doi:10.4028/www.scientific.net/AEF.5.259

Chen, C. J. (1993). *Introduction to scanning tunneling microscopy* (Vol. 227). New York: Oxford University Press.

Cheng, J., Shen, J., & Xiao, F. (2011). Moisture susceptibility of warm-mix asphalt mixtures containing nanosized hydrated lime. *Journal of Materials in Civil Engineering, 23*(11), 1552–1559. doi:10.1061/(ASCE)MT.1943-5533.0000308

Chen, J., Kou, S., & Poon, C. (2012). Hydration and properties of nano-TiO_2 blended cement composites. *Cement and Concrete Composites, 34*(5), 642–649.

Chen, J., & Poon, C. (2009). Photocatalytic construction and building materials: From fundamentals to applications. *Building and Environment, 44*(9), 1899–1906.

Chen, S. J., Collins, F. G., Macleod, A. J. N., Pan, Z., Duan, W. H., & Wang, C. M. (2011). Carbon nanotube–cement composites: A retrospect. *The IES Journal Part A: Civil & Structural Engineering, 4*(4), 254–265.

Chong, K. P. (2002). Research and Challenges in Nanomechanics, 90-minute Nanotechnology Webcast. *ASME*.

Choolaei, M., Rashidi, A. M., Ardjmand, M., Yadegari, A., & Soltanian, H. (2012). The effect of nanosilica on the physical properties of oil well cement. *Materials Science and Engineering A, 538*, 288–294. doi:10.1016/j.msea.2012.01.045

Chung, D. D. L. (1998). Self-monitoring structural materials. *Materials Science and Engineering R Reports, 22*(2), 57–78. doi:10.1016/S0927-796X(97)00021-1

Compilation of References

Claxton, N. S., Fellers, T. J., & Davidson, M. W. (2006). Laser scanning confocal microscopy. Tallahassee, FL: Department of Optical Microscopy and Digital Imaging, Florida State University. Retrieved from http://www.olympusconfocal.com/theory/LSCMIntro.pdf

Cobb, M. D., & Macoubrie, J. (2004). Public perceptions about nanotechnology: Risks, benefits and trust. *Journal of Nanoparticle Research*, *6*(4), 395–405. doi:10.1007/s11051-004-3394-4

Coffey, T., Urquhart, S. G., & Ade, H. (2002). Characterization of the effects of soft X-ray irradiation on polymers. *Journal of Electron Spectroscopy and Related Phenomena*, *122*(1), 65–78. doi:10.1016/S0368-2048(01)00342-5

Coleman, J. N., Khan, U., Blau, W. J., & Gun'ko, Y. K. (2006). Small but strong: A review of the mechanical properties of carbon nanotube–polymer composites. *Carbon*, *44*(9), 1624–1652. doi:10.1016/j.carbon.2006.02.038

Collepardi, M., Olagot, J. O., Skarp, U., & Troli, R. (2002). Influence of amorphous colloidal silica on the properties of self-compacting concretes. In *Proceedings of the International Conference in Concrete Constructions - innovations and developments in concrete materials and constructions*. Dundee, UK: ICE.

Constantinides, G., & Ulm, F. (2004). The effect of two types of C-S-H on the elasticity of cement-based materials: Results from nanoindentation and micromechanical modeling. *Cement and Concrete Research*, *34*(1), 67–80.

Constantinides, G., & Ulm, F. (2007). The nanogranular nature of C-S-H. *Journal of the Mechanics and Physics of Solids*, *55*(1), 64–90.

Constantinides, G., Ulm, F.-J., & Van Vliet, K. (2003). On the use of nanoindentation for cementitious materials. *Materials and Structures*, *36*(3), 191–196. doi:10.1007/BF02479557

Crewe, A. V., Wall, J., & Langmore, J. (1970). Visibility of single atoms. *Science*, *168*(3937), 1338–1340. doi:10.1126/science.168.3937.1338 PMID:17731040

Crooks, G. E. (1998). Nonequilibrium measurements of free energy differences for microscopically reversible Markovian systems. *Journal of Statistical Physics*, *90*(5-6), 1481–1487. doi:10.1023/A:1023208217925

Dabrowski, A. (2001). Adsorption--from theory to practice. *Advances in Colloid and Interface Science*, *93*(1-3), 135–224. doi:10.1016/S0001-8686(00)00082-8 PMID:11591108

Dabrunz, A., Duester, L., Prasse, C., Seitz, F., Rosenfeldt, R., Schilde, C., & Schulz, R. et al. (2011). Biological surface coating and molting inhibition as mechanisms of TiO 2 nanoparticle toxicity in Daphnia magna. *PLoS ONE*, *6*(5), e20112. doi:10.1371/journal.pone.0020112 PMID:21647422

Dai, Y., Sun, M., Liu, C., & Li, Z. (2010). Electromagnetic wave absorbing characteristics of carbon black cement-based composites. *Cement and Concrete Composites*, *32*(7), 508–513.

Dangers Come in Small Particles. (n.d.). *Hazards Magazine*. Retrieved 13 September, 2014 from http://www.hazards.org/nanotech/safety.htm

Das, B. B., & Mitra, A. (2014). Nanomaterials for Construction Engineering-A Review. *International Journal of Materials. Mechanics and Manufacturing*, *2*(1), 41–46. doi:10.7763/IJMMM.2014.V2.96

De Paola, N., Hirose, T., Mitchell, T., Di Toro, G., Viti, C., & Shimamoto, T. (2011). Fault lubrication and earthquake propagation in carbonate rocks. In R. Borja (Ed.), *Multiscale and Multiphysics Processes in Geomechanics* (pp. 153–156). Springer Berlin Heidelberg. doi:10.1007/978-3-642-19630-0_39

Degussa. (2008). *Zinc Oxide, Cerium Oxide, Indium TinOxide, just part of the growing nano product range from Degussa*. Retrieved April 20, 2015 from http://www.azonano.com/Details.asp?ArticleID=1603

DeJong, M. J., & Ulm, F. (2007). The nanogranular behavior of C-S-H at elevated temperatures (up to 700°C). *Cement and Concrete Research*, *37*(1), 1–12.

Deliyanni, E. A., Bakoyannakis, D. N., Zouboulis, A. I., & Matis, K. A. (2003). Sorption of As(V) ions by akaganeite-type nanocrystals. *Chemosphere*, *50*(1), 155–163. doi:10.1016/S0045-6535(02)00351-X PMID:12656241

Department of Physics, The Chinese University of Hong Kong. (n.d.). Retrieved from http://www.hk-phy.org/atomic_world/tem/tem02_e.html

Department of Physics, University of California at Santa Barbara. (n.d.). Retrieved from http://web.physics.ucsb.edu/~hhansma/biomolecules.htm

Derbyshire, H., & Miller, E. R. (1981). The photodegradation of wood during solar irradiation. *Holz als Roh- und Werkstoff*, *39*(8), 341–350. doi:10.1007/BF02608404

Derbyshire, H., Miller, E. R., Sell, J., & Turkulin, H. (1995). Assessment of wood photodegradation by microtensile testing. *Drvna Industrija*, *46*(3), 123–132.

Deshpande, A., Bao, W., Miao, F., Lau, C. N., & LeRoy, B. J. (2009). Spatially resolved spectroscopy of monolayer graphene on SiO_2. *Physical Review B: Condensed Matter and Materials Physics*, *79*(20), 205411. doi:10.1103/PhysRevB.79.205411

Dhawan, A., & Sharma, V. (2010). Toxicity assessment of nanomaterials: Methods and challenges. *Analytical and Bioanalytical Chemistry*, *398*(2), 589–605. doi:10.1007/s00216-010-3996-x PMID:20652549

Dhillon, A., & Kumar, D. (2015). Development of a nanoporous adsorbent for the removal of health-hazardous fluoride ions from aqueous systems. *Journal of Materials Chemistry A*, *3*(8), 4215–4228. doi:10.1039/C4TA06147K

Dhir, R. K., Newlands, M. D., & Csetenyi, L. J. (2005). Introduction. In *Proceedings of the International Conference – Application of Technology in Concrete Design*.

Dipanjan Mukherjee. (2014). *Low Cost Light Weight Concrete Making By Using Waste Materials*. Global Journal of Engineering Science and Research Management.

Compilation of References

Do, M. H., Phan, N. H., Nguyen, T. D., Pham, T. T. S., Nguyen, V. K., Vu, T. T. T., & Nguyen, T. K. P. (2011). Activated carbon/Fe_3O_4 nanoparticle composite: Fabrication, methyl orange removal and regeneration by hydrogen peroxide. *Chemosphere*, *85*(8), 1269–1276. doi:10.1016/j.chemosphere.2011.07.023 PMID:21840037

Domone, P., & Illston, J. (Eds.). (2010). *Construction materials: their nature and behaviour*. CRC Press.

Dresselhaus, M. S., & Dresselhaus, G. (2001). Ph. in Carbon Nanotubes. Topics Appl. Phys, 80, 287–327.

Drexler, K. E. (1981). Molecular engineering: An approach to the development of general capabilities for molecular manipulation. *Proceedings of the National Academy of Sciences of the United States of America*, *78*(9), 5275–5278. doi:10.1073/pnas.78.9.5275 PMID:16593078

Drexler, K. E. (1996). *Engines of Creation*. Fourth Estate.

Dubinin, M. M., & Radushkevich, L. V. (1960). The equation of the characteristic curve of the activated charcoal. In Proceedings Academy of Sciences USSR (vol. 55,pp. 331–337).

Du, H., Quek, S. T., & Dai Pang, S. (2013, April). Smart multifunctional cement mortar containing graphite nanoplatelet. In *Proceedings of SPIE Smart Structures and Materials + Nondestructive Evaluation and Health Monitoring*. Orlando, FL: International Society for Optics and Photonics.

Dunn, D. N., & Hull, R. (1999). Reconstruction of three-dimensional chemistry and geometry using focused ion beam microscopy. *Applied Physics Letters*, *75*(21), 3414–3416. doi:10.1063/1.125311

Dunphy Guzman, K. A., Taylor, M. R., & Banfield, J. F. (2006). Environmental risks of nanotechnology: *National nanotechnology initiative funding, 2000-2004*. Environmental Science & Technology, *40*(5), 1401–1407. PMID:16568748

Du, W. L., Xu, Z. R., Han, X. Y., Xu, Y. L., & Miao, Z. G. (2008). Preparation, characterization and adsorption properties of chitosan nanoparticles for eosin Y as a model anionic dye. *Journal of Hazardous Materials*, *153*(1-2), 152–156. doi:10.1016/j.jhazmat.2007.08.040 PMID:17890000

Dweck, J., Ferreira da Silva, P. F., Büchler, P. M., & Cartledge, F. K. (2002). Study by thermogravimetry of the evolution of ettringite phase during type II Portland cement hydration. *Journal of Thermal Analysis and Calorimetry*, *69*(1), 179–186. doi:10.1023/A:1019950126184

Ebbesen, T. W. (1996). Carbon nanotubes. *Physics Today*, *49*(6), 26–35. doi:10.1063/1.881603

Edvardsen, C. (1999). Water permeability and autogenous healing of cracks in concrete. *ACI Materials Journal*, *96*(4).

Ershadi, V., Ebadi, T., Rabani, A., Ershadi, L., & Soltanian, H. (2011). The effect of nanosilica on cement matrix permeability in oil well to decrease the pollution of receptive environment. *Int. J. Environ. Sci. Develop*, *2*, 128–132. doi:10.7763/IJESD.2011.V2.109

Eskandarian, L., Arami, M., & Pajootan, E. (2014). Evaluation of Adsorption Characteristics of Multiwalled Carbon Nanotubes Modified by a Poly (propylene imine) Dendrimer in Single and Multiple Dye Solutions: Isotherms, Kinetics, and Thermodynamics. *Journal of Chemical & Engineering Data*, *59*(2), 444–454. doi:10.1021/je400913z

European Commission Technical Guidance Document (TGD) on Risk Assessment. (2003). *European Commission Joint Research Center (ECJRC)*. Retrieved from http://ecb.jrc.ec.europa.eu/tgd/

Fanella, D. A., & Naaman, A. E. (1985). Stress–strain properties of fiber reinforced mortar in compression. *Journal of the American Concrete Institute*, *82*(4), 475–483.

Fan, J. J., Tang, J. Y., Cong, L. Q., & Mcolm, I. J. (2004). Influence of synthetic nano-ZrO2 powder on the strength property of portland cement. *Jianzhu Cailiao Xuebao*, *7*(4), 462–467.

Farre, M., Sanchis, J., & Barcelo, D. (2011). Analysis and assessment of the occurrence, the fate and the behavior of nanomaterials in the environment. *Trends in Analytical Chemistry*, *30*(2), 517–527. doi:10.1016/j.trac.2010.11.014

Federici, G., Shaw, B. J., & Handy, R. D. (2007). Toxicity of titanium dioxide nanoparticles to rainbow trout (Oncorhynchus mykiss): Gill injury, oxidative stress, and other physiological effects. *Aquatic Toxicology (Amsterdam, Netherlands)*, *84*(4), 415–430. doi:10.1016/j.aquatox.2007.07.009 PMID:17727975

Fei, J., Cui, Y., Zhao, J., Gao, L., Yang, Y., & Li, J. (2011). Large-scale preparation of 3D self-assembled iron hydroxide and oxide hierarchical nanostructures and their applications for water treatment. *Journal of Materials Chemistry*, *21*(32), 11742–11746. doi:10.1039/c1jm11950h

Fendler, J. H. (1987). Atomic and molecular clusters in membrane mimetic chemistry. *Chemical Reviews*, *87*(5), 877–899. doi:10.1021/cr00081a002

Feng, Y., Gong, J. L., Zeng, G. M., Niu, Q. Y., Zhang, H. Y., Niu, C. G., & Yan, M. et al. (2010). Adsorption of Cd (II) and Zn (II) from aqueous solutions using magnetic hydroxyapatite nanoparticles as adsorbents. *Chemical Engineering Journal*, *162*(2), 487–494. doi:10.1016/j.cej.2010.05.049

Ferrari, A. C., Meyer, J. C., Scardaci, V., Casiraghi, C., Lazzeri, M., Mauri, F., & Geim, A. K. et al. (2006). Raman spectrum of graphene and graphene layers. *Physical Review Letters*, *97*(18), 187401. doi:10.1103/PhysRevLett.97.187401 PMID:17155573

Feser, M., Hornberger, B., Jacobsen, C., De Geronimo, G., Rehak, P., Holl, P., & Strüder, L. (2006). Integrating Silicon detector with segmentation for scanning transmission X-ray microscopy. *Nuclear Instruments & Methods in Physics Research. Section A, Accelerators, Spectrometers, Detectors and Associated Equipment*, *565*(2), 841–854. doi:10.1016/j.nima.2006.05.086

Feynman, R. P. (1960). There's plenty of room at the bottom. *Engineering and Science*, *23*(5), 22–36.

Feynman, R. P. (1961). *There's plenty of room at the bottom*. New York: Reinhold.

Compilation of References

Fiore, V., Scalici, T., Di Bella, G., & Valenza, A. (2015). A review on basalt fibre and its composites. *Composites. Part B, Engineering*, *74*, 74–94. doi:10.1016/j.compositesb.2014.12.034

Flores, I., Sobolev, K., Torres-Martinez, L. M., Cuellar, E. L., Valdez, P. L., & Zarazua, E. (2010). Performance of cement systems with nano-SiO2 particles produced by using the sol-gel method. *Transportation Research Record*, *2141*, 10–14. doi:10.3141/2141-03

Foo, K. Y., & Hameed, B. H. (2010). Insights into the modeling of adsorption isotherm systems. *Chemical Engineering Journal*, *156*(1), 2–10. doi:10.1016/j.cej.2009.09.013

Freitag, W., & Stoye, D. (Eds.). (2008). *Paints, coatings and solvents*. John Wiley & Sons.

Freundlich, H.M.F. (1906). Over the adsorption in solution. *Journal of Physical Chemistry*, *57*(A), 385–471.

Fubini, B., Ghiazza, M., & Fenoglio, I. (2010). Physico-chemical features of engineered nanoparticles relevant to their toxicity. *Nanotoxicology*, *4*(4), 347–363. doi:10.3109/17435390.2010.509519 PMID:20858045

Fujishima, A., Zhang, X., & Tryk, D. A. (2008). TiO 2 photocatalysis and related surface phenomena. *Surface Science Reports*, *63*(12), 515–582. doi:10.1016/j.surfrep.2008.10.001

Gaitero, J. J., Campillo, I., & Guerrero, A. (2008). Reduction of the calcium leaching rate of cement paste by addition of silica nanoparticles. *Cement and Concrete Research*, *38*(8-9), 1112–1118.

Galao, O., Baeza, F. J., Zornoza, E., & Garcés, P. (2013). Strain and damage sensing properties on multifunctional cement composites with CNF admixture. *Cement and Concrete Composites*, *46*, 90–98.

Galooyak, S. S., Dabir, B., Nazarbeygi, A. E., & Moeini, A. (2010). Rheological properties and storage stability of bitumen/SBS/montmorillonite composites. *Construction & Building Materials*, *24*(3), 300–307. doi:10.1016/j.conbuildmat.2009.08.032

Ganguli, A. K., Vaidya, S., & Ahmad, T. (2008). Synthesis of nanocrystalline materials through reverse micelles:A versatile methodology for synthesis of complex metal oxides. *Bulletin of Materials Science*, *31*(3), 415–419. doi:10.1007/s12034-008-0065-6

Gao, D., Sturm, M., & Mo, Y. L. (2009). Electrical resistance of carbon-nanofiber concrete. *Smart Materials and Structures*, *18*(9), 095039.

Gao, X. F., Lo, Y. T., & Tam, C. M. (2002). Investigation of micro-cracks and microstructure of high performance lightweight aggregate concrete. *Building and Environment*, *37*(5), 485–489. doi:10.1016/S0360-1323(01)00051-8

Gateshki, M., Petkov, V., Williams, G., Pradhan, S. K., & Ren, Y. (2005). Atomic-scale structure of nanocrystalline ZrO_2 prepared by high-energy ball milling. *Physical Review B: Condensed Matter and Materials Physics*, *71*(22), 224107–224109. doi:10.1103/PhysRevB.71.224107

Gautam, R. K., Gautam, P. K., Banerjee, S., Soni, S., Singh, S. K., & Chattopadhyaya, M. C. (2015). Removal of Ni (II) by magnetic nanoparticles. *Journal of Molecular Liquids*, *204*(4), 60–69. doi:10.1016/j.molliq.2015.01.038

Gautam, R. K., Mudhoo, A., & Chattopadhyaya, M. C. (2013). Kinetic, equilibrium, thermodynamic studies and spectroscopic analysis of Alizarin Red S removal by mustard husk. *Journal of Environmental Chemical Engineering*, *1*(4), 1283–1291. doi:10.1016/j.jece.2013.09.021

Gay, C., & Sanchez, F. (2010). Performance of Carbon Nanofiber-Cement Composites with a High-Range Water Reducer. *Transportation Research Record: Journal of the Transportation Research Board*, *2142*(1), 109-113.

Gay, C., & Sanchez, F. (2010). Performance of carbon nanofiber-cement composites with a high-range water reducer. *Transportation Research Record*, *2142*, 109–113. doi:10.3141/2142-16

Ge, F., Li, M. M., Ye, H., & Zhao, B. X. (2012). Effective removal of heavy metal ions Cd^{2+}, Zn^{2+}, Pb^{2+}, Cu^{2+} from aqueous solution by polymer-modified magnetic nanoparticles. *Journal of Hazardous Materials*, *211–212*(4), 366–372.

Ghafari, E., Costa, H., Julio, E., Portugal, A., & Duraes, L. (2014). The effect of nanosilica addition on flowability, strength and transport properties of ultra high performance concrete. *Materials & Design*, *59*, 1–9.

Ghasemi, E., & Sillanpaa, M. (2015). Magnetic hydroxyapatite nanoparticles: An efficient adsorbent for the separation and removal of nitrate and nitrite ions from environmental samples. *Journal of Separation Science*, *38*(1), 164–169. doi:10.1002/jssc.201400928 PMID:25376506

Ghavami, K. (2005). Bamboo as reinforcement in structural concrete elements. *Cement and Concrete Composites*, *27*(6), 637–649. doi:10.1016/j.cemconcomp.2004.06.002

Ghile, D. B. (2006). *Effects of nanoclay modification on rheology of bitumen and on performance of asphalt mixtures*. Delft, The Netherlands: Delft University of Technology.

Giner, V. T., Baeza, F. J., Ivorra, S., Zornoza, E., & Galao, Ó. (2012). Effect of steel and carbon fiber additions on the dynamic properties of concrete containing silica fume. *Materials & Design*, *34*, 332–339. doi:10.1016/j.matdes.2011.07.068

Glenn, J. C. (2006). Nanotechnology: Future military environmental health considerations. *Technological Forecasting and Social Change*, *73*(2), 128–137. doi:10.1016/j.techfore.2005.06.010

Glezer, A. M. (2011). Structural Classification of Nanomaterials. *Russian Metallurgy (Metally)*, *2011*(4), 263–269. doi:10.1134/S0036029511040057

Goddard, W. A. III, Brenner, D. W., Lyshevski, S. E., & Iafrate, G. J. (2004). Properties of High-Volume Fly Ash Concrete Incorporating Nano-SiO2. *Cement and Concrete Research*, *34*(6), 1043–1049. doi:10.1016/j.cemconres.2003.11.013

Goddard, W. A. III, Brenner, D., Lyshevski, S. E., & Iafrate, G. J. (Eds.). (2007). *Handbook of nanoscience, engineering, and technology*. CRC Press.

Compilation of References

Godt, J., Scheidig, F., Siestrup, C. G., Esche, V., Brandenburg, P., Reich, A., & Groneberg, D. A. (2006). The toxicity of cadmium and resulting hazards for human health. *Journal of Occupational Medicine and Toxicology (London, England)*, *1*(22), 1–6. PMID:16961932

Goh, S. W., Akin, M., You, Z., & Shi, X. (2011). Effect of deicing solutions on the tensile strength of micro-or nano-modified asphalt mixture. *Construction & Building Materials*, *25*(1), 195–200. doi:10.1016/j.conbuildmat.2010.06.038

Gold nanoparticles show promise for early detection of heart attacks. (n.d.). New York University Polytechnic School of Engineering. Retrieved June 20, 2015 from, http://www.eurekalert.org/pub_releases/2015-01/nyup-gns011515.php

Gong, H., Zhang, Y., Quan, J., & Che, S. (2011). Preparation and properties of cement based piezoelectric composites modified by CNTs. *Current Applied Physics*, *11*(3), 653–656.

Gong, J., Chen, L., Zeng, G., Long, F., Deng, J., Niu, Q., & He, X. (2012). Shellac-coated iron oxide nanoparticles for removal of cadmium(II) ions from aqueous solution. *Journal of Environmental Sciences (China)*, *24*(7), 1165–1173. doi:10.1016/S1001-0742(11)60934-0 PMID:23513435

Gopal, K., Tripathy, S. S., Bersillon, J. L., & Dubey, S. P. (2007). Chlorination byproducts, their toxicodynamics and removal from drinking water. *Journal of Hazardous Materials*, *140*(1-2), 1–6. doi:10.1016/j.jhazmat.2006.10.063 PMID:17129670

Gordon, T., Chen, L., Fine, J., Schlesinger, R., Su, W., Kimmel, T., & Amdur, M. (1992). Pulmonary effects of inhaled zinc-oxide in human-subjects, guinea-pigs, rats, and rabbits. *American Industrial Hygiene Association Journal*, *53*(8), 503–509. doi:10.1080/15298669291360030 PMID:1509990

Gorkem, C., & Sengoz, B. (2009). Predicting stripping and moisture induced damage of asphalt concrete prepared with polymer modified bitumen and hydrated lime. *Construction & Building Materials*, *23*(6), 2227–2236. doi:10.1016/j.conbuildmat.2008.12.001

Granier, J. J., & Pantoya, M. L. (2004). Laser ignition of nanocomposite thermites. *Combustion and Flame*, *138*(4), 373–382. doi:10.1016/j.combustflame.2004.05.006

Green, M., & Howman, E. (2005). Semiconductor quantum dots and free radical induced DNA nicking. *Chemical Communications (Cambridge)*, *1*(1), 121–123. doi:10.1039/b413175d PMID:15614393

Grodzinski, P., Silver, M., & Molnar, L. K. (2006). *Nanotechnology for cancer diagnostics: promises and challenges*. Academic Press.

Grove, J., Vanikar, S., & Crawford, G. (2010). Nanotechnology new tools to address old problems. *Transportation Research Record*, *2141*, 47–51. doi:10.3141/2141-09

Groves, G. W., Sueur, P. J., & Sinclair, W. (1986). Transmission Electron Microscopy and Microanalytical Studies of Ion-Beam-Thinned Sections of Tricalcium Silicate Paste. *Journal of the American Ceramic Society*, *69*(4), 353–356.

Grujicic, M., Cao, G., & Gersten, B. (2003). Enhancement of field emission in carbon nanotubes through adsorption of polar molecules. *Applied Surface Science, 206*(1), 167–177.

Guo, Z. (2013). *Study on the Electromagnetic Wave Absorbing Properties of Multi-walled Carbon Nanotube/Cement Composites.* (Master's Dissertation). Dalian University of Technology, Dalian, China.

Gutmann, P. F. (1987). Bubble characteristics as they pertain to compressive strength and freeze-thaw durability. In *MRS Proceedings* (Vol. 114, p. 271). Cambridge University Press.

Habel, K., Viviani, M., Denarié, E., & Brühwiler, E. (2006). Development of the mechanical properties of an Ultra-High Performance Fiber Reinforced Concrete (UHPFRC). *Cement and Concrete Research, 36*(7), 1362–1370. doi:10.1016/j.cemconres.2006.03.009

Habermehl-Cwirzen, K., Penttala, V., & Cwirzen, A. (2008). Surface decoration of carbon nanotubes and mechanical properties of cement/carbon nanotube composites. *Advances in Cement Research, 20*(2), 65–73.

Haddad, R. H., Al-Rousan, R., Ghanma, L., & Nimri, Z. (2015). Modifying CFRP-concrete bond characteristics from pull-out testing. *Magazine of Concrete Research, 67*(13), 707–717. doi:10.1680/macr.14.00271

Haider, M., Braunshausen, G., & Schwan, E. (1995). Correction of the spherical aberration of a 200 kV TEM by means of a hexapole-corrector. *Optik (Stuttgart), 99*(4), 167–179.

Hamilton, S. J. (2004). Review of selenium toxicity in the aquatic food chain. *The Science of the Total Environment, 326*(1-3), 1–31. doi:10.1016/j.scitotenv.2004.01.019 PMID:15142762

Hammel, E., Tang, X., Trampert, M., Schmitt, T., Mauthner, K., Eder, A., & Pötschke, P. (2004). Carbon nanofibers for composite applications. *Carbon, 42*(5-6), 1153–1158. doi:10.1016/j.carbon.2003.12.043

Han, B., Yu, X., & Ou, J. (2014). *Self-sensing concrete in smart structures.* Waltham, MA: Butterworth Heinemann (Elsevier).

Han, B., & Ou, J. (2007). Embedded piezoresistive cement-based stress/strain sensor. *Sensors and Actuators. A, Physical, 138*(2), 294–298.

Han, B., Sun, S., Ding, S., Zhang, L., Yu, X., & Ou, J. (2015). Review of nanocarbon-engineered multifunctional cementitious composites. *Composites. Part A, Applied Science and Manufacturing, 70*, 69–81.

Han, B., Yang, Z., Shi, X., & Yu, X. (2013). Transport Properties of Carbon-Nanotube/Cement Composites. *Journal of Materials Engineering and Performance, 22*(1), 184–189.

Han, B., Yu, X., & Kwon, E. (2009). A self-sensing carbon nanotube/cement composite for traffic monitoring. *Nanotechnology, 20*(44), 445501.

Compilation of References

Han, B., Yu, X., & Ou, J. (2011). Multifunctional and smart carbon nanotube reinforced cement-based materials. In K. Gopalakrishnan, B. Birgisson, P. Taylor, & N. O. Attoh-Okine (Eds.), *Nanotechnology in civil infrastructure* (pp. 1–47). Berlin, Germany: Springer.

Han, B., Zhang, K., Burnham, T., Kwon, E., & Yu, X. (2013). Integration and road tests of a self-sensing CNT concrete pavement system for traffic detection. *Smart Materials and Structures*, *22*(1), 15020.

Han, B., Zhang, K., Yu, X., Kwon, E., & Ou, J. (2012). Electrical characteristics and pressure-sensitive response measurements of carboxyl MWNT/cement composites. *Cement and Concrete Composites*, *34*(6), 794–800.

Hannesson, G., Kuder, K., Shogren, R., & Lehman, D. (2012). The influence of high volume of fly ash and slag on the compressive strength of self-consolidating concrete. *Construction & Building Materials*, *30*, 161–168.

Hansen, S. F. (2009). *Regulation and risk assessment of nanomaterials: too little, too late*. Department of Environmental Engineering, Technical University of Denmark. Retrieved August 14, 2015 from http://www2.er.dtu.dk/publications/fulltext/2009/ENV2009-069.pdf

Han, Y., & Yan, W. (2014). Bimetallic nickel-iron nanoparticles for groundwater decontamination: Effect of groundwater constituents on surface deactivation. *Water Research*, *66*, 149–159. doi:10.1016/j.watres.2014.08.001 PMID:25201338

Haruehansapong, S., Pulngern, T., & Chucheepsakul, S. (2014). Effect of the particle size of nanosilica on the compressive strength and the optimum replacement content of cement mortar containing nano-SiO$_2$. *Construction & Building Materials*, *50*, 471–477.

Hasan, U., Chegenizadeh, A., Budihardjo, M. A., & Nikraz, H. (2015). *Experimental evaluation of construction waste and ground granulated blast furnace slag as alternative soil stabilisers*. Manuscript submitted for publication.

Hasan, U., Chegenizadeh, A., Budihardjo, M., & Nikraz, H. (2015). A review of the stabilisation techniques on expansive soils. *Australian Journal of Basic & Applied Sciences*, *9*(7), 541–548.

Hashmi, S. (2014). *Comprehensive Materials Processing*. Elsevier Science.

Hassan, M. M., Dylla, H., Mohammad, L. N., & Rupnow, T. (2010). Evaluation of the durability of titanium dioxide photocatalyst coating for concrete pavement. *Construction & Building Materials*, *24*(8), 1456–1461. doi:10.1016/j.conbuildmat.2010.01.009

Hassan, T., & Rizkalla, S. (2003). Investigation of bond in concrete structures strengthened with near surface mounted carbon fiber reinforced polymer strips. *Journal of Composites for Construction*, *7*(3), 248–257. doi:10.1061/(ASCE)1090-0268(2003)7:3(248)

He, X., & Shi, X. (2008). Chloride Permeability and Microstructure of Portland Cement Mortars Incorporating Nanomaterials. *Transportation Research Record: Journal of the Transportation Research Board*, *2070*(1), 13-21.

Head, M. K., Wong, H. S., & Buenfeld, N. R. (2006). Characterisation of 'Hadley' grains by confocal microscopy. *Cement and Concrete Research*, *36*(8), 1483–1489. doi:10.1016/j.cemconres.2005.12.020

Health and Food Safety, Scientific Committees. (n.d.). European Comission. Retrieved August 20, 2015, from http://ec.europa.eu/health/scientific_committees/?opinions_layman/en/nanotechnologies/1-2/

He, C. N., Zhao, N. Q., Shi, C. S., & Song, S. Z. (2009). Fabrication of carbon nanomaterials by chemical vapor deposition. *Journal of Alloys and Compounds*, *484*(1-2), 6–11. doi:10.1016/j.jallcom.2009.04.088

Hermann, P., Hoehl, A., Patoka, P., Huth, F., Rühl, E., & Ulm, G. (2013). Near-field imaging and nano-Fourier-transform infrared spectroscopy using broadband synchrotron radiation. *Optics Express*, *21*(3), 2913–2919. doi:10.1364/OE.21.002913 PMID:23481749

High Strength Concrete. (n.d.). Retrieved May 04, 2015 from, http://www.concrete.org.uk/fingertips_nuggets.asp?cmd=display&id=528

Hinds, W. C. (2012). *Aerosol technology: properties, behavior, and measurement of airborne particles*. John Wiley & Sons.

Hirth, J. P., & Lothe, J. (1982). *Theory of Dislocations*. Krieger Publishing Company.

Hodge, A. T. (1981). Vitruvius, lead pipes and lead poisoning. *American Journal of Archaeology*, 486-491. Retrieved June 02, 2015 from, http://www.bbc.co.uk/schools/primaryhistory/indus_valley/technology_and_jobs/

Holzer, L., Muench, B., Wegmann, M., Gasser, P., & Flatt, R. J. (2006). FIB-Nanotomography of Particulate Systems—Part I: Particle Shape and Topology of Interfaces. *Journal of the American Ceramic Society*, *89*(8), 2577–2585. doi:10.1111/j.1551-2916.2006.00974.x

Horsfall, M., & Spiff, A. I. (2005). Equilibrium sorption study of Al^{3+}, Co^{2+} and Ag^{2+} in aqueous solutions by fluted pumpkin (Telfairia occidentalis HOOK) waste biomass. *Acta Chimica Slovenica*, *52*(2), 174–181.

Hoshino, A., Hanaki, K., Suzuki, K., & Yamamoto, K. (2004). Applications of t-lymphoma labeled with fluorescent quantum dots to cell tracing markers in mouse body. *Biochemical and Biophysical Research Communications*, *314*(1), 46–53. doi:10.1016/j.bbrc.2003.11.185 PMID:14715244

Hosseini, P., Booshehrian, A., & Farshchi, S. (2010). Influence of nano-SiO2 addition on microstructure and mechanical properties of cement mortars for ferrocement. *Transportation Research Record*, *2141*, 15–20. doi:10.3141/2141-04

Hosseini, P., Booshehrian, A., & Madari, A. (2011). Developing Concrete Recycling Strategies by Utilization of Nano-SiO_2 Particles. *Waster and Biomass Valorization*, *2*(3), 347–355.

Compilation of References

Hou, T. C., & Lynch, J. P. (2005). *Conductivity-based strain monitoring and damage characterization of fiber reinforced cementitious structural components*. Paper presented at the Proceedings of SPIE - The International Society for Optical Engineering. doi:10.1117/12.599955

Hou, D., Zhu, Y., Lu, Y., & Li, Z. (2014). Mechanical properties of calcium silicate hydrate (C–S–H) at nano-scale: A molecular dynamics study. *Materials Chemistry and Physics*, *146*(3), 503–511.

Hou, P., Cheng, X., Qian, J., Zhang, R., Cao, W., & Shah, S. P. (2015). Characteristics of surface-treatment of nano-SiO2 on the transport properties of hardened cement pastes with different water-to-cement ratios. *Cement and Concrete Composites*, *55*, 26–33. doi:10.1016/j.cemconcomp.2014.07.022

Hou, P., Kawashima, S., Kong, D., Corr, D. J., Qian, J., & Shah, S. P. (2013). Modification effects of colloidal nanoSiO2 on cement hydration and its gel property. *Composites. Part B, Engineering*, *45*(1), 440–448. doi:10.1016/j.compositesb.2012.05.056

Hou, P., Kawashima, S., Wang, K., Corr, D. J., Qian, J., & Shah, S. P. (2013). Effects of colloidal nanosilica on rheological and mechanical properties of fly ash-cement mortar. *Cement and Concrete Composites*, *35*(1), 12–22.

Hou, X., Zhou, F., Yu, B., & Liu, W. (2007). Superhydrophobic zinc oxide surface by differential etching and hydrophobic modification. *Materials Science and Engineering A*, *452*, 732–736. doi:10.1016/j.msea.2006.11.057

Howser, R. N., Dhonde, H. B., & Mo, Y. L. (2011). Self-sensing of carbon nanofiber concrete columns subjected to reversed cyclic loading. *Smart Materials and Structures*, *20*(8), 085031.

Ho, Y. S., & McKay, G. (1999). Pseudo-second order model for sorption processes. *Process Biochemistry*, *34*(5), 451–465. doi:10.1016/S0032-9592(98)00112-5

Huang, S. (2012). *Multifunctional Graphite Nanoplatelets (GNP) Reinforced Cementitious Composites*. (Master's Dissertation). National University of Singapore.

Huang, H. L., Lin, C. C., & Hsu, K. (2015). Comparison of resistance improvement to fungal growth on green and conventional building materials by nano-metal impregnation. *Building and Environment*, *93*(P2), 119–127. doi:10.1016/j.buildenv.2015.06.016

Huang, S. H., & Chen, D. H. (2009). Rapid removal of heavy metal cations and anions from aqueous solutions by an amino-functionalized magnetic nano-adsorbent. *Journal of Hazardous Materials*, *163*(1), 174–179. doi:10.1016/j.jhazmat.2008.06.075 PMID:18657903

Huang, W., Wang, S., Zhu, Z., Li, L., Yao, X., Rudolph, V., & Haghseresht, F. (2008). Phosphate removal from wastewater using red mud. *Journal of Hazardous Materials*, *158*(1), 35–42. doi:10.1016/j.jhazmat.2008.01.061 PMID:18314264

Huang, X., Jain, P. K., El-Sayed, I. H., & El-Sayed, M. A. (2007). Gold nanoparticles: Interesting optical properties and recent applications in cancer diagnostics and therapy. *Nanomedicine (London)*, *2*(5), 681–693. doi:10.2217/17435889.2.5.681 PMID:17976030

Hui, Y., Zhanping, Y., Liang, L., Huei, L. C., David, W., Khin, Y. Y., & Wei, G. S. et al. (2012). Properties and Chemical Bonding of Asphalt and Asphalt Mixtures Modified with Nanosilica. *Journal of Materials in Civil Engineering.* doi:10.1061/(ASCE)MT.1943-5533.0000690

Hull, M. S., Quadros, M. E., Born, R., Provo, J., Lohani, V. K., & Mahajan, R. L. (2014). Sustainable Nanotechnology: A Regional Perspective. In M. S. Hull & D. M. Bowman (Eds.), *Nanotechnology Environmental Health and Safety* (2nd ed.; pp. 395–424). Oxford, UK: William Andrew Publishing.

Hunashyal, A. M., Sundeep, G. V., Quadri, S. S., & Banapurmath, N. R. (2011). Experimental investigations to study the effect of carbon nanotubes reinforced in cement-based matrix composite beams. *Proceedings of the Institution of Mechanical Engineers. Part N, Journal of Nanoengineering and Nanosystems, 225*(1), 17–22.

Hund-Rinke, K., & Simon, M. (2006). Ecotoxic effect of photocatalytic active nanoparticles (TiO2) on algae and daphnids (8 pp). *Environmental Science and Pollution Research International, 13*(4), 225–232. doi:10.1065/espr2006.06.311 PMID:16910119

Hunt. (2000). Material Evidence Conserving historic building fabric. *New Heritage Materials.*

Hüsken, G., Hunger, M., & Brouwers, H. (2009). Experimental study of photocatalytic concrete products for air purification. *Building and Environment, 44*(12), 2463–2474.

Huth, F., Govyadinov, A., Amarie, S., Nuansing, W., Keilmann, F., & Hillenbrand, R. (2012). Nano-FTIR Absorption Spectroscopy of Molecular Fingerprints at 20 nm Spatial Resolution. *Nano Letters, 12*(8), 3973–3978. doi:10.1021/nl301159v PMID:22703339

Hyeon, T. (2003). Chemical synthesis of magnetic nanoparticles. *Chemical Communications (Cambridge),* (8): 927–934. doi:10.1039/b207789b PMID:12744306

Ibrahim, R. K., Hamid, R., & Taha, M. R. (2012). Fire resistance of high-volume fly ash mortars with nanosilica addition. *Construction & Building Materials, 36*, 779–786.

Iijima, S. (1991). Helical microtubules of graphitic carbon. *Nature, 354*(6348), 56–58. doi:10.1038/354056a0

Inkson, B. J., Steer, T., Möbus, G., & Wagner, T. (2001). Subsurface nanoindentation deformation of Cu–Al multilayers mapped in 3D by focused ion beam microscopy. *Journal of Microscopy, 201*(2), 256–269. doi:10.1046/j.1365-2818.2001.00767.x PMID:11207928

Institute of Experimental and Applied Physics, University of Kiel. (n.d.). Retrieved from http://www.ieap.uni-kiel.de/surface/ag-kipp/stm/stm.htm

International Programme on Chemical Safety. (2004). *IPCS Risk Assessment Terminology.* World Health Organization. Retrieved from http://www.who.int/ipcs/methods/ harmonization/areas/ipcsterminologyparts1and2.pdf

Ishigami, M., Chen, J. H., Cullen, W. G., Fuhrer, M. S., & Williams, E. D. (2007). Atomic structure of graphene on SiO_2. *Nano Letters, 7*(6), 1643–1648. doi:10.1021/nl070613a PMID:17497819

Compilation of References

Islam, N., & Miyazaki, K. (2010). An empirical analysis of nanotechnology research domains. *Technovation, 30*(4), 229–237. doi:10.1016/j.technovation.2009.10.002

Ivorra, S., Garcés, P., Catalá, G., Andión, L. G., & Zornoza, E. (2010). Effect of silica fume particle size on mechanical properties of short carbon fiber reinforced concrete. *Materials & Design, 31*(3), 1553–1558. doi:10.1016/j.matdes.2009.09.050

Jafari, M., & Esna-ashari, M. (2012). Effect of waste tire cord reinforcement on unconfined compressive strength of lime stabilized clayey soil under freeze–thaw condition. *Cold Regions Science and Technology, 82*(0), 21–29. doi:10.1016/j.coldregions.2012.05.012

Jahromi, S. G., Andalibizade, B., & Vossough, S. (2010). Engineering properties of nanoclay modified asphalt concrete mixtures. *The Arabian Journal for Science and Engineering, 35*(1B), 89–103.

Jahromi, S. G., & Khodaii, A. (2009). Effects of nanoclay on rheological properties of bitumen binder. *Construction & Building Materials, 23*(8), 2894–2904. doi:10.1016/j.conbuildmat.2009.02.027

Jain, S., & Jayaram, R. V. (2010). Removal of basic dyes from aqueous solution by low-cost adsorbent: Wood apple shell (Feronia acidissima). *Desalination, 250*(3), 921–927. doi:10.1016/j.desal.2009.04.005

Jalal, M., Pouladkhan, A., Harandi, O. F., & Jafari, D. (2015). Comparative study on effects of Class F fly ash, nano silica and silica fume on properties of high performance self compacting concrete. *Construction & Building Materials, 94*, 90–104. doi:10.1016/j.conbuildmat.2015.07.001

Jalal, M., Ramezanianpour, A. A., & Pool, M. K. (2013). Split tensile strength of binary blended self compacting concrete containing low volume fly ash and TiO_2 nanoparticles. *Composites. Part B, Engineering, 55*, 324–337.

Janssen, R. A. J., & Stouwdam, J. W. (2008). Red, green, and blue quantum dot LEDs with solution processable ZnO nanocrystal electron injection layers. *Journal of Materials Chemistry, 18*(16), 1889–1894. doi:10.1039/b800028j

Jaques, P. A., & Kim, C. S. (2000). Measurement of total lung deposition of inhaled ultrafine particles in healthy men and women. *Inhalation Toxicology, 12*(8), 715–731. doi:10.1080/08958370050085156 PMID:10880153

Jennings, H. M. (2003). Colloid model of C-S-H and implications to the problem of creep and shrinkage. *Materials and Structures, 37*(265), 59–70.

Jirsak, O., & Dao, T. A. (2009). Production, properties and end-uses of nanofibres. *Proc. of the Nanotechnology in Construction, 3*, 95–100. doi:10.1007/978-3-642-00980-8_11

Ji, T. (2005). Preliminary study on the water permeability and microstructure of concrete incorporating nano-SiO_2. *Cement and Concrete Research, 35*(10), 1943–1947. doi:10.1016/j.cemconres.2005.07.004

Ji, T., Huang, Y. Z., & Zheng, Z. Q. (2003). Primary investigation of physics and mechanics properties of nano-concrete. *Concrete (London)*, *3*(48), 13–14.

Jo, B. W., Chakraborty, S., & Kim, K. H. (2014). Investigation on the effectiveness of chemically synthesized nano cement in controlling the physical and mechanical performances of concrete. *Construction & Building Materials*, *70*, 1–8.

Jo, B., Chakraborty, S., Kim, K. H., & Lee, Y. S. (2014). Effectiveness of the Top-Down Nanotechnology in the Production of Ultrafine Cement. *Journal of Nanomaterials*, *2014*, 1–9.

Jo, B., Kim, C., Tae, G., & Park, J. (2007). Characteristics of cement mortar with nano-SiO_2 particles. *Construction & Building Materials*, *21*(6), 1351–1355.

Jones, J. E. (1924, October 1). *On the determination of molecular fields. II. From the equation of state of a gas.* Paper presented at the Royal Society of London. Series A: Mathematical, Physical and Engineering Sciences, London, UK.

Jonkers, H. M., Thijssen, A., Muyzer, G., Copuroglu, O., & Schlangen, E. (2010). Application of bacteria as self-healing agent for the development of sustainable concrete. *Ecological Engineering*, *36*(2), 230–235. doi:10.1016/j.ecoleng.2008.12.036

Jose, M. V., Steinert, B. W., Thomas, V., Dean, D. R., Abdalla, M. A., Price, G., & Janowski, G. M. (2007). Morphology and mechanical properties of Nylon 6/MWNT nanofibers. *Polymer*, *48*(4), 1096–1104. doi:10.1016/j.polymer.2006.12.023

Joy, D. C. (1997). Scanning electron microscopy for materials characterization. *Current Opinion in Solid State and Materials Science*, *2*(4), 465–468. doi:10.1016/S1359-0286(97)80091-5

Kadivar, Barkhordari, & Kadivar. (2011). Nanotechnology in Geotechnical Engineering. *Advanced Materials Research*, *261-263*, 524-528. doi: 10.4028/www.scientific.net/AMR.261-263.524

Kadu, B. S., & Chikate, R. C. (2013). Improved adsorptive mineralization capacity of Fe–Ni sandwiched montmorillonite nanocomposites towards magenta dye. *Chemical Engineering Journal*, *228*, 308–317. doi:10.1016/j.cej.2013.04.103

Kalaitzidis, S., & Christanis, K. (2003). Scanning electron microscope studies of the Philippi peat (NE Greece): Initial aspects. *International Journal of Coal Geology*, *54*(1), 69–77. doi:10.1016/S0166-5162(03)00020-X

Karn, B., Roco, M., Masciangioli, T., & Savage, N. (2003, May). *Nanotechnology and the Environment.* National Nanotechnology Initiative Workshop.

Kaur, R., Hasan, A., Iqbal, N., Alam, S., Saini, M. K., & Raza, S. K. (2014). Synthesis and surface engineering of magnetic nanoparticles for environmental cleanup and pesticide residue analysis: A review. *Journal of Separation Science*, *37*(14), 1805–1825. doi:10.1002/jssc.201400256 PMID:24777942

Kawashima, S., Hou, P., Corr, D. J., & Shah, S. P. (2013). Modification of cement-based materials with nanoparticles. *Cement and Concrete Composites*, *36*, 8–15.

Compilation of References

Kelsall, R. W., Hamley, I. W., & Geoghegan, M. (2004). *Nanoscale science and technology.* Chisester, UK: John Wiley and Sons, Ltd.

Kerienė, J., Kligys, M., Laukaitis, A., Yakovlev, G., Špokauskas, A., & Aleknevičius, M. (2013). The influence of multi-walled carbon nanotubes additive on properties of non-autoclaved and autoclaved aerated concretes. *Construction & Building Materials, 49,* 527–535.

Khan, M. S. (2011). Nanotechnology in transportation: Evolution of a revolutionary technology. *TR News,* (277), 3-8.

Khanna, V., Bakshi, B. R., & Lee, L. J. (2008). Carbon nanofiber production. *Journal of Industrial Ecology, 12*(3), 394–410. doi:10.1111/j.1530-9290.2008.00052.x

Khitab, A., & Arshad, M. T. (2014). Nano Construction Materials, *Rev. Adv. Mat. Sci., 3,* 131–138.

Khoshakhlagh, A., Nazari, A., & Khalaj, G. (2012). Effects of Fe_2O_3 Nanoparticles on Water Permeability and Strength Assessments of High Strength Self-Compacting Concrete. *Journal of Materials Science and Technology, 28*(1), 73–82.

Kim, H. K., Nam, I. W., & Lee, H. K. (2012). Microstructure and mechanical/EMI shielding characteristics of CNT/cement composites with various silica fume contents. *UKC 2012 on Science Technology, and Entrepreneurship,* 8-11.

Kim, Y. T., Han, J. H., Hong, B. H., & Kwon, Y. U. (2010). Electrochemical Synthesis of CdSe quantum-Dot arrays on a graphene basal plane using mesoporous silica thin-film templates. *Advanced Materials, 22*(4), 515–518. doi:10.1002/adma.200902736

Kim, Y., & Zimmerman, S. C. (1998). Applications of dendrimers in bio-organic chemistry. *Current Opinion in Chemical Biology, 2*(6), 733–742. doi:10.1016/S1367-5931(98)80111-7 PMID:9914193

Kjellsen, K. O., Jennings, H. M., & Lagerblad, B. (1996). Evidence of hollow shells in the microstructure of cement paste. *Cement and Concrete Research, 26*(4), 593–599.

Knol, A. B., de Hartog, J. J., Boogaard, H., Slottje, P., van der Sluijs, J. P., Lebret, E., & Brunekreef, B. et al. (2009). Expert elicitation on ultrafine particles: Likelihood of health effects and causal pathways. *Particle and Fibre Toxicology, 6*(1), 1. doi:10.1186/1743-8977-6-19 PMID:19630955

Köksal, F., Altun, F., Yiğit, I., & Şahin, Y. (2008). Combined effect of silica fume and steel fiber on the mechanical properties of high strength concretes. *Construction & Building Materials, 22*(8), 1874–1880. doi:10.1016/j.conbuildmat.2007.04.017

Koleva, D. A. (2008). *Nano-materials with tailored properties for self healing of corrosion damages in reinforced concrete, IOP self healing materials.* SenterNovem, TheNetherlands.

Kong, D., Du, X., Wei, S., Zhang, H., Yang, Y., & Shah, S. P. (2012). Influence of nano-silica agglomeration on microstructure and properties of the hardened cement-based materials. *Construction & Building Materials, 37,* 707–715.

Konsta-Gdoutos, M. S., Metaxa, Z. S., & Shah, S. P. (2010). Multi-scale mechanical and fracture characteristics and early-age strain capacity of high performance carbon nanotube/cement nanocomposites. *Cement and Concrete Composites*, *32*(2), 110–115.

Koo, Y., Littlejohn, G., Collins, B., Yun, Y., Shanov, V. N., Schulz, M., & Sankar, J. et al. (2014). Synthesis and characterization of Ag–TiO 2–CNT nanoparticle composites with high photocatalytic activity under artificial light. *Composites. Part B, Engineering*, *57*, 105–111. doi:10.1016/j.compositesb.2013.09.004

Kostoff, R. N., Koytcheff, R. G., & Lau, C. G. (2007). Global nanotechnology research literature overview. *Technological Forecasting and Social Change*, *74*(9), 1733–1747. doi:10.1016/j.techfore.2007.04.004

Kramer, B. (2003). *Advances in Solid State Physics* (Vol. 43). Springer. doi:10.1007/b12017

Krivanek, O. L., Dellby, N., & Lupini, A. R. (1999). Towards sub-electron beams. *Ultramicroscopy*, *78*(1-4), 1–11. doi:10.1016/S0304-3991(99)00013-3

Krivanek, O. L., Nellist, P. D., Dellby, N., Murfitt, M. F., & Szilagyi, Z. (2003). Towards sub-0.5 Å electron beams. *Ultramicroscopy*, *96*(3), 229–237. doi:10.1016/S0304-3991(03)00090-1 PMID:12871791

Kuempel, E. D. (2006). *Titanium dioxide 93*. International Agency for Research on Cancer.

Kuennen, T. (2003). Correcting Problem Concrete Pavements. *Better Roads*, *73*, 32–38.

Kuhlbusch, T., Asbach, C., Fissan, H., Gohler, D., & Stintz, M. (2011). Nanoparticle exposure at nanotechnology workplaces: A review. *Particle and Fibre Toxicology*, *8*(1), 22. doi:10.1186/1743-8977-8-22 PMID:21794132

Kumar, A. P., Depan, D., Singh Tomer, N., & Singh, R. P. (2009). Nanoscale particles for polymer degradation and stabilization—Trends and future perspectives. *Progress in Polymer Science*, *34*(6), 479–515. doi:10.1016/j.progpolymsci.2009.01.002

Kumar, S., Kolay, P., Malla, S., & Mishra, S. (2012). Effect of Multiwalled Carbon Nanotubes on Mechanical Strength of Cement Paste. *Journal of Materials in Civil Engineering*, *24*(1), 84–91. doi:10.1061/(ASCE)MT.1943-5533.0000350

Kyzas, G. Z., & Kostoglou, M. (2014). Green Adsorbents for Wastewaters: A Critical Review. *Materials (Basel)*, *7*(1), 333–364. doi:10.3390/ma7010333

Lademann, J., Weigmann, H. J., Rickmeyer, C., Barthelmes, H., Schaefer, H., Mueller, G., & Sterry, W. (1998). Penetration of titanium dioxide microparticles in a sunscreen formulation into the horny layer and the follicular orifice. *Skin Pharmacology and Applied Skin Physiology*, *12*(5), 247–256. doi:10.1159/000066249 PMID:10461093

Lagergren, S. (1898). About the theory of so-called adsorption of soluble substances, *Kungliga Svenska Vetenskapsakademiens. Handlingar*, *24*(4), 1–39.

Compilation of References

Lai, H., Chen, W., & Chiang, L. (2000). Free radical scavenging activity of fullerenol on the ischemia-reperfusion intestine in dogs. *World Journal of Surgery*, *24*(4), 450–454. doi:10.1007/s002689910071 PMID:10706918

Lam, C. W., James, J. T., McCluskey, R., & Hunter, R. L. (2004). Pulmonary toxicity of single-wall carbon nanotubes in mice 7 and 90 days after intratracheal instillation. *Toxicological Sciences*, *77*(1), 126–134. doi:10.1093/toxsci/kfg243 PMID:14514958

Land, G., & Stephan, D. (2012). The influence of nano-silica on the hydration of ordinary Portland cement. *Journal of Materials Science*, *47*(2), 1011–1017. doi:10.1007/s10853-011-5881-1

Langmuir, I. (1916). The constitution and fundamental properties of solids and liquids. *Journal of the American Chemical Society*, *38*(11), 2221–2295. doi:10.1021/ja02268a002

Lasagni, F., Lasagni, A., Holzapfel, C., Mücklich, F., & Degischer, H. P. (2006). Three Dimensional Characterization of Unmodified and Sr-Modified Al-Si Eutectics by FIB and FIB EDX Tomography. *Advanced Engineering Materials*, *8*(8), 719–723. doi:10.1002/adem.200500276

Lazzara, G., & Milioto, S. (2010). Dispersions of nanosilica in biocompatible copolymers. *Polymer Degradation & Stability*, *95*(4), 610–617. doi:10.1016/j.polymdegradstab.2009.12.007

Lee, C.-J., Scheufele, D. A., & Lewenstein, B. V. (2005). Public attitudes toward emerging technologies. *Science Communication*, *7*(2), 240–267. doi:10.1177/1075547005281474

Lee, J., Mahendra, S., & Alvarez, P. J. (2010). Nanomaterials in the construction industry: A review of their applications and environmental health and safety considerations. *ACS Nano*, *4*(7), 3580–3590. doi:10.1021/nn100866w PMID:20695513

Lee, S., Seo, Y., & Kim, Y. R. (2010). Effect of hydrated lime on dynamic modulus of asphalt-aggregate mixtures in the state of North Carolina. *KSCE Journal of Civil Engineering*, *14*(6), 829–837. doi:10.1007/s12205-010-0944-4

Lee, W., Kang, S. H., Kim, J. Y., Kolekar, G. B., Sung, Y. E., & Han, S. H. (2009). TiO2 nanotubes with a ZnO thin energy barrier for improved current efficiency of CdSe quantum-dot-sensitized solar cells. *Nanotechnology*, *20*(33), 335706. doi:10.1088/0957-4484/20/33/335706

Lefeuvre, S., Federova, E., Gomonova, O., & Tao, J. (2010). Microwave Sintering of Micro-and Nano-Sized Alumina Powder. *Advances in Modeling of Microwave sintering*, 8-9.

Lefeuvre, S., & Gomonova, O. (2012). Modeling at the Nano Level: Application to Physical Processes. In S. M. Musa (Ed.), *Computational Finite Element Methods in Nanotechnology* (pp. 559–583). Taylor & Francis. doi:10.1201/b13002-17

Lesko, S., Lesniewska, E., Nonat, A., Mutin, J.-C., & Goudonnet, J.-P. (2001). Investigation by atomic force microscopy of forces at the origin of cement cohesion. *Ultramicroscopy*, *86*(1), 11–21. doi:10.1016/S0304-3991(00)00091-7 PMID:11215612

Leydekker, S. (2008). *Nanomaterials in architecture, interior architecture and design*. Birkhäuser Verlag AG, BaselBoston-Berlin.

Li, W., Li, Y., & Wang, X. (2010). *Notice of retraction nano-TiO$_2$ modified super-hydrophobic surface coating for anti-ice and anti-flashover of insulating layer.* Paper presented at the Power and Energy Engineering Conference (APPEEC).

Li, B., & Zhong, W. H. (2011). Review on polymer/graphite nanoplatelet nanocomposites. *Journal of Materials Science, 46*(17), 5595–5614. doi:10.1007/s10853-011-5572-y

Liberti, L., & Kucherenko, S. (2005). Comparison of deterministic and stochastic approaches to global optimization. *International Transactions in Operational Research, 12*(3), 263–285. doi:10.1111/j.1475-3995.2005.00503.x

Li, C., & Chou, T. (2008). Modeling of damage sensing in fiber composites using carbon nanotube networks. *Composites Science and Technology, 68*(15), 3373–3379.

Li, G. (2004). Properties of high-volume fly ash concrete incorporating nano-SiO$_2$. *Cement and Concrete Research, 34*(6), 1043–1049.

Li, G. Y., Wang, P. M., & Zhao, X. (2007). Pressure-sensitive properties and microstructure of carbon nanotube reinforced cement composites. *Cement and Concrete Composites, 29*(5), 377–382. doi:10.1016/j.cemconcomp.2006.12.011

Li, G., Wang, P., & Zhao, X. (2005). Mechanical behavior and microstructure of cement composites incorporating surface-treated multi-walled carbon nanotubes. *Carbon, 43*(6), 1239–1245.

Li, H., Xiao, H. G., & Ou, J. P. (2004). A study on mechanical and pressure-sensitive properties of cement mortar with nanophase materials. *Cement and Concrete Research, 34*(3), 435–438.

Li, H., Xiao, H., Guan, X., Wang, Z., & Yu, L. (2014). Chloride diffusion in concrete containing nano-TiO$_2$ under coupled effect of scouring. *Composites. Part B, Engineering, 56*, 698–704.

Li, H., Xiao, H., & Ou, J. (2006). Effect of compressive strain on electrical resistivity of carbon black-filled cement-based composites. *Cement and Concrete Composites, 28*(9), 824–828.

Li, H., Xiao, H., & Ou, J. (2008). Electrical property of cement-based composites filled with carbon black under long-term wet and loading condition. *Composites Science and Technology, 68*(9), 2114–2119.

Li, H., Xiao, H., Yuan, J., & Ou, J. (2004). Microstructure of cement mortar with nano-particles. *Composites. Part B, Engineering, 35*(2), 185–189.

Li, H., Zhang, M. H., & Ou, J. P. (2007). Flexural fatigue performance of concrete containing nano-particles for pavement. *International Journal of Fatigue, 29*(7), 1292–1301. doi:10.1016/j.ijfatigue.2006.10.004

Li, H., Zhang, M., & Ou, J. (2006). Abrasion resistance of concrete containing nano-particles for pavement. *Wear, 260*(11-12), 1262–1266.

Compilation of References

Li, L., Fan, M., Brown, R. C., & Leeuwen, J. V. (2006). Synthesis, Properties, and Environmental Applications of Nanoscale Iron-Based Materials: A Review. *Critical Reviews in Environmental Science and Technology*, *36*(5), 405–431. doi:10.1080/10643380600620387

Lim, S. C., Kim, K. S., Jeong, S. Y., Cho, S., Yoo, J. E., & Lee, Y. H. (2005). Nanomanipulator-assisted fabrication and characterization of carbon nanotubes inside scanning electron microscope. *Micron (Oxford, England)*, *36*(5), 471–476. doi:10.1016/j.micron.2005.03.005 PMID:15896968

Lim, S., & Mondal, P. (2014). Micro- and nano-scale characterization to study the thermal degradation of cement-based materials. *Materials Characterization*, *92*, 15–25.

Lim, T.-C. (2007). Long range relationship between Morse and Lennard–Jones potential energy functions. *Molecular Physics*, *105*(8), 1013–1018. doi:10.1080/00268970701261449

Linch, K. D. (2002). Respirable concrete dust--silicosis hazard in the construction industry. *Applied Occupational and Environmental Hygiene*, *17*(3), 209–221. doi:10.1080/104732202753438298 PMID:11871757

Lin, D. F., Lin, K. L., Chang, W. C., Luo, H. L., & Cai, M. Q. (2008). Improvements of nano-SiO_2 on sludge/fly ash mortar. *Waste Management (New York, N.Y.)*, *28*(6), 1081–1087.

Lin, D., & Xing, B. (2007). Phytotoxicity of nanoparticles: Inhibition of seed germination and root growth. *Environmental Pollution*, *150*(2), 243–250. doi:10.1016/j.envpol.2007.01.016 PMID:17374428

Lin, S., Reppert, J., Hu, Q., Hudson, J. S., Reid, M. L., Ratnikova, T. A., & Ke, P. C. et al. (2009). Uptake, translocation, and transmission of carbon nanomaterials in rice plants. *Small*, *5*(10), 1128–1132. PMID:19235197

Little, D. N., Males, E. H., Prusinski, J. R., & Stewart, B. (2000). *Cementitious stabilization*. Transportation in the New Millennium.

Liu, J. F., Zhao, Z. S., & Jiang, G. B. (2008). Coating Fe_3O_4 Magnetic Nanoparticles with Humic Acid for High Efficient Removal of Heavy Metals in Water. *Environmental Science & Technology*, *42*(18), 6949–6954. doi:10.1021/es800924c PMID:18853814

Liu, Q., Tong, T., Liu, S., Yang, D., & Yu, Q. (2014). Investigation of using hybrid recycled powder from demolished concrete solids and clay bricks as a pozzolanic supplement for cement. *Construction & Building Materials*, *73*, 754–763. doi:10.1016/j.conbuildmat.2014.09.066

Liu, X., Chen, L., Liu, A., & Wang, X. (2012). Effect of Nano-$CaCO_3$ on Properties of Cement Paste. *Energy Procedia*, *16*, 991–996.

Liu, Z., Cai, W., He, L., Nakayama, N., Chen, K., Sun, X., & Dai, H. et al. (2006). In vivo biodistribution and highly efficient tumour targeting of carbon nanotubes in mice. *Nature Nanotechnology*, *2*(1), 47–52. doi:10.1038/nnano.2006.170 PMID:18654207

Liu, Z., Fujita, N., Miyasaka, K., Han, L., Stevens, S. M., Suga, M., & Terasaki, O. (2013). A review of fine structures of nanoporous materials as evidenced by microscopic methods. *Microscopy*, *62*(1), 109–146. doi:10.1093/jmicro/dfs098 PMID:23349242

Li, W., Buschhorn, S. T., Schulte, K., & Bauhofer, W. (2011). The imaging mechanism, imaging depth, and parameters influencing the visibility of carbon nanotubes in a polymer matrix using an SEM. *Carbon*, *49*(6), 1955–1964. doi:10.1016/j.carbon.2010.12.069

Li, X., Cai, W., An, J., Kim, S., Nah, J., Yang, D., & Ruoff, R. S. et al. (2009). Large-area synthesis of high-quality and uniform graphene films on copper foils. *Science*, *324*(5932), 1312–1314. doi:10.1126/science.1171245 PMID:19423775

Li, Z. J., Wang, L., Yuan, L. Y., Xiao, C. L., Mei, L., Zheng, L. R., & Shi, W. Q. et al. (2015). Efficient removal of uranium from aqueous solution by zero-valent iron nanoparticle and its graphene composite. *Journal of Hazardous Materials*, *290*, 26–33. doi:10.1016/j.jhazmat.2015.02.028 PMID:25734531

Li, Z., Wang, H., He, S., Lu, Y., & Wang, M. (2006). Investigations on the preparation and mechanical properties of the nano-alumina reinforced cement composite. *Materials Letters*, *60*(3), 356–359.

López-Arce, P., Garcia-Guinea, J., Gracia, M., & Obis, J. (2003). Bricks in historical buildings of Toledo City: Characterisation and restoration. *Materials Characterization*, *50*(1), 59–68. doi:10.1016/S1044-5803(03)00101-3

Lovern, S. B., & Klaper, R. (2006). Daphnia magna mortality when exposed to titanium dioxide and fullerene (C60) nanoparticles. *Environmental Toxicology and Chemistry*, *25*(4), 1132–1137. doi:10.1897/05-278R.1 PMID:16629153

Lovern, S. B., Strickler, J. R., & Klaper, R. (2007). Behavioral and physiological changes in Daphnia magna when exposed to nanoparticle suspensions (titanium dioxide, nano-C60, and C60HxC70Hx). *Environmental Science & Technology*, *41*(12), 4465–4470. doi:10.1021/es062146p PMID:17626453

Lovric, J., Bazzi, H., Cuie, Y., Fortin, G., Winnik, F., & Maysinger, D. (2005). Differences in subcellular distribution and toxicity of green and red emitting CdTe quantum dots. *Journal of Molecular Medicine (Berlin, Germany)*, *83*(5), 377–385. doi:10.1007/s00109-004-0629-x PMID:15688234

Ltifi, M., Guefrech, A., Mounanga, P., & Khelidj, A. (2011). *Experimental study of the effect of addition of nano-silica on the behaviour of cement mortars.* Paper presented at the Procedia Engineering. doi:10.1016/j.proeng.2011.04.148

Lucas, S. S., Ferreira, V. M., & de Aguiar, J. L. B. (2013). Incorporation of titanium dioxide nanoparticles in mortars -Influence of microstructure in the hardened state properties and photocatalytic activity. *Cement and Concrete Research*, *43*, 112–120.

Compilation of References

Ludvig, P., Calixto, J. M., Ladeira, L. O., & Gaspar, I. C. P. (2011). Using Converter Dust to Produce Low Cost Cementitious Composites by in situ Carbon Nanotube and Nanofiber Synthesis. *Materials (Basel)*, *4*(12), 575–584.

Luo, J., Duan, Z., & Li, H. (2009). The influence of surfactants on the processing of multi-walled carbon nanotubes in reinforced cement matrix composites.[a]. *Physica Status Solidi*, *206*(12), 2783–2790.

Lv, S., Ma, Y., Qiu, C., Sun, T., Liu, J., & Zhou, Q. (2013). Effect of graphene oxide nanosheets of microstructure and mechanical properties of cement composites. *Construction & Building Materials*, *49*, 121–127.

Mackie, W. A., Morrissey, J. L., Hinrichs, C. H., & Davis, P. R. (1992). Field emission from hafnium carbide. *Journal of Vacuum Science & Technology. A, Vacuum, Surfaces, and Films*, *10*(4), 2852–2856. doi:10.1116/1.577719

Maher, M., Uzarowski, L., Moore, G., & Aurilio, V. (2006, November). Sustainable Pavements-Making the Case for Longer Design Lives for Flexible Pavements. In *Proceedings of the Fifty-First Annual Conference of the Canadian Technical Asphalt Association (CTAA)*.

Majeed, Z. H., Taha, M. R., & Jawad, I. T. (2014). Stabilization of Soft Soil Using Nanomaterials. *Research Journal of Applied Sciences. Engineering and Technology*, *8*(4), 503–509.

Makar, J. M., Margeson, J. C., & Luh, J. (2005). *Carbon nanotube/cement composites-early results and potential applications.* Paper presented at 3rd International Conference on Construction Materials: Performance, Innovations and Structural Implications, Vancouver, Canada.

Makar, J., Margeson, J., & Luh, J. (2005). *Carbon nanotube/cement composites-early results and potential applications.* Paper presented at the 3rd International Conference on Construction Materials: Performance, Innovations and Structural Implications, Vancouver, Canada.

Makar, J. M., & Beaudoin, J. J. (2003). Carbon nanotubes and their application in the construction industry.*Proceedings, 1 Inter. Symp. on Nanotechnology in Constructionst*, (pp. 331-341).

Makar, J. M., & Chan, G. W. (2009). Growth of Cement Hydration Products on Single-Walled Carbon Nanotubes. *Journal of the American Ceramic Society*, *92*(6), 1303–1310.

Makar, J. M., & Chan, G. W. (2009). Growth of Cement Hydration Products on Single-Walled Carbon Nanotubes. *Journal of the American Ceramic Society*, *92*(6), 1303–1310. doi:10.1111/j.1551-2916.2009.03055.x

Makar, J., Margeson, J., & Luh, J. (2005). Carbon nanotube/cement composites - early results and potential applications. *NRCC-47643, 3rd International Conference on Construction Materials: Performance, Innovations and Structural Implications*, (pp. 1-10).

Malik, M. A., Wani, M. Y., & Hashim, M. A. (2014). Microemulsion method: A novel route to synthesize organic and inorganic nanomaterials: 1st Nano Update. *Arabian Journal of Chemistry*, *5*(4), 397–417. doi:10.1016/j.arabjc.2010.09.027

Mamalis, A. G., Vogtländer, L. O. G., & Markopoulos, A. (2004). Nanotechnology and nanostructured materials: Trends in carbon nanotubes. *Precision Engineering*, *28*(1), 16–30. doi:10.1016/j.precisioneng.2002.11.002

Mane, V. S., & Vijay Babu, P. V. (2011). Studies on the adsorption of Brilliant Green dye from aqueous solution onto low-cost NaOH treated saw dust. *Desalination*, *273*(2-3), 321–329. doi:10.1016/j.desal.2011.01.049

Mann, A. K. P., & Skrabalak, S. E. (2011). Synthesis of single-crystalline nanoplates by spray pyrolysis: A metathesis route to Bi2WO6. *Chemistry of Materials*, *23*(4), 1017–1022. doi:10.1021/cm103007v

Manoharan, D., Loganathan, A., Kurapati, V., & Nesamony, V. J. (2015). Unique sharp photoluminescence of size-controlled sonochemically synthesized zirconia nanoparticles. *Ultrasonics Sonochemistry*, *23*, 174–184. doi:10.1016/j.ultsonch.2014.10.004 PMID:25453213

Mao, Q., Zhao, B., Sheng, D., Yang, Y., & Shui, Z. (1995). Resistance Changement of Compression Sensible Cement Specimen Under Different Stresses.*Workshop on Smart Materials*, (pp. 35-39).

Masahito, S., Kashiwagi, Y., Li, Y., Arstila, K., Richard, O., Cott, D. J., & Vereecken, P. M. et al. (2011). Measuring the electrical resistivity and contact resistance of vertical carbon nanotube bundles for application as interconnects. *Nanotechnology*, *22*(8), 085302. doi:10.1088/0957-4484/22/8/085302 PMID:21242623

Materazzi, A. L., Ubertini, F., & D'Alessandro, A. (2013). Carbon nanotube cement-based transducers for dynamic sensing of strain. *Cement and Concrete Composites*, *37*, 2–11.

Maynard, A. D. (2006). *A research strategy for addressing risk. In Nanotechnology*. Woodrow Wilson International Center for Scholars.

Maynard, A. D., Warheit, D. B., & Philbert, M. A. (2010). The new toxicology of sophisticated materials: Nanotoxicology and beyond. *Toxicological Sciences*. PMID:21177774

Mayo, J. T., Yavuz, C., Yean, S., Cong, L., Shipley, H., Yu, W., & Colvin, V. L. et al. (2007). The effect of nanocrystalline magnetite size on arsenic removal. *Science and Technology of Advanced Materials*, *8*(1-2), 71–75. doi:10.1016/j.stam.2006.10.005

McCarthy, M. J. &. D. (1999). Towards maximising the use of fly ash as a binder. *Fuel*, *78*(2), 121–132.

McQuarrie, D. A. (2008). *Quantum chemistry*. University Science Books.

Meng, T., Yu, Y., Qian, X., Zhan, S., & Qian, K. (2012). Effect of nano-TiO_2 on the mechanical properties of cement mortar. *Construction & Building Materials*, *29*, 241–245.

Meridian Institute. (2006). *Overview and comparison of conventional water treatment technologies and nano based water treatment technologies. In Global Dialogue on nanotechnology and the poor: opportunities and risks*. Chennai, India: Meridian Institute.

Compilation of References

Merritt, F. S., & Ricketts, J. T. (2001). *Building design and construction handbook* (Vol. 13). New York, NY: McGraw-Hill.

Metaxa, Z. S., Konsta-Gdoutos, M. S., & Shah, S. P. (2010). Carbon Nanofiber-Reinforced cement-based materials. *Transportation Research Record*, *2142*, 114–118. doi:10.3141/2142-17

Meyer, E. (1992). Atomic force microscopy. *Progress in Surface Science*, *41*(1), 3–49. doi:10.1016/0079-6816(92)90009-7

Meyer, J. C., Girit, C. O., Crommie, M. F., & Zettl, A. (2008). Imaging and dynamics of light atoms and molecules on graphene. *Nature*, *454*(7202), 319–322. doi:10.1038/nature07094 PMID:18633414

Minsky, M. (1988). Memoir on inventing the confocal scanning microscope. *Scanning*, *10*(4), 128–138. doi:10.1002/sca.4950100403

Mohan, D., & Pittman, C. U. Jr. (2007). Arsenic removal from water/wastewater using adsorbents—A critical review. *Journal of Hazardous Materials*, *142*(1-2), 1–53. doi:10.1016/j.jhazmat.2007.01.006 PMID:17324507

Mohapatra, M., & Anand, S. (2010). Synthesis and applications of nano-structured iron oxides/hydroxides – a review. *International Journal of Engineering Science and Technology*, *2*(8), 127–146.

Mohapatra, M., Anand, S., Mishra, B. K., Giles, D. E., & Singh, P. (2009). Review of fluoride removal from drinking water. *Journal of Environmental Management*, *91*(1), 67–77. doi:10.1016/j.jenvman.2009.08.015 PMID:19775804

Mojtaba, G., Morteza, M. S., Majid, T., Jalal, K. R., & Reza, T. (2012). Modification of stone matrix asphalt with nano-SiO2. *J. Basic Appl. Sci. Res*, *2*(2), 1338–1344.

Mokerov, V. G., Fedorov, Y. V., Velikovski, L. E., & Scherbakova, M. Y. (2001). New quantum dot transistor. *Nanotechnology*, *12*(4), 552–555. doi:10.1088/0957-4484/12/4/336

Monteiro, P. J., Clodic, L., Battocchio, F., Kanitpanyacharoen, W., Chae, S. R., Ha, J., & Wenk, H. R. (2013). Incorporating carbon sequestration materials in civil infrastructure: A micro and nano-structural analysis. *Cement and Concrete Composites*, *40*, 14–20. doi:10.1016/j.cemconcomp.2013.03.013

Morse, G. (2004). You Don't Have a Nanostrategy? *Harvard Business Review*, *82*(2), 32.

Morse, P. M. (1929). Diatomic Molecules According to the Wave Mechanics. II. Vibrational Levels. *Physical Review*, *34*(1), 57–64. doi:10.1103/PhysRev.34.57

Morsy, M. S., & Aglan, H. A. (2007). Development and characterization of nanostructured-perlite-cementitious surface compounds. *Journal of Materials Science*, *42*(24), 10188–10195.

Morsy, M. S., Alsayed, S. H., & Aqel, M. (2011). Hybrid effect of carbon nanotube and nano-clay on physico-mechanical properties of cement mortar. *Construction & Building Materials*, *25*(1), 145–149.

Moussavi, G., & Mahmoudi, M. (2009). Removal of azo and anthraquinone reactive dyes from industrial wastewaters using MgO nanoparticles. *Journal of Hazardous Materials*, *168*(2-3), 806–812. doi:10.1016/j.jhazmat.2009.02.097 PMID:19303210

Mo, Y. L., & Roberts, R. H. (2013). Carbon Nanofiber Concrete for Damage Detection of Infrastructure. In R. Maguire (Ed.), *Advances in Nanofibers* (pp. 125–143). Rijeka, Croatia: InTech.

Münch, B., Gasser, P., Holzer, L., & Flatt, R. (2006). FIB-Nanotomography of Particulate Systems—Part II: Particle Recognition and Effect of Boundary Truncation. *Journal of the American Ceramic Society*, *89*(8), 2586–2595. doi:10.1111/j.1551-2916.2006.01121.x

Munoz, J. F., Sanfilippo, J. M., Tejedor, M. I., Anderson, M. A., & Cramer, S. M. (2008). *Preliminary study of the effects of nanoporous films in ITZ properties of concrete*. Poster presented at 87th Annual Transportation Research Board meeting, TRB 2008, Washington, DC.

Murata, Y., Tawara, H., Obata, H., & Takeuchi, K. (1999). Air purifying pavement: Development of photocatalytic concrete blocks. *J Adv Oxidat Technol*, *4*(2), 227–230.

Murray, A. R., Kisin, E., Kommineni, C., Kagan, V. E., Castranova, V., & Shvedova, A. A. (2007). Single-walled carbon nanotubes induce oxidative stress and inflammation in skin. *The Toxicologist*, *96*, A1406.

Musa, S. M. (2012). *Computational Finite Element Methods in Nanotechnology*. Taylor & Francis. doi:10.1201/b13002

Musso, S., Tulliani, J., Ferro, G., & Tagliaferro, A. (2009). Influence of carbon nanotubes structure on the mechanical behavior of cement composites. *Composites Science and Technology*, *69*(11-12), 1985–1990.

Nakaso, K., Han, B., Ahn, K. H., Choi, M., & Okuyama, K. (2003). Synthesis of non-agglomerated nanoparticles by an electrospray assisted chemical vapor deposition (ES-CVD) method. *Aerosol Science*, *34*(7), 869–881. doi:10.1016/S0021-8502(03)00053-3

Nanocor Technical. (n.d.). *Nanoclay Structures*. Retrieved 12 November 2014, from http://www.nanocor.com/nano_struct.asp

Nanomaterials in Maintenance Work: Occupational Risks and Prevention. (n.d.). *European Agency for safety and health at work*. Retrieved July 11, 2015 from https://osha.europa.eu/en/publications/e-facts/e-fact-74-nanomaterials-in-maintenance-work-occupational-risks-and-prevention

NanoScienceWorks. (n.d.). *Types of Carbon Nanotubes*. Retrieved 04/03/2015, 2014, from http://www.nanoscienceworks.org/Members/siebo/657px-types_of_carbon_nanotubes.png/view

Nanotechnologies, Nanoparticles: What Hazards – What Risks? (2006). Ministry of Ecology and Sustainable Development (MOE & SD). Retrieved from http://www.developpement-durable.gouv.fr/IMG/pdf/CPP_NanotechnologiesNanoparticles.pdf

Nasibulin, A. G., Tapper, U., Tian, Y., Penttala, V., Karppinen, M. J., Kauppinen, E. I. S., & Malm, J. E. M. (2009). A novel cement-based hybrid material. *New Journal of Physics*, *11*(2), 23013.

Compilation of References

Nasibulina, L. I., Anoshkin, I. V., Semencha, A. V., Tolochko, O. V., Malm, J. E., Karppinen, M. J., & Kauppinen, E. I. (2012). Carbon nanofiber/clinker hybrid material as a highly efficient modificator of mortar mechanical properties. *Materials Physics and Mechanics, 13,* 77–84.

National Research Council. (2006). *Geological and geotechnical engineering in the new millennium: Opportunities for research and technological innovation.* Washington, DC: The National Academies Press.

Nawrocki, J., & Kasprzyk-Hordern, B. (2010). The efficiency and mechanisms of catalytic ozonation. *Applied Catalysis B: Environmental, 99*(1-2), 27–42. doi:10.1016/j.apcatb.2010.06.033

Nayak, B. B., Behera, D., & Mishra, B. K. (2010). Synthesis of silicon carbide dendrite by the arc plasma process and observation of nanorod bundles in the dendrite arm. *Journal of the American Ceramic Society, 93*(10), 3080–3083. doi:10.1111/j.1551-2916.2010.04060.x

Nazari, A., Rafieipour, M. H., & Riahi, S. (2011). The effects of CuO nanoparticles on properties of self compacting concrete with GGBFS as binder. *Materials Research-Ibero-American Journal of Materials, 14*(3), 307–316.

Nazari, A., & Riahi, S. (2010). The effect of TiO_2 nanoparticles on water permeability and thermal and mechanical properties of high strength self-compacting concrete. *Materials Science and Engineering A, 528*(2), 756–763.

Nazari, A., & Riahi, S. (2011). The Effects of TiO_2 Nanoparticles on Flexural Damage of Self-compacting Concrete. *International Journal of Damage Mechanics, 20*(7), 1049–1072.

Nazari, A., & Riahi, S. (2011a). The effects of SiO_2 nanoparticles on physical and mechanical properties of high strength compacting concrete. *Composites. Part B, Engineering, 42*(3), 570–578.

Nazari, A., & Riahi, S. (2011b). Abrasion resistance of concrete containing SiO_2 and Al_2O_3 nanoparticles in different curing media. *Energy and Building, 43*(10), 2939–2946.

Nazari, A., & Riahi, S. (2011c). Compressive strength and abrasion resistance of concrete containing SiO_2 and Cr_2O_3 nanoparticles in different curing media. *Magazine of Concrete Research, 64*(2), 177–188.

Nazari, A., & Riahi, S. (2011d). Al_2O_3 nanoparticles in concrete and different curing media. *Energy and Building, 43*(6), 1480–1488.

Nazari, A., & Riahi, S. (2011e). Improvement compressive strength of concrete in different curing media by Al_2O_3 nanoparticles. *Materials Science and Engineering A, 528*(3), 1183–1191.

Nazari, A., & Riahi, S. (2011g). TiO_2 nanoparticles effects on physical, thermal and mechanical properties of self compacting concrete with ground granulated blast furnace slag as binder. *Energy and Building, 43*(4), 995–1002.

Nazari, A., & Riahi, S. (2011h). Computer-aided design of the effects of Fe_2O_3 nanoparticles on split tensile strength and water permeability of high strength concrete. *Materials & Design, 32*(7), 3966–3979.

Nazari, A., & Riahi, S. (2011i). The effects of zinc dioxide nanoparticles on flexural strength of self-compacting concrete. *Composites. Part B, Engineering, 42*(2), 167–175.

Nazari, A., & Riahi, S. (2011j). Effects of CuO nanoparticles on compressive strength of self-compacting concrete. *Sadhana, 36*(3), 371–391.

Nazari, A., Riahi, S., Riahi, S., Shamekhi, S. F., & Khademno, A. (2010). Benefits of Fe_2O_3 nanoparticles in concrete mixing matrix. *J Am Sci, 6*(4), 102–106.

Nazari, A., Riahi, S., Riahi, S., Shamekhi, S. F., & Khademno, A. (2010). Mechanical properties of cement mortar with Al_2O_3 nanoparticles. *Journal of American Science, 6*(4), 94–97.

Nazari, A., Sh, R., Sh, R., Shamekhi, S. F., & Khademno, A. (2010a). Influence of Al_2O_3 nanoparticles on the compressive strength and workability of blended concrete. *Journal of American Science, 6*(5), 6–9.

Nazari, A., Sh, R., Sh, R., Shamekhi, S. F., & Khademno, A. (2010b). Benefits of Fe_2O_3 nanoparticles in concrete mixing matrix. *Journal of American Science, 6*(4), 102–106.

Nazari, A., Sh, R., Sh, R., Shamekhi, S. F., & Khademno, A. (2010c). The effects of incorporation Fe_2O_3 nanoparticles on tensile and flexural strength of concrete. *Journal of American Science, 6*(4), 90–93.

Neaspec Gmb, H. (n.d.). *Nano-FTIR – Nanoscale Infrared Spectroscopy with a thermal source*. Retrieved 03/03/2015, 2015, from http://www.neaspec.com/application/nano-ftir-nanoscale-infrared-spectroscopy-with-a-thermal-source/

Neitola, R., Ruuska, H., & Pakkanen, T. A. (2005). Ab Initio Studies on Nanoscale Friction between Graphite Layers: Effect of Model Size and Level of Theory. *The Journal of Physical Chemistry B, 109*(20), 10348–10354. doi:10.1021/jp044065q PMID:16852254

Nielsen, E., Ostergaard, G., & Larsen, J. (2007). *Toxicological Risk Assessment of Chemicals: A Practical Guide* (pp. 2–3). New York, NY: Informa Healthcare.

Nielsen, E., Ostergaard, G., & Larsen, J. C. (2008). *Toxicological risk assessment of chemicals: A practical guide*. CRC Press. doi:10.1201/9781420006940

Niroumanda, H., Zainb, M. F. M., & Alhosseini, S. N. (2013). The Influence of Nano-Clays on Compressive Strength of Earth Bricks as Sustainable Materials.*Procedia - Social and Behavioral Sciences*. doi:10.1016/j.sbspro.2013.08.945

Nochaiya, T., & Chaipanich, A. (2011). Behavior of multi-walled carbon nanotubes on the porosity and microstructure of cement-based materials. *Applied Surface Science, 257*(6), 1941–1945.

Noorvand, H., Abang Ali, A. A., Demirboga, R., Farzadnia, N., & Noorvand, H. (2013). Incorporation of nano TiO2 in black rice husk ash mortars. *Construction & Building Materials, 47*, 1350–1361.

Nordan, M. M. (2005). *Nanotechnology: where does the US stand*. Lux Research Incorporated.

Compilation of References

Oberdörster, E. (2004). Manufactured nanomaterials (fullerenes, C60) induce oxidative stress in juvenile largemouth bass. *Environmental Health Perspectives, 112*(10), 1058–1062. doi:10.1289/ehp.7021 PMID:15238277

Oberdörster, E. (2004). *Toxicity of C60 Fullerenes to Two Aquatic Species: Daphnia and Largemouth Bass*. Anaheim, CA: American Chemical Society.

Oberdörster, G., Oberdörster, E., & Oberdörster, J. (2005). Nanotoxicology: An emerging discipline evolving from studies of ultrafine particles. *Environmental Health Perspectives, 113*(7), 823–839. doi:10.1289/ehp.7339 PMID:16002369

Oberdürster, G. (2000). Toxicology of ultrafine particles: in vivo studies. *Philosophical Transactions of the Royal Society of London. Series A: Mathematical, Physical and Engineering Sciences, 358*(1775), 2719-2740.

Office of the Press Secretary. (2000). National Nanotechnology Initiative - Leading To The Next Industrial Revolution. *Microscale Thermophysical Engineering, 4*(3), 205–212. doi:10.1080/10893950050148160

Ohara, S., Sato, K., Tan, Z., Shimoda, H., Ueda, M., & Fukui, T. (2010). Novel mechanochemical synthesis of fine $FeTiO_3$ nanoparticles by a high-speed ball-milling process. *Journal of Alloys and Compounds, 504*(1), L17–L19. doi:10.1016/j.jallcom.2010.05.090

Okada, T., Kawashima, K., Nakata, Y., & Ning, X. (2005). Synthesis of ZnO nanorods by laser ablation of ZnO and Zn targets in He and O2 background gas. *Japanese Journal of Applied Physics, Part 1: Regular Papers and Short Notes and Review Papers, 44*(1 B), 688-691.

Oliver, W. C., & Pharr, G. M. (2004). Measurement of hardness and elastic modulus by instrumented indentation: Advances in understanding and refinements to methodology. *Journal of Materials Research, 19*(01), 3–20. doi:10.1557/jmr.2004.19.1.3

Oltulu, M., & Şahin, R. (2011). Single and combined effects of nano-SiO_2, nano-Al_2O_3 and nano-Fe_2O_3 powders on compressive strength and capillary permeability of cement mortar containing silica fume. *Materials Science and Engineering A, 528*(22-23), 7012–7019.

Oltulu, M., & Şahin, R. (2013). Effect of nano-SiO_2, nano-Al_2O_3 and nano-Fe_2O_3 powders on compressive strengths and capillary water absorption of cement mortar containing fly ash: A comparative study. *Energy and Building, 58*, 292–301.

Oltulu, M., & Şahin, R. (2014). Pore structure analysis of hardened cement mortars containing silica fume and different nano-powders. *Construction & Building Materials, 53*, 658–664.

Ou, J., & Han, B. (2009). Piezoresistive cement-based strain sensors and self-sensing concrete components. *Journal of Intelligent Material Systems and Structures, 20*(3), 329–336.

Owlad, M., Aroua, M. K., Daud, W. A. W., & Baroutian, S. (2009). Removal of Hexavalent Chromium-Contaminated Water and Wastewater: A Review. *Water, Air, and Soil Pollution, 200*(1-4), 59–77. doi:10.1007/s11270-008-9893-7

Özen, H. (2011). Rutting evaluation of hydrated lime and SBS modified asphalt mixtures for laboratory and field compacted samples. *Construction & Building Materials, 25*(2), 756–765. doi:10.1016/j.conbuildmat.2010.07.010

Öztürk, A., & Malkoc, E. (2014). Adsorptive potential of cationic Basic Yellow 2 (BY2) dye onto natural untreated clay (NUC) from aqueous phase: Mass transfer analysis, kinetic and equilibrium profile. *Applied Surface Science, 299*, 105–115. doi:10.1016/j.apsusc.2014.01.193

Pacewska, B., Bukowska, M., Wilińska, I., & Swat, M. (2002). Modification of the properties of concrete by a new pozzolan—a waste catalyst from the catalytic process in a fluidized bed. *Cement and Concrete Research, 32*(1), 145–152. doi:10.1016/S0008-8846(01)00646-9

Pacheco-Torgal, F., & Jalali, S. (2011). Nanotechnology: Advantages and drawbacks in the field of construction and building materials. *Construction & Building Materials, 25*(2), 582–590. doi:10.1016/j.conbuildmat.2010.07.009

Pan, B., & Xing, B. (2008). Adsorption mechanisms of organic chemicals on carbon nanotubes. *Environmental Science & Technology, 42*(24), 9005–9013. doi:10.1021/es801777n PMID:19174865

Papakonstantinou, C. G. (2005). Protective coatings with nano constituent materials.*2nd International Symposium on Nanotechnology in Construction*, Bilbao, Spain

Papanikolaou, N. C., Hatzidaki, E. G., Belivanis, S., Tzanakakis, G. N., & Tsatsakis, A. M. (2005). Lead toxicity update. A brief review. *Medical Science Monitor, 11*(10), 329–336. PMID:16192916

Park, J. W., Benz, C. C., & Martin, F. J. (2004). Future directions of liposome-and immunoliposome-based cancer therapeutics.[). WB Saunders.]. *Seminars in Oncology, 31*, 196–205. doi:10.1053/j.seminoncol.2004.08.009 PMID:15717745

Paul, B., Parashar, V., & Mishra, A. (2015). Graphene in the Fe_3O_4 nano-composite switching the negative influence of humic acid coating into an enhancing effect in the removal of arsenic from water. *Environmental Science and Water Research Technology, 1*(1), 77–83. doi:10.1039/C4EW00034J

Pedersen, J. A., Yeager, M. A., & Suffet, I. H. (2003). Xenobiotic Organic Compounds in Runoff from Fields Irrigated with Treated Wastewater. *Journal of Agricultural and Food Chemistry, 51*(5), 1360–1372. doi:10.1021/jf025953q PMID:12590482

Pelekani, C., & Snoeyink, V. L. (2001). A kinetic and equilibrium study of competitive adsorption between atrazine and congo red dye on activated carbon: The importance of pore size distribution. *Carbon, 39*(1), 25–37. doi:10.1016/S0008-6223(00)00078-6

Pendleton, P., & Wu, S. H. (2003). Kinetics of dodecanoic acid adsorption from caustic solution by activated carbon. *Journal of Colloid and Interface Science, 266*(2), 245–250. doi:10.1016/S0021-9797(03)00575-7 PMID:14527446

Peyvandi, A., Sbia, L. A., Soroushian, P., & Sobolev, K. (2013). Effect of the cementitious paste density on the performance efficiency of carbon nanofiber in concrete nanocomposite. *Construction & Building Materials, 48*, 265–269.

Compilation of References

Peyvandi, A., Soroushian, P., Abdol, N., & Balachandra, A. M. (2013). Surface-modified graphite nanomaterials for improved reinforcement efficiency in cementitious paste. *Carbon*, *63*, 175–186.

Peyvandi, A., Soroushian, P., Balachandra, A. M., & Sobolev, K. (2013). Enhancement of the durability characteristics of concrete nanocomposite pipes with modified graphite nanoplatelets. *Construction & Building Materials*, *47*, 111–117.

Peyvandi, A., Soroushian, P., & Jahangirnejad, S. (2013). Enhancement of the structural efficiency and performance of concrete pipes through fiber reinforcement. *Construction & Building Materials*, *45*, 36–44. doi:10.1016/j.conbuildmat.2013.03.084

Picraux, S. T. (2014). *Nanotechnology*. Retrieved 9 November 2014, from http://www.britannica.com/EBchecked/topic/962484/nanotechnology/236436/Properties-at-the-nanoscale

Pierce, J. (2004). Safe as sunshine? *The Engineer*. Retrieved August 23, 2015 from http://www2.er.dtu.dk/publications/fulltext/2007/MR2007-239.pdf

Pillay, K., Cukrowsk, E. M., & Coville, N. J. (2009). Multi-walled carbon nanotubes as adsorbents for the removal of parts per billion levels of hexavalent chromium from aqueous solution. *Journal of Hazardous Materials*, *166*(2-3), 1067–1075. doi:10.1016/j.jhazmat.2008.12.011 PMID:19157694

Plassard, C., Lesniewska, E., Pochard, I., & Nonat, A. (2004). Investigation of the surface structure and elastic properties of calcium silicate hydrates at the nanoscale. *Ultramicroscopy*, *100*(3), 331–338. doi:10.1016/j.ultramic.2003.11.012 PMID:15231326

Plassard, C., Lesniewska, E., Pochard, I., & Nonat, A. (2005). Nanoscale experimental investigation of particle interactions at the origin of the cohesion of cement. *Langmuir*, *21*(16), 7263–7270. doi:10.1021/la050440+ PMID:16042451

Pokropivny, V. V., & Skorokhod, V. V. (2007). Classification of nanostructures by dimensionality and concept of surface forms engineering in nanomaterial science. *Materials Science and Engineering C*, *27*(5), 990–993. doi:10.1016/j.msec.2006.09.023

Poland, C., Duffin, R., Kinloch, I., Maynard, A., Wallace, W., Seaton, A., & Donaldson, K. et al. (2008). Carbon nanotubes introduced into the abdominal cavity of mice show asbestos-like pathogenicity in a pilot study. *Nature Nanotechnology*, *3*(7), 423–428. doi:10.1038/nnano.2008.111 PMID:18654567

Poon, B., Rittel, D., & Ravichandran, G. (2008). An analysis of nanoindentation in linearly elastic solids. *International Journal of Solids and Structures*, *45*(24), 6018–6033. doi:10.1016/j.ijsolstr.2008.07.021

Portland Cement Association. (n.d.). *Concrete Pavement*. Retrieved 13/03/2015, 2015, from http://www.cement.org/cement-concrete-basics/products/concrete-pavement

Pospíchal, O., Kucharczyková, B., Misák, P., & Vymazal, T. (2010). Freeze-thaw resistance of concrete with porous aggregate. *Procedia Engineering*, *2*(1), 521–529. doi:10.1016/j.proeng.2010.03.056

Pottier, A., Chanéac, C., Tronc, E., Mazerolles, L., & Jolivet, J. P. (2001). Synthesis of brookite TiO2 nanoparticlesby thermolysis of TiCl4 in strongly acidic aqueous media. *Journal of Materials Chemistry, 11*(4), 1116–1121. doi:10.1039/b100435m

Pradhan, D., & Leung, K. T. (2008). Vertical growth of two-dimensional zinc oxide nanostructures on ITO-coated glass: Effects of deposition temperature and deposition time. *The Journal of Physical Chemistry C, 112*(5), 1357–1364. doi:10.1021/jp076890n

Puvaneshwari, N., Muthukrishnan, J., & Gunasekaran, P. (2006). Toxicity assessment and microbial degradation of azo dyes. *Indian Journal of Experimental Biology, 44*(8), 618–626. PMID:16924831

Pyrgiotakis, G., Blattmann, C. O., Sotiris, P., & Philip, D. (2013). Nanoparticle-nanoparticle interactions in biological media by Atomic Force Microscopy. *Langmuir, 29*(36), 11385–11395. doi:10.1021/la4019585 PMID:23978039

Qian, K. L., Meng, T., Qian, X. Q., & Zhan, S. L. (2009). Research on some properties of fly ash concrete with nano-$CaCO_3$ middle slurry. *Key Engineering Materials, 405*, 186–190.

Qingwen, L., Zuyuan, H., Tokunaga, T., & Hotate, K. (2011, Aug. 28). *An ultra-high-resolution large-dynamic-range fiber optic static strain sensor using Pound-Drever-Hall technique.* Paper presented at the Quantum Electronics Conference & Lasers and Electro-Optics (CLEO/IQEC/PACIFIC RIM).

Quercia, G., Spiesz, P., Husken, G., & Brouwers, J. (2012). *Effects of amorphous nano-silica additions on mechanical and durability performance of SCC mixtures.* Paper presented at the International congress on durability of concrete, Trondheim, Norway.

Qu, X. L., Brame, J., Li, Q., & Alvarez, J. J. P. (2013b). Nanotechnology for a safe and sustainable water supply: Enabling integrated water treatment and reuse. *Accounts of Chemical Research, 46*(3), 834–843. doi:10.1021/ar300029v PMID:22738389

Qu, X., Alvarez, J. J. P., & Li, Q. (2013a). Applications of nanotechnology in water and wastewater treatment. *Water Research, 47*(12), 3931–3946. doi:10.1016/j.watres.2012.09.058 PMID:23571110

Rachel's Environment and Health News. (2003). Retrieved from http://www.rachel.org/files/rachel/Rachels_Environment_Health_News_2362.pdf

Rakesh, K., Renu, M., & Arun, K. M. (2011). *Opportunities & Challenges for Use of Nanotechnology in Cement-Based Materials.* New Delhi: Rigid Pavements Division, Central Road Research Institute (CRRI).

Ramamoorthy, S. K., Skrifvars, M., & Persson, A. (2015). A review of natural fibers used in biocomposites: Plant, animal and regenerated cellulose fibers. *The Polish Review, 55*(1), 107–162. doi:10.1080/15583724.2014.971124

Rana, A. K., Rana, S. B., Kumari, A., & Kiran, V. (2009). Significance of nanotechnology in construction engineering. *International Journal of Recent Trends in Engineering, 1*(4), 46–48.

Compilation of References

Rapaport, D. C. (2004). *The art of molecular dynamics simulation*. Cambridge university press. doi:10.1017/CBO9780511816581

Ravishankar, T. N., Ramakrishnappa, T., Nagaraju, G., & Rajanaika, H. (2014). Synthesis and Characterization of CeO_2 Nanoparticles via Solution Combustion Method for Photocatalytic and Antibacterial Activity Studies. *Chemistry Open*, *4*(2), 146–154. PMID:25969812

Raziyeh, S., Arami, M., Mahmoodi, N. M., Bahrami, H., & Khorramfar, S. (2010). Novel biocompatible composite (Chitosan–zinc oxide nanoparticle): Preparation,characterization and dye adsorption properties. *Colloids and Surfaces. B, Biointerfaces*, *80*(1), 86–93. doi:10.1016/j.colsurfb.2010.05.039 PMID:20566273

Redlich, O., & Peterson, D. L. (1959). A useful adsorption isotherm. *Journal of Physical Chemistry*, *63*(6), 1024–1026. doi:10.1021/j150576a611

Renwick, L. C., Brown, D., Clouter, A., & Donaldson, K. (2004). Increased inflammation and altered macrophage chemotactic responses caused by two ultrafine particles types. *Occupational and Environmental Medicine*, *61*(5), 442–447. doi:10.1136/oem.2003.008227 PMID:15090666

Research, C. A. C. (2011). The Effects of ZnO_2 Nanoparticles on Strength Assessments and Water Permeability of Concrete in Different Curing Media. *Materials Research-Ibero-American Jouranl of Materials*, *14*(2), 178–188.

Revel, G. M., Martarelli, M., Bengochea, M. Á., Gozalbo, A., Orts, M. J., Gaki, A., & Emiliani, M. et al. (2013). Nanobased coatings with improved NIR reflecting properties for building envelope materials: Development and natural aging effect measurement. *Cement and Concrete Composites*, *36*(1), 128–135. doi:10.1016/j.cemconcomp.2012.10.002

Reynolds, G. H. (2003). Nanotechnology and regulatory policy: Three futures. *Harvard Journal of Law & Technology*, *17*, 179.

Riahi, S., & Nazari, A. (2011a). Compressive strength and abrasion resistance of concrete containing SiO_2 and CuO nanoparticles in different curing media. *Science China-Technological Sciences*, *54*(9), 2349–2357.

Riahi, S., & Nazari, A. (2011b). Physical, mechanical and thermal properties of concrete in different curing media containing ZnO_2 nanoparticles. *Energy and Building*, *43*(8), 1977–1984.

Richardson, I. G. (1999). The nature of C-S-H in hardened cements. *Cement and Concrete Research*, *29*(8), 1131–1147.

Richardson, I. G., & Groves, G. W. (1992). Microstructure and microanalysis of hardened cement pastes involving ground granulated blast-furnace slag. *Journal of Materials Science*, *27*(22), 6204–6212.

Roco, M. C., & Bainbridge, W. S. (2005). Societal implications of nanoscience and nanotechnology: Maximizing human benefit. *Journal of Nanoparticle Research*, *7*(1), 1–13. doi:10.1007/s11051-004-2336-5

Roduner, E. (2006). *Nanoscopic Materials Size-Dependent Phenomena* (1st ed.). Stuttgart, Germany: RSC Publishing.

Rose, H. (1990). Outline of a spherically corrected semiaplanatic medium-voltage transmission electron-microscope. *Optik (Stuttgart)*, *85*(1), 19–24.

Roy, A., Adhikari, B., & Majumder, S. B. (2013). Equilibrium, Kinetic, and Thermodynamic Studies of Azo Dye Adsorption from Aqueous Solution by Chemically Modified Lignocellulosic Jute Fiber. *Industrial & Engineering Chemistry Research*, *52*(19), 6502–6512. doi:10.1021/ie400236s

Rudin, T., & Pratsinis, S. E. (2012). Homogeneous Iron Phosphate Nanoparticles by Combustion of Sprays. *Industrial & Engineering Chemistry Research*, *51*(23), 7891–7900. doi:10.1021/ie202736s PMID:23407874

Saafi, M. (2009). Wireless and embedded carbon nanotube networks for damage detection in concrete structures. *Nanotechnology*, *20*(39), 395502.

Sáez De Ibarra, Y., Gaitero, J. J., Erkizia, E., & Campillo, I. (2006). Atomic force microscopy and nanoindentation of cement pastes with nanotube dispersions. *Physica Status Solidi (a)*, *203*(6), 1076-1081.

Sáez De Ibarra, Y., Gaitero, J. J., Erkizia, E., & Campillo, I. (2006). Atomic force microscopy and nanoindentation of cement pastes with nanotube dispersions. *Physica Status Solidi (A). Applications and Materials Science*, *203*(6), 1076–1081.

Safiuddin, M., Gonzalez, M., Cao, J., & Tighe, S. L. (2014). State-of-the-art report on use of nano-materials in concrete. *International Journal of Pavement Engineering*, *15*(10), 940–949. doi:10.1080/10298436.2014.893327

Sahm, T., Mädler, L., Gurlo, A., Barsan, N., Pratsinis, S. E., & Weimar, U. (2004). Flame spray synthesis of tin dioxide nanoparticles for gas sensing. *Sensors and Actuators. B, Chemical*, *98*(2-3), 148–153. doi:10.1016/j.snb.2003.10.003

Sahoo, S. K., Parveen, S., & Panda, J. J. (2007). The present and future of nanotechnology in human health care. *Nanomedicine; Nanotechnology, Biology, and Medicine*, *3*(1), 20–31. doi:10.1016/j.nano.2006.11.008 PMID:17379166

Said, A. M., Zeidan, M. S., Bassuoni, M. T., & Tian, Y. (2012). Properties of concrete incorporating nano-silica. *Construction & Building Materials*, *36*, 838–844.

Saini, R., Saini, S., & Sharma, S. (2010). Nanotechnology: The future medicine. *Journal of Cutaneous and Aesthetic Surgery*, *3*(1), 32.

Saito, R., Dresselhaus, G., & Dresselhaus, M. S. (1998). *Physical Properties of Carbon Nanotubes*. World Scientific.

Sajanlal, P. R., Sreeprasad, T. S., Samal, A. K., & Pradeep, T. (2011). *Anisotropic nanomaterials: structure, growth, assembly, and functions*. doi: 10.3402/nr.v2i0.5883

Compilation of References

Sakamoto, T., Cheng, Z., Takahashi, M., Owari, M., & Nihei, Y. (1998). Development of an ion and electron dual focused beam apparatus for three-dimensional microanalysis. *Japanese Journal of Applied Physics*, *37*(4R), 2051–2056. doi:10.1143/JJAP.37.2051

Salam, M. A., Gabal, M. A., & Obaid, A. Y. (2012). Preparation and characterization of magnetic multi-walled carbon nanotubes/ferrite nanocomposite and its application for the removal of aniline from aqueous solution. *Synthetic Metals*, *161*(23), 2651–2658. doi:10.1016/j.synthmet.2011.09.038

Salemi, N., & Behfarnia, K. (2013). Effect of nano-particles on durability of fiber-reinforced concrete pavement. *Construction & Building Materials*, *48*(0), 934–941. doi:10.1016/j.conbuildmat.2013.07.037

Salerno, M., Landoni, P., & Verganti, R. (2008). Designing foresight studies for Nanoscience and Nanotechnology (NST) future developments. *Technological Forecasting and Social Change*, *75*(8), 1202–1223. doi:10.1016/j.techfore.2007.11.011

Salman, S. A., Usami, T., Kuroda, K., & Okido, M. (2014). Synthesis and characterization of cobalt nanoparticles using hydrazine and citric acid. Journal of Nanotechnology. Article ID 525193, 6 pages.

Salvetat, J. P., Bonard, J. M., Thomson, N. B., Kulik, A. J., Forró, L., Benoit, W., & Zuppiroli, L. (1999). Mechanical properties of carbon nanotubes. *Applied Physics. A, Materials Science & Processing*, *69*(3), 255–260. doi:10.1007/s003390050999

Sanchez, F., & Sobolev, K. (2010). Nanotechnology in concrete - A review. *Construction & Building Materials*, *24*(11), 2060–2071. doi:10.1016/j.conbuildmat.2010.03.014

Sandyadav. (2012). Retrieved September 02, 2015 from http://commons.wikimedia.org/wiki/File%3AWhite_Marble_Rocks_at_Bhedaghat.jpg

Sawada, H., Sasaki, T., Hosokawa, F., Yuasa, S., Terao, M., Kawazoe, M., Nakamichi, T., Kaneyama, T., Kondo, Y., Kimoto, K. & Suenaga, K. (2009). Correction of higher order geometrical aberration by triple 3-fold astigmatism field. *Journal of Electron Microscopy*, 1-7.

Saxl, O. (2001). *Opportunities for Industry in the Application of Nanotechnology*. The Institute of Nanotechnology.

Sayes, C., Marchione, A., Reed, K., & Warheit, D. (2007). Comparative pulmonary toxicity assessments of C60 water suspensions in rats: Few differences in fullerene toxicity in vivo in contrast to in vitro profiles. *Nano Letters*, *7*(8), 2399–2406. doi:10.1021/nl0710710 PMID:17630811

Scarfato, P., Di Maio, L., Fariello, M. L., Russo, P., & Incarnato, L. (2012). Preparation and evaluation of polymer/clay nanocomposite surface treatments for concrete durability enhancement. *Cement and Concrete Composites*, *34*(3), 297–305. doi:10.1016/j.cemconcomp.2011.11.006

Schaffer, M., Wagner, J., Schaffer, B., Schmied, M., & Mulders, H. (2007). Automated three-dimensional X-ray analysis using a dual-beam FIB. *Ultramicroscopy*, *107*(8), 587–597. doi:10.1016/j.ultramic.2006.11.007 PMID:17267131

Schierz, A., & Zänker, H. (2009). Aqueous suspensions of carbon nanotubes: Surface oxidation, colloidal stability and uranium sorption. *Environmental Pollution, 157,* 1088-1094.

Schneider, T. (2007). Evaluation and control of occupational health risks from nanoparticles. Tema Nord. Nordic Council of Ministers, Copenhagen, 2007:581.

Schodek, D. L., Ferreira, P., & Ashby, M. F. (2009). *Nanomaterials, nanotechnologies and design: an introduction for engineers and architects.* Butterworth-Heinemann.

Schulte, P., Geraci, C., Zumwalde, R., Hoover, M., & Kuempel, E. (2008). Occupational risk management of engineered nanoparticles. *Journal of Occupational and Environmental Hygiene, 5*(4), 239–249. doi:10.1080/15459620801907840 PMID:18260001

Scott-Fordsmand, J., Krogh, P., Schaefer, M., & Johansen, A. (2008). The toxicity testing of double-walled nanotubescontaminated food to Eisenia veneta earthworms. *Ecotoxicology and Environmental Safety, 71*(3), 616–619. doi:10.1016/j.ecoenv.2008.04.011 PMID:18514310

Scrivener, K. L., & Kirkpatrick, R. J. (2008). Innovation in use and research on cementitious material. *Cement and Concrete Research, 38*(2), 128–136.

Selvam, R. P., Subramani, V. J., Murray, S., & Hall, K. (2009). *Potential application of nanotechnology on cement based materials.* Potential Application of Nanotechnology on Cement Based Materials.

Senff, L., Labrincha, J. A., Ferreira, V. M., Hotza, D., & Repette, W. L. (2009). Effect of nano-silica on rheology and fresh properties of cement pastes and mortars. *Construction & Building Materials, 23*(7), 2487–2491.

Şengül, H., Theis, T., & Ghosh, S. (2008). Towards sustainable nano products: An overview of nanomanufacturing methods. *Journal of Industrial Ecology, 12*(3), 329–359.

Shah, A. H., Sharma, U. K., Roy, D. A., & Bhargava, P. (2013). Spalling behaviour of nano SiO_2 high strength concrete at elevated temperatures. In *Proceedings of MATEC Web of Conferences.* London, UK: EDP Sciences.

Shah, S. P., Konsta-Gdoutos, M. S., & Metaxa, Z. S. (2010). Exploration of fracture characteristics, nanoscale properties and nanostructure of cementitious matrices with carbon nanotubes and carbon nanofibers. In *Proceedings of the 7th international conference on fracture mechanics of concrete and concrete structures.* Seoul, Korea: Korea Concrete Institute.

Shaikh, F. U. A., & Supit, S. W. M. (2014). Mechanical and durability properties of high volume fly ash (HVFA) concrete containing calcium carbonate ($CaCO_3$) nanoparticles. *Construction & Building Materials, 70,* 309–321.

Shaikh, F. U. A., & Supit, S. W. M. (2015). Chloride induced corrosion durability of high volume fly ash concretes containing nano particles. *Construction & Building Materials, 99,* 208–225. doi:10.1016/j.conbuildmat.2015.09.030

Compilation of References

Sharma, Y. C., & Uma. (2010). Optimization of Parameters for Adsorption of Methylene Blue on a Low-Cost Activated Carbon. *J. Chem. Eng. Data, 55*(11), 435–439.

Sharma, A., Wang, J., & Gutierrez, M. S. (2007). *Nanoscale Simulations of Rock and Clay Minerals*. Advances in Measurement and Modeling of Soil Behavior.

Sharma, Y. C., Srivastava, V., Upadhyay, S. N., & Weng, C. H. (2008). Alumina Nanoparticles for the Removal of Ni(II) from Aqueous Solutions. *Industrial & Engineering Chemistry Research, 47*(21), 8095–8100. doi:10.1021/ie800831v

Sheela, T., Nayaka, Y. A., Viswanatha, R., Basavanna, S., & Venkatesha, T. G. (2012). Kinetics and thermodynamics studies on the adsorption of Zn (II), Cd (II) and Hg (II) from aqueous solution using zinc oxide nanoparticles. *Powder Technology, 217*, 163–170. doi:10.1016/j.powtec.2011.10.023

Shekari, A. H., & Razzaghi, M. S. (2011). Influence of Nano Particles on Durability and Mechanical Properties of High Performance Concrete. *Procedia Engineering, 14*, 3036–3041.

Shen, J., Li, Z., Wu, Y., Zhang, B., & Li, F. (2015). Dendrimer-based preparation of mesoporous alumina nanofibers by electrospinning and their application in dye adsorption. *Chemical Engineering Journal, 264*, 48–55. doi:10.1016/j.cej.2014.11.069

Shih, J. Y., Chang, T. P., & Hsiao, T. C. (2006). Effect of nanosilica on characterization of Portland cement composite. *Materials Science and Engineering A-structural Materials Properties Microstructure and Processing, 424*(1-2), 266–274.

Shi, T., Schins, R., Knaapen, A., Kuhlbusch, T., Pitz, M., Heinrich, J., & Borm, P. (2003). Hydroxyl radical generation by electron paramagnetic resonance as a new method to monitor ambient particulate matter composition. *Journal of Environmental Monitoring, 5*(4), 550–556. doi:10.1039/b303928p PMID:12948226

Shon, H., Phuntsho, S., Okour, Y., Cho, D. L., Kim, K. S., Li, H. J., & Kim, J. H. et al. (2008). Visible light responsive titanium dioxide (TiO_2). *Journal of the Korean Industrial and Engineering Chemistry, 19*(1), 1–16.

Shukla, P., Bhatia, V., Gaur, V., Basniwal, R. K., Singh, B. K., & Jain, V. K. (2012). Multiwalled carbon nanotubes reinforced portland cement composites for smoke detection. *Solid State Phenomena, 185*, 21–24.

Siddiqi, Z. A., & Ashraf, M. (2009). *Steel Structures* (2nd ed.). Lahore: Help Civil Engineering Publisher.

Siddique, R., & Mehta, A. (2014). Effect of carbon nanotubes on properties of cement mortars. *Construction & Building Materials, 50*(0), 116–129. doi:10.1016/j.conbuildmat.2013.09.019

Siegel, R. W., Hu, E., & Roco, M. C. (1999, September). Nanostructure science and technology: a worldwide study. *IWGN*.

Siegrist, M., & Cvetkovich, G. (2000). Perception of hazards: The role of social trust and knowledge. *Risk Analysis, 20*(5), 713–719. doi:10.1111/0272-4332.205064 PMID:11110217

Siegrist, M., Keller, C., Kastenholz, H., Frey, S., & Wiek, A. (2007). Laypeople's and experts' perception of nanotechnology hazards. *Risk Analysis, 27*(1), 59–69. doi:10.1111/j.1539-6924.2006.00859.x PMID:17362400

Siever, R. (1979). Plate-tectonic controls on diagenesis. *The Journal of Geology, 87*(2), 127–155. doi:10.1086/628405

Silva, A. J., Varesche, M. B., Foresti, E., & Zaiat, M. (2002). Sulphate removal from industrial wastewater using a packed-bed anaerobic reactor. *Process Biochemistry, 37*(9), 927–935. doi:10.1016/S0032-9592(01)00297-7

Sindu, B. S., Sasmal, S., & Gopinath, S. (2014). A multi-scale approach for evaluating the mechanical characteristics of carbon nanotube incorporated cementitious composites. *Construction & Building Materials, 50*, 317–327.

Singh. (2005). Engineering Materials (5th ed.). Academic Press.

Singh, A. P., Gupta, B. K., Mishra, M., Chandra, A., Mathur, R. B., & Dhawan, S. K. (2013). Multiwalled carbon nanotube/cement composites with exceptional electromagnetic interference shielding properties. *Carbon, 56*(0), 86–96.

Singh, A. P., Mishra, M., Chandra, A., & Dhawan, S. K. (2011). Graphene oxide/ferrofluid/cement composites for electromagnetic interference shielding application. *Nanotechnology, 22*(46), 465701.

Singh, J. P., Dixit, G., Srivastava, R. C., Agrawal, H. M., Reddy, V. R., & Gupta, A. (2012). Observation of bulk like magnetic ordering below the blocking temperature in nanosized zinc ferrite. *Journal of Magnetism and Magnetic Materials, 324*(16), 2553–2559. doi:10.1016/j.jmmm.2012.03.045

Singh, M., Kumar, M., Štěpánek, F., Ulbrich, P., Svoboda, P., Santava, E., & Singla, M. L. (2011). Liquid-Phase Synthesis of Nickel Nanoparticles stabilized by PVP and study of their structural and magnetic properties. *Advanced Material Letters, 2*(6), 409–414. doi:10.5185/amlett.2011.4257

Sips, R. (1948). Combined form of Langmuir and Freundlich equations. *The Journal of Chemical Physics, 16*(5), 490–495. doi:10.1063/1.1746922

Skinner, L. B., Chae, S. R., Benmore, C. J., Wenk, H. R., & Monteiro, P. J. M. (2010). Nanostructure of calcium silicate hydrates in cements. *Physical Review Letters, 104*(19), 195502. doi:10.1103/PhysRevLett.104.195502

Skorokhod, V., Ragulya, A., & Uvarova, I. (2001). *Physico-chemical Kinetics in Nanostructured Systems*. Kyiv: Academperiodica.

Smith, M. R. (Ed.). (1999). *Stone: Building stone, rock fill and armourstone in construction*. Geological Society of London.

Compilation of References

Snell, L. M., & Snell, B. G. (2002). Oldest concrete street in the United States. *Concrete International-Detroit, 24*(3), 72–74.

Sobolev, K., Flores, I., Hermosillo, R., & Torres-Martínez, L. M. (2006). Nanomaterials and Nanotechnology for High-Performance Cement Composites. *ACI Session on Nanotechnology of Concrete: Recent Developments and Future Perspectives*, 91–118.

Sobolev, K., Flores, I., Hermosillo, R., & Torres-Martínez, L. M. (2008). *Nanomaterials and nanotechnology for high-performance cement composites.* Paper presented at the American Concrete Institute, ACI Special Publication.

Sobolev, K., & Gutiérrez, M. F. (2005). How nanotechnology can change the concrete world. *American Ceramic Society Bulletin, 84*(10), 14.

So, M. R., & Voter, A. F. (2000). Temperature-accelerated dynamics for simulation of infrequent events. *The Journal of Chemical Physics, 112*(21), 9599–9606. doi:10.1063/1.481576

Srivastava, V., Sharma, Y. C., & Sillanpää, M. (2015 a). Green synthesis of magnesium oxide nanoflower and its application for the removal of divalent metallic species from synthetic wastewater. *Ceramics International, 44*(5), 6702–6709. doi:10.1016/j.ceramint.2015.01.112

Srivastava, V., Sharma, Y. C., & Sillanpää, M. (2015 b). Application of nano-magnesso ferrite (n-MgFe2O4) for the removal Co^{2+} ions from synthetic wastewater: Kinetic, equilibrium and thermodynamic studies. *Applied Surface Science, 338*, 42–54. doi:10.1016/j.apsusc.2015.02.072

Stankiewicz, A., Szczygieł, I., & Szczygieł, B. (2013). Self-healing coatings in anti-corrosion applications. *Journal of Materials Science, 48*(23), 8041–8051. doi:10.1007/s10853-013-7616-y

Steinfeldt, M., Von Gleich, A., Petschow, U., Haum, R., Chudoba, T., & Haubold, S. (2004). Nachhaltigkeitseffekte durch Herstellung und Anwendung nanotechnischer Produkte. Institut für ökologische Wirtschaftsforschung GmbH, Berlin, Germany.

Stevens, S. M., Jansson, K., Xiao, C., Asahina, S., Klingstedt, M., Grüner, D., & Europe, J. E. O. L. (2009). An appraisal of high resolution scanning electron microscopy applied to porous materials. *JEOL News, 44*(1), 17.

Steyn, W. J. M. (2008). *Development of auto-luminescent surfacing for concrete pavements.* Paper presented at Transportation Research Board, Washington, DC.

Steyn. (2008). Research and Application of Nanotechnology in Transportation. *27th Southern African Transport Conference*.

Stone, V., (2010). *Engineered nanoparticles: review of health and environmental safety (ENRHES).* Edinburgh, FP7.

Sumer, M. (2012). Compressive strength and sulfate resistance properties of concretes containing Class F and Class C fly ashes. *Construction & Building Materials, 34*, 531–536.

Sunter, C. (1996). *The high road: Where are we now.* Cape Town, South Africa: Tafelberg.

Supit, S. W., Shaikh, F. U., Singh, L. P., & Ahalawat, S. K. B. S. (2014). Effect of nano-CaCO$_3$ on compressive strength development of high volume fly ash mortars and concretes. *Journal of Advanced Concrete Technology, 12*, 178–186.

Sutherland, W. J., Bardsley, S., Bennun, L., Clout, M., Côté, I. M., Depledge, M. H., & Gibbons, D. W. et al. (2011). Horizon scan of global conservation issues for 2011. *Trends in Ecology & Evolution, 26*(1), 10–16. doi:10.1016/j.tree.2010.11.002 PMID:21126797

Swamy, R. N., Ali, S. A., & Theodorakopoulos, D. D. (1983, September). Early strength fly ash concrete for structural applications. *ACI Journal Proceedings, 80*(5), 414–423.

Syed. (1970). *Materials of Construction*. Academic Press.

Tabor, D. (1996). Indentation hardness: Fifty years on a personal view. *Philosophical Magazine A, 74*(5), 1207–1212. doi:10.1080/01418619608239720

Taha, M. R. (2009). Geotechnical Properties of Soil-Ball Milled Soil Mixtures. In Z. Bittnar, P. M. Bartos, J. Němeček, V. Šmilauer, & J. Zeman (Eds.), *Nanotechnology in Construction 3* (pp. 377–382). Springer Berlin Heidelberg. doi:10.1007/978-3-642-00980-8_51

Taha, M. R., & Ying, T. (2010). Effects of carbon nanotube on kaolinite: Basic geotechnical behavior. *World Journal Of Engineering, 7*(2), 472–473.

Takeda, K., Suzuki, K., Ishihara, A., Kubo-Irie, M., Fujimoto, R., Tabata, M., & Sugamata, M. et al. (2009). Nanoparticles transferred from pregnant mice to their offspring can damage the genital and cranial nerve systems. *Journal of Health Science, 55*(1), 95–102. doi:10.1248/jhs.55.95

Taniguchi, N. (1974). On the basic concept of nanotechnology. *Proceedings of the International Conference on Production Engineering*.

Tani, T., Lutz, M., & Pratsinis, S. E. (2002). Homogeneous ZnO nanoparticles by flame spray pyrolysis. *Journal of Nanoparticle Research, 4*(4), 337–343. doi:10.1023/A:1021153419671

Taplin, J. H. (1959). A method for following the hydration reaction in portland cement paste. *Australian Journal of Applied Science, 10*(3), 329–345.

Taylor, H. F. W. (1997). Cement Chemistry (2nd ed.). London, UK: Thomas Telford. doi:10.1680/cc.25929

Tchebinyayeff, M. R. (1981). *U.S. Patent No. 4,306,587*. Washington, DC: U.S. Patent and Trademark Office.

Tegart, G. (2009). Energy and nanotechnologies: Priority areas for Australia's future. *Technological Forecasting and Social Change, 76*(9), 1240–1246. doi:10.1016/j.techfore.2009.06.010

Tempkin, M. I., & Pyzhev, V. (1940). Kinetics of ammonia synthesis on promoted iron catalyst. *Acta Physicochimica U.R.S.S., 12*, 327–356.

Compilation of References

Templeton, R., Ferguson, P., Washburn, K., Scrivens, W., & Chandler, G. (2006). Life-cycle effects of single-walled carbon nanotubes (SWNTs) on an estuarine meiobenthic copepod. *Environmental Science & Technology*, *40*(23), 7387–7393. doi:10.1021/es060407p PMID:17180993

Teng, M., Qiao, J., Li, F., & Bera, P. K. (2012). Electrospun mesoporous carbon nanofibers produced from phenolic resin and their use in the adsorption of large dye molecules. *Carbon*, *50*(8), 2877–2886. doi:10.1016/j.carbon.2012.02.056

Teoh, W. Y., Amal, R., & Lutz, M. (2010). Flame spray pyrolysis: An enabling technology for nanoparticles design and fabrication. *Nanoscale*, *2*(8), 1324–1347. doi:10.1039/c0nr00017e PMID:20820719

Thiruchitrambalam, M., Palkar, V. R., & Gopinathan, V. (2004). Hydrolysis of aluminium metal and sol-gel processing of nano alumina. *Materials Letters*, *58*(24), 3063–3066. doi:10.1016/j.matlet.2004.05.043

Thomas, J. J., & Jennings, H. M. (2006). A colloidal interpretation of chemical aging of the C-S-H gel and its effects on the properties of cement paste. *Cement and Concrete Research*, *36*(1), 30–38.

Thomas, M. D., & Bamforth, P. B. (1999). Modelling chloride diffusion in concrete: Effect of fly ash and slag. *Cement and Concrete Research*, *29*(4), 487–495.

Thorek, D. L., Chen, A. K., Czupryna, J., & Tsourkas, A. (2006). Super paramagnetic iron oxide nanoparticle probes for molecular imaging. *Annals of Biomedical Engineering*, *34*(1), 23–38. doi:10.1007/s10439-005-9002-7 PMID:16496086

Tian, P. (2008). Molecular dynamics simulations of nanoparticles. *Annual Reports Section "C"(Physical Chemistry)*, *104*, 142-164.

Tian, P., Han, X. Y., Ning, G., Fang, H., Ye, J., Gong, W., & Lin, Y. (2013). Synthesis of Porous Hierarchical MgO and Its Superb Adsorption Properties. *Applied Materials and Interfaces*, *5*(23), 12411–12418. doi:10.1021/am403352y PMID:24224803

Tiwari, D. K., Behari, J., & Sen, P. (2008). Application of Nanoparticles in Waste Water Treatment. *World Applied Sciences Journal*, *3*(3), 417–433.

Toksha, B. G., Shirsath, S. E., Patange, S. M., & Jadhav, K. M. (2008). Structural investigations and magnetic properties of cobalt ferrite nanoparticles prepared by sol–gel auto combustion method. *Solid State Communications*, *147*(11), 479–483. doi:10.1016/j.ssc.2008.06.040

Tomiyasu, B., Fukuju, I., Komatsubara, H., Owari, M., & Nihei, Y. (1998). High spatial resolution 3D analysis of materials using gallium focused ion beam secondary ion mass spectrometry (FIB SIMS). *Nuclear Instruments & Methods in Physics Research. Section B, Beam Interactions with Materials and Atoms*, *136*, 1028–1033. doi:10.1016/S0168-583X(97)00790-8

Toth, J. (1981). A uniform interpretation of gas/solid adsorption. *Journal of Colloid and Interface Science*, *79*(1), 85–95. doi:10.1016/0021-9797(81)90050-3

Tripathi, S., Sonkar, S. K., & Sarkar, S. (2011). Growth stimulation of gram (Cicer arietinum) plant by water soluble carbon nanotubes. *Nanoscale*, *3*(3), 1176–1181. doi:10.1039/c0nr00722f PMID:21253651

Trouiller, B., Reliene, R., Westbrook, A., Solaimani, P., & Schiestl, R. H. (2009). Titanium dioxide nanoparticles induce dna damage and genetic instability in vivo in mice. *Cancer Research*, *69*(22), 8784–8789. doi:10.1158/0008-5472.CAN-09-2496 PMID:19887611

Tsuchiya, T., Oguri, I., Yamakoshi, Y., & Miyata, N. (1996). Novel harmful effects of fullerene on mouse embryos in vitro and in vivo. *FEBS Letters*, *393*(1), 139–145. doi:10.1016/0014-5793(96)00812-5 PMID:8804443

Tyliszczak, T., & Chou, K. W. (n.d.). *STXM in nanoscience ver2*. Lawrence Berkeley National Laboratory. Retrieved from http://pdf.internationalx.net/STXM-in-nanoscience-ver2---Lawrence-Berkeley-National-Laboratory-download-w18866.html#

Tyliszczak, T., Warwick, T., Kilcoyne, A. L. D., Fakra, S., Shuh, D. K., Yoon, T. H., ... Acremann, Y. (2004). Soft X-ray scanning transmission microscope working in an extended energy range at the advanced light source. In Synchrotron Radiation Instrumentation (Vol. 705, pp. 1356-1359).

U.S. Environmental Protection Agency Nanotechnology White Paper. (2007). Retrieved May 17, 2015 from http://www.epa.gov/osa/pdfs/nanotech/epa-nanotechnology-whitepaper-0207.pdf

Uchic, M. D., Holzer, L., Inkson, B. J., Principe, E. L., & Munroe, P. (2007). Three-dimensional microstructural characterization using focused ion beam tomography. *MRS Bulletin*, *32*(05), 408–416. doi:10.1557/mrs2007.64

Uskokovic, V. (2007). Nanotechnologies: What we do not know. *Technology in Society.*, *29*(1), 43–61. doi:10.1016/j.techsoc.2006.10.005

Valentini, L., Biagiotti, J., Kenny, J. M., & Santucci, S. (2003). Morphological characterization of single-walled carbon nanotubes-PP composites. *Composites Science and Technology*, *63*(8), 1149–1153. doi:10.1016/S0266-3538(03)00036-8

Varghese, P. C. (2010). *Building Materials*. PHI Learning Private Limited.

Videa, J. R. P., Zhao, L., Morenoc, M. L. L., de la Rosa, G., Hong, J., & Torresdey, J. L. G. (2011). Nanomaterials and the environment: A review for the biennium 2008–2010. *Journal of Hazardous Materials*, *186*(1), 1–15. PMID:21134718

Virji, M. A., & Stefaniak, A. B. (2014). 8.06 - A Review of Engineered Nanomaterial Manufacturing Processes and Associated Exposures. In *Comprehensive Materials Processing* (pp. 103–125). Oxford, UK: Elsevier. doi:10.1016/B978-0-08-096532-1.00811-6

Walters, D. A., Ericson, L. M., Casavant, M. J., Liu, J., Colbert, D. T., Smith, K. A., & Smalley, R. E. (1999). Elastic strain of freely suspended single-wall carbon nanotube ropes. *Applied Physics Letters*, *74*(25), 3803–3805. doi:10.1063/1.124185

Compilation of References

Wang, M. J., Reznek, S. R., Kutsovsky, Y., & Mahmud, K. (2002). *Elastomeric compounds with improved wet skid resistance and methods to improve wet skid resistance.* US Patent 6469089. Retrieved from www.nanoandme.org/nano-products/paints-and-coatings

Wang, B., Feng, W., Wang, T., Jia, G., Wang, M., Shi, J., & Chai, Z. et al. (2006). Acute toxicity of nano- and micro-scale zinc powder in healthy adult mice. *Toxicology Letters, 161*(2), 115–123. doi:10.1016/j.toxlet.2005.08.007 PMID:16165331

Wang, B., Wang, L., & Lai, F. C. (2008). Freezing resistance of HPC with nano-SiO_2. *Journal of Wuhan University of Technology-Mater, 23*(1), 85–88.

Wang, J. N., Su, L. F., & Wu, Z. P. (2008). Growth of highly compressed and regular coiled carbon nanotubes by a spray-pyrolysis method. *Crystal Growth & Design, 8*(5), 1741–1747. doi:10.1021/cg700671p

Wang, L., Ding, T., & Wang, P. (2009). Influence of carbon black concentration on piezoresistivity for carbon-black-filled silicone rubber composite. *Carbon, 47*(14), 3151–3157.

Wang, L., & Yamauchi, Y. (2009). Facile synthesis of three-dimensional dendritic platinum nanoelectrocatalyst. *Chemistry of Materials, 21*(15), 3562–3569. doi:10.1021/cm901161g

Wang, Y., Zhao, X., Du, J., & Lan, S. (2008). Study on improving mechanical property and pressure sensibility of cement-based composites with nano-sized carbon black. *New Building Materials, 35*(12), 6–9.

Wang, Z. L., Liu, Y., & Zhang, Z. (2002). *Handbook of nanophase and nanostructured materials: Characterization* (Vol. 2). Kluwer Academic/Plenum.

Wang, Z., Han, E., & Ke, W. (2006). Effect of nanoparticles on the improvement in fire-resistant and anti-ageing properties of flame-retardant coating. *Surface and Coatings Technology, 200*(20–21), 5706–5716. doi:10.1016/j.surfcoat.2005.08.102

Wansom, S., Kidner, N. J., Woo, L. Y., & Mason, T. O. (2006). AC-impedance response of multi-walled carbon nanotube/cement composites. *Cement and Concrete Composites, 28*(6), 509–519.

Warneke, B. A., Scott, M. D., Leibowitz, B. S., Zhou, L., Bellew, C. L., Chediak, J. A., ... Pister, K. S. (2002). *An autonomous 16 mm 3 solar-powered node for distributed wireless sensor networks.* Paper presented at the Sensors. doi:10.1109/ICSENS.2002.1037346

Weber, W. J. Jr, & Morris, J. C. (1963). Kinetics of adsorption on carbon from solution. *Journal of Sanitation Engineering Division American Society of Civil Engineering, 89*(1), 31–60.

Wiesner, M. R., Lowry, G. V., Alvarez, P., Dionysiou, D., & Biswas, P. (2006). Assessing the risks of manufactured nanomaterials. *Environmental Science & Technology, 40*(14), 4336–4345. doi:10.1021/es062726m PMID:16903268

Wille, K., Naaman, A. E., El-Tawil, S., & Parra-Montesinos, G. J. (2012). Ultra-high performance concrete and fiber reinforced concrete: Achieving strength and ductility without heat curing. *Materials and Structures/Materiaux et Constructions, 45*(3), 309-324.

Williamson, R. B. (1972). Solidification of Portland cement. *Progress in Materials Science, 15*(3), 189–286.

Wu, R., Qu, J., He, H., & Yu, Y. (2004). Removal of azo-dye Acid Red B (ARB) by adsorption and catalytic combustion using magnetic $CuFe_2O_4$ powder. *Applied Catalysis B: Environmental, 48*(1), 49–56. doi:10.1016/j.apcatb.2003.09.006

Wynand, J. M. (2009). Potential Applications of Nanotechnology in Pavement Engineering. *Journal of Transportation Engineering, 135*(10), 764–772. doi:10.1061/(ASCE)0733-947X(2009)135:10(764)

Xia, H., Feng, J., Wang, H., Lai, M. O., & Lu, L. (2010). MnO2 nanotube and nanowire arrays by electrochemical deposition for supercapacitors. *Journal of Power Sources, 195*(13), 4410–4413. doi:10.1016/j.jpowsour.2010.01.075

Xiao, F., Amirkhanian, A. N., & Amirkhanian, S. N. (2011). Influence of Carbon Nanoparticles on the Rheological Characteristics of Short-Term Aged Asphalt Binders. *Journal of Materials in Civil Engineering, 23*(4), 423–431. doi:10.1061/(ASCE)MT.1943-5533.0000184

Xiao, H., Li, H., & Ou, J. (2011). Self-monitoring Properties of Concrete Columns with Embedded Cement-based Strain Sensors. *Journal of Intelligent Material Systems and Structures, 22*(2), 191–200.

Xu, H., Zeiger, B. W., & Suslick, K. S. (2013). Sonochemical synthesis of nanomaterials. *Chemical Society Reviews, 42*(7), 2555–2567. doi:10.1039/C2CS35282F PMID:23165883

Xu, K., Cao, P., & Heath, J. R. (2009). Scanning tunneling microscopy characterization of the electrical properties of wrinkles in exfoliated graphene monolayers. *Nano Letters, 9*(12), 4446–4451. doi:10.1021/nl902729p PMID:19852488

Yakovlev, G., Kerienė, J., Gailius, A., & Girnienė, I. (2006). Cement based foam concrete reinforced by carbon nanotubes. *Materials Science, 12*(2), 147–151.

Yang, C., Hazeghi, A., Takei, K., Hong-Yu, C., Chan, P. C. H., Javey, A., & Wong, H. S. P. (2010, 6-8 Dec. 2010). *Graphitic interfacial layer to carbon nanotube for low electrical contact resistance.* Paper presented at the Electron Devices Meeting (IEDM), 2010 IEEE International.

Yang, J., & Tighe, S. (2013). A review of advances of Nanotechnology in asphalt mixtures. *Procedia: Social and Behavioral Sciences, 96*, 1269–1276. doi:10.1016/j.sbspro.2013.08.144

Yang, J., Zeng, Q., Peng, L., Lei, M., Song, H., Tie, B., & Gu, J. (2013). La-EDTA coated Fe_3O_4 nanomaterial: Preparation and application in removal of phosphate from water. *Journal of Environmental Sciences (China), 25*(2), 413–418. doi:10.1016/S1001-0742(12)60014-X PMID:23596964

Yang, K., Wu, W., Jing, Q., & Zhu, L. (2008). Aqueous Adsorption of Aniline, Phenol, and their Substitutes by Multi-Walled Carbon Nanotubes. *Environmental Science & Technology, 42*(21), 7931–7936. doi:10.1021/es801463v PMID:19031883

Compilation of References

Yang, L., & Watts, D. J. (2005). Particle surface characteristics may play an important role in phytotoxicity of alumina nanoparticles. *Toxicology Letters*, *158*(2), 122–132. doi:10.1016/j.toxlet.2005.03.003 PMID:16039401

Yang, T., Keller, B., & Magyari, E. (2002). AFM investigation of cement paste in humid air at different relative humidities. *Journal of Physics-London-D Applied Physics*, *35*(8), L25–L28. doi:10.1088/0022-3727/35/8/101

Yang, Y., Lepech, M. D., Yang, E. H., & Li, V. C. (2009). Autogenous healing of engineered cementitious composites under wet–dry cycles. *Cement and Concrete Research*, *39*(5), 382–390. doi:10.1016/j.cemconres.2009.01.013

Yao, W., Zuo, J., & Wu, K. (2013). Microstructure and thermoelectric properties of carbon nanotube-carbon fiber/cement composites. *Journal of Functional Materials*, *13*.

Yao, H., You, Z., Li, L., Goh, S. W., Lee, C. H., Yap, Y. K., & Shi, X. (2013). Rheological properties and chemical analysis of nanoclay and carbon microfiber modified asphalt with Fourier transform infrared spectroscopy. *Construction & Building Materials*, *38*(0), 327–337. doi:10.1016/j.conbuildmat.2012.08.004

Yao, N., & Wang, Z. L. (2005). *Handbook of Microscopy for Nanotechnology*. Springer. doi:10.1007/1-4020-8006-9

Yazdanbakhsh, A., Grasley, Z., Tyson, B., & Abu Al-Rub, R. K. (2010). Distribution of carbon nanofibers and nanotubes in cementitious composites. *Transportation Research Record*, *2142*, 89–95. doi:10.3141/2142-13

Yeganeh, J. K., Sadeghi, M., & Kourki, H. (2008). Recycled HIPS and nanoclay in improvement of cement mortar properties. *Malaysian Polym J*, *3*(2), 32–38.

Ye, Q., Zhang, Z. N., Kong, D. Y., Chen, R. S., & Ma, C. C. (2003). Comparison of properties of high-strength concrete with nano-SiO2 and silica fume added. *Journal of Building Materials*, *6*(4), 281–285.

Ye, Q., Zhang, Z., Kong, D., & Chen, R. (2007). Influence of nano-SiO_2 addition on properties of hardened cement paste as compared with silica fume. *Construction & Building Materials*, *21*(3SI), 539–545.

Yilbas, B. S. (2014). Introduction to Nano- and Microscale Processing – Modeling. In M. S. J. Hashmi (Ed.), *Comprehensive Materials Processing* (pp. 1–2). Oxford, UK: Elsevier. doi:10.1016/B978-0-08-096532-1.00700-7

Yin, L., Cheng, Y., Espinasse, B., Colman, B. P., Auffan, M., Wiesner, M., & Bernhardt, E. S. et al. (2011). More than the ions: The effects of silver nanoparticles on Lolium multiflorum. *Environmental Science & Technology*, *45*(6), 2360–2367. doi:10.1021/es103995x PMID:21341685

Yogesh, K. K., Muralidhara, H. B., Nayaka, Y. A., Balasubramanyam, J., & Hanumanthappa, H. (2013). Low-cost synthesis of metal oxide nanoparticles and their application in adsorption of commercial dye and heavy metal ion in aqueous solution. *Powder Technology*, *246*, 125–136.

Yoo, H., Lee, S., Kang, D., Kim, T., Gweon, D., Lee, S., & Kim, K. (2006). Confocal Scanning Microscopy. *International Journal of Precision Engineering and Manufacturing, 7*(4), 3–7.

Yousefi, A., Allahverdi, A., & Hejazi, P. (2013). Effective dispersion of nano-TiO_2 powder for enhancement of photocatalytic properties in cement mixes. *Construction & Building Materials, 41*, 224–230.

You, Z., Mills-Beale, J., Foley, J. M., Roy, S., Odegard, G. M., Dai, Q., & Goh, S. W. (2011). Nanoclay-modified asphalt materials: Preparation and characterization. *Construction & Building Materials, 25*(2), 1072–1078. doi:10.1016/j.conbuildmat.2010.06.070

Yu, J. Y., Feng, P. C., Zhang, H. L., & Wu, S. P. (2009). Effect of organomontmorillonite on aging properties of asphalt. *Construction & Building Materials, 23*(7), 2636–2640. doi:10.1016/j.conbuildmat.2009.01.007

Yu, J., Zeng, X., Wu, S., Wang, L., & Liu, G. (2007). Preparation and properties of montmorillonite modified asphalts. *Materials Science and Engineering A, 447*(1–2), 233–238. doi:10.1016/j.msea.2006.10.037

Zaafarani, N., Raabe, D., Singh, R. N., Roters, F., & Zaefferer, S. (2006). Three-dimensional investigation of the texture and microstructure below a nanoindent in a Cu single crystal using 3D EBSD and crystal plasticity finite element simulations. *Acta Materialia, 54*(7), 1863–1876. doi:10.1016/j.actamat.2005.12.014

Zapata, L. E., Portela, G., Suárez, O. M., & Carrasquillo, O. (2013). Rheological performance and compressive strength of superplasticized cementitious mixtures with micro/nano-SiO2 additions. *Construction & Building Materials, 41*, 708–716. doi:10.1016/j.conbuildmat.2012.12.025

Zare-Shahabadi, A., Shokuhfar, A., & Ebrahimi-Nejad, S. (2010). Preparation and rheological characterization of asphalt binders reinforced with layered silicate nanoparticles. *Construction & Building Materials, 24*(7), 1239–1244. doi:10.1016/j.conbuildmat.2009.12.013

Zhang, B., Xi, M., Zhang, D., Zhang, H., & Zhang, B. (2009). The effect of styrene-butadiene-rubber/montmorillonite modification on the characteristics and properties of asphalt. *Construction & Building Materials, 23*(10), 3112–3117. doi:10.1016/j.conbuildmat.2009.06.011

Zhang, G., Ren, Z., Zhang, X., & Chen, J. (2013). Nanostructured iron(III)-copper(II) binary oxide: A novel adsorbent for enhanced arsenic removal from aqueous solutions. *Water Research, 47*(12), 4022–4031. doi:10.1016/j.watres.2012.11.059 PMID:23571113

Zhang, J., Morsdorf, L., & Tasan, C. C. (2016). Multi-probe microstructure tracking during heat treatment without an in-situ setup: Case studies on martensitic steel, dual phase steel and β-Ti alloy. *Materials Characterization, 111*, 137–146. doi:10.1016/j.matchar.2015.11.019

Zhang, M. H., & Islam, J. (2012). Use of nano-silica to reduce setting time and increase early strength of concretes with high volumes of fly ash or slag. *Construction & Building Materials, 29*, 573–580.

Compilation of References

Zhang, M. H., Islam, J., & Peethamparan, S. (2012). Use of nano-silica to increase early strength and reduce setting time of concretes with high volumes of slag. *Cement and Concrete Composites, 34*(5), 650–662.

Zhang, M., & Li, H. (2011). Pore structure and chloride permeability of concrete containing nano-particles for pavement. *Construction & Building Materials, 25*(2), 608–616.

Zhang, W. (2003). Nanoscale iron particles for environmental remediation: An overview. *Journal of Nanoparticle Research, 5*(3/4), 323–332. doi:10.1023/A:1025520116015

Zhang, X., Li, Q., Holesinger, T. G., Arendt, P. N., Huang, J., Kirven, P. D., & Zhao, Y. et al. (2007). Ultrastrong, stiff, and lightweight Carbon-Nanotube fibers. *Advanced Materials, 19*(23), 4198–4201.

Zhang, Y., Brar, V. W., Wang, F., Girit, C., Yayon, Y., Panlasigui, M., & Crommie, M. F. et al. (2008). Giant phonon-induced conductance in scanning tunnelling spectroscopy of gate-tunable graphene. *Nature Physics, 4*(8), 627–630. doi:10.1038/nphys1022

Zhao, H., Zhang, Y., Bradford, P. D., Zhou, Q., Jia, Q., Yuan, F.-G., & Zhu, Y. (2010). Carbon nanotube yarn strain sensors. *Nanotechnology, 21*(30), 305502. doi:10.1088/0957-4484/21/30/305502 PMID:20610871

Zhao, J., Wang, Z., White, J. C., & Xing, B. (2014). Graphene in the Aquatic Environment: Adsorption, Dispersion, Toxicity and Transformation. *Environmental Science & Technology, 48*(17), 9995–10009. doi:10.1021/es5022679 PMID:25122195

Zhao, X., Wang, J., Wu, F., Wang, T., Cai, Y., Shi, Y., & Jiang, G. (2010). Removal of fluoride from aqueous media by Fe_3O_4@$Al(OH)_3$ magnetic nanoparticles. *Journal of Hazardous Materials, 173*(1-3), 102–109. doi:10.1016/j.jhazmat.2009.08.054 PMID:19747775

Zheng, L., O'connell, M., Doorn, S., Liao, X., Zhao, Y., Akhadov, E., & Dye, R. et al. (2004). Ultralong single-wall carbon nanotubes. *Nature Materials, 3*(10), 673–676. doi:10.1038/nmat1216 PMID:15359345

Zheng, Y., Li, N., & Zhang, W.-D. (2012). Preparation of nanostructured microspheres of Zn–Mg–Al layered double hydroxides with high adsorption property. *Colloids and Surfaces. A, Physicochemical and Engineering Aspects, 415*, 195–201. doi:10.1016/j.colsurfa.2012.10.014

Zhong, B. Q., & Zhu, Q. (2002). Study on the application of polypropylene fibre concrete in water and hydropower projects. *Water Res. Plann. Design, 1*, 54–58.

Zhong, Y., Liu, L., Wikman, S., Cui, D., & Shen, Z. (2016). Intragranular cellular segregation network structure strengthening 316L stainless steel prepared by selective laser melting. *Journal of Nuclear Materials, 470*, 170–178. doi:10.1016/j.jnucmat.2015.12.034

Zhu, J., Lu, Z., Aruna, S. T., Aurbach, D., & Gedanken, A. (2000). Sonochemical Synthesis of SnO_2 Nanoparticles and Their Preliminary Study as Li Insertion Electrodes. *Chemistry of Materials, 12*(9), 2557–2566. doi:10.1021/cm990683l

Zhu, S., Oberdorster, E., & Haasch, M. (2006). Toxicity of an engineered nanoparticle (Fullerene, C60) in two aquatic species, daphnia and fathead minnow. *Marine Environmental Research, 62*, S5–S9. doi:10.1016/j.marenvres.2006.04.059 PMID:16709433

Zhu, W., Bartos, P. J. M., & Porro, A. (2004). Application of nanotechnology in construction. *Materials and Structures, 37*(9), 649–658. doi:10.1007/BF02483294

Zuyuan, H., Liu, Q., & Tokunaga, T. (2013). *Sensing the Earth Crustal Deformation with Fiber Optics.* Paper presented at the Frontiers in Optics.

Zuyuan, H., Qingwen, L., & Tokunaga, T. (2012, 6-11 May 2012). *Realization of nano-strain-resolution fiber optic static strain sensor for geo-science applications.* Paper presented at the Conference on Lasers and Electro-Optics (CLEO).

About the Contributors

Anwar Khitab, PhD, is the head of the civil engineering department at Mirpur University of Science and Technology, Mirpur, Pakistan. He is a graduate of National Institute of Applied Sciences in Toulouse, France. He also attended Paul Sabatier University, where he was awarded a master's degree in civil engineering. Professionally, he started out in 1997 as an assistant manager in National Engineering and Scientific Commission (NESCOM) in Islamabad, Pakistan. While employed there, he was promoted to manager in 1999 and then general manager in 2005. He left NESCOM in 2009 and joined Preston University Islamabad and in 2010, he joined The University of Wah Cant in Pakistan. During his carrier, he won fellowships from Pakistan Atomic Energy Commission, Ministry of Science and Technology Government of Pakistan, Embassy of France in Pakistan and Ministry of Education, Government of France . He wrote a book on construction materials in 2012, published by the Allied Books, Lahore. He also wrote many research articles on construction materials which have been publish in various impact factor journals.

Waqas Anwar, MSc, is a lecturer in the Civil Engineering Department at Mirpur University of Science & Technology, Pakistan. He holds a Master's degree in Civil Engineering from The University of Nottingham, England. Currently he is a PhD scholar in the field of Civil Engineering. Mr. Waqas joined the academic career since January 2014. In his academic career, he remained involved in many research works and also published some research papers with good impact factors. Besides academic experience, he also has good engineering field experience in government as well as private firms including PWD, Descon and Xinjiang Beixin. He is an Associate member of ASCE and is also playing vital role in different academia societies.

* * *

Imtiaz Ahmed studied Civil Engineering at Undergraduate and Post-Graduate level from a very reputable institute in Pakistan. His areas of interest is Transportation Engineering.

About the Contributors

Naveed Ahmad is currently working as a faculty member at the Department of Civil Engineering, University of Engineering and Technology Taxila, Pakistan.

Sushmita Banerjee completed her MSc in Environmental Science from University of Allahabad, India. Now she is doing her D.Phil. Her area of research is wastewater treatment using low cost agricultural wastes and nanoengineered materials.

Mahesh Chandra Chattopadhyaya is an eminent researcher in the field of nano particle synthesis, wastewater treatment, electrochemical sensors and solid oxide fuel cell. He has significant contribution in the related areas.

Amin Chegenizadeh is a Researcher at the Department of Civil Engineering, Faculty of Science and Engineering, Curtin University.

Siqi Ding received the B.S. degree in material science and technology from Dalian Jiaotong University, Dalian, China in 2013. He is currently pursuing the M.S. degree at the School of Civil Engineering, Dalian University of Technology, Dalian, China. His current research interests include smart materials and structures, sensors, and traffic detection.

Elaheh Esmaeili is an Assistant Professor Dept. of Chemical Engineering Birjand University of Technology.

Pavan Kumar Gautam did his M.Sc. Environmental Science from Baba Saheb Bhim Rao Ambbedkar University, Lucknow, India. Then he joined D.Phil. in the Department of Chemistry, University of Allahabad.

Ravindra Kumar Gautam did his post graduation in Environmental Science from University of Allahabad, India in 2009. He did a post graduate diploma in Disaster Management from Indira Gandhi National Open University, New Delhi in 2010. Thereafter, he worked for one year in National Environmental Engineering Research Institute, Council of Scientific & Industrial Research (NEERI-CSIR), Nagpur, India. He qualified for CSIR-UGC National Eligibility Test for Junior Research Fellowship. He has published 57 research papers including original research articles, reviews, books, book chapters, and conference proceedings. He has written a book entitled, "Environmental Magnetism: Fundamentals and Applications" (ISBN-10: 3659209090 | ISBN-13: 978-3659209093) which was published by LAP Lambert Academic Publishing, Saarbrucken, Germany. Yet, 5 research articles and 3 books are in pipeline. He is selected as a Fellow of the Indian Chemical Society and Life Member of the Indian Science Congress Association in 2013. He is a member of

About the Contributors

the editorial boards of International Journal of Nanoscience and Nanoengineering, American Journal of Environmental Engineering and Science, and International Journal of Environmental Monitoring and Protection. He also serves as a reviewer for more than 20 journals of international repute. Currently, he is engaged in doctoral work in the Department of Chemistry, University of Allahabad, Allahabad 211002, India. His areas of interests are adsorption and nanomaterials, and their analogues for water/wastewater remediation.

Baoguo Han received his PhD in the field of smart materials and structures from the Harbin Institute of Technology, China, in 2005. He is a professor of civil engineering in Dalian University of Technology, China. His main research interests include cement and concrete materials, smart materials and structures, sensors, structural health monitoring, traffic detection and nanotechnology. He is the author of 1 book, 7 book chapters and more than 50 published papers. He also was awarded the New Century Excellent Talents in University by the Ministry of Education of China and the first prize of Natural Science by the Ministry of Education of China as the 3rd participant.

Israr Ul Haq did his B.Sc in Civil Engineering from Balochistan University of Engineering and Technology Khuzdar, Pakistan and M.Sc in Structural Engineering from National University of Sciences and Technology (NUST) Islamabad, Pakistan. He has three (03) years' experience of field work with NESPAK Private Limited, Pakistan as Junior Engineer and one (01) year field experience with DHA Islamabad, Phase 1 as Site Engineer. Currently he is in teaching field and working as Lecturer in Civil Engineering Department of Mirpur University of Science and Technology (MUST) Mirpur, Azad Kashmir for more than three (03) years.

Muhammad Hassan has been working as a faculty member in Civil Engineering Department of Mirpur University of Science and Technology, Mirpur AJK, Pakistan.

Umair Hasan received his MPhil in Civil Engineering from Curtin University in 2015. He has a wide range of experience in the construction industry and is currently pursuing PhD at Department of Civil Engineering, Faculty of Science and Engineering, Curtin University and writes and researches wide range of topics related to geotechnical engineering, nanotechnology and project management.

Muhammad Umer Arif Khan is a Lecturer in the Civil Engineering Department, MUST.

Iman Mansouri is an Assistant Professor Dept. of Civil Engineering Birjand University of Technology.

Imran Mehmood is a faculty member in CED, MUST Mirpur, AJK Pakistan.

Hamid Nikraz is a Professor at the Department of Civil Engineering, Faculty of Science and Engineering, Curtin University.

Jinping Ou received the Ph.D. degree from the Harbin Institute of Technology, China, in 1987. He is currently a Professor with the Harbin Institute of Technology, and the Dalian University of Technology. He has authored four books and over 200 published papers. His main research interests include structural damage, reliability and health monitoring, structural vibration and control, and smart material and structures.

Ahmed Sharif received his first class honours B.Sc. Engineering degree in Metallurgical Engineering from the Bangladesh University of Engineering and Technology (BUET), Dhaka, Bangladesh in 1998 and also the M. Sc. Engineering degree in 2001 with record CGPA. He completed his Ph. D. in 2005 at City University of Hong Kong (CityU). He began his academic career in BUET as a Lecturer from 1999 and was promoted to Full Professor in 2014. He began his research activities during his final year of undergraduate study in 1997. From this work, his research output was first disseminated in 1999 at a conference in Ireland. Since then he has published steadily in different international conferences and peer-reviewed journals. His publication record, 40 peer-reviewed (SCI listed) journal papers with h-index of 19 reflects his potentiality and leadership qualities in research. A part of his PhD research has led to the award of an international prize the "IEEE CPMT Young Scientist Award" for his paper presentation in an IEEE conference held in Japan in 2004. This demonstrates an international recognition of his research output by contemporary researchers worldwide. Currently he is serving as the Head of the Department of Materials and Metallurgical Engineering, BUET, Dhaka.

Shengwei Sun is a PhD candidate in Smart/intelligent Concrete Materials and Structure, High Performance Concrete Materials and Structure, Harbin Institute of Technology.

Xun Yu received a Ph.D. in mechanical engineering from the University of Minnesota-Twin Cities in 2006. He then joined the Department of Mechanical and Industrial Engineering at the University of Minnesota-Duluth, where he worked as an assistant professor from 2006-2010 and associate professor from 2010-2011. He

About the Contributors

was an associate professor and associate chair for academic affairs in the Mechanical and Energy Engineering of the University of North Texas from 2011- 2015. He is an associate professor and chair of Department of Mechanical Engineering of New York Institute of Technology from 2015.

Liqing Zhang is a PhD candidate in civil engineer of Dailian University of Technology, China. Her main research interests include cement and concrete materials, smart materials, and nanotechnology.

Index

A

AFM 80, 90-93, 102, 152, 154, 156, 177, 221, 224
Anthropogenic 144, 178
asphalt 161, 169, 177-178, 187-191, 197-203, 206
atomic force 32, 76, 80, 90, 92, 108, 138, 152, 167-168, 172, 224, 248
Atterberg's limits 161, 164, 169

B

bricks 1-2, 6-12, 14, 26, 107, 203
Building Materials 1-2, 27, 65-66, 68-70, 72-73, 75-79, 84, 106-107, 109, 131-132, 134-135, 137-139, 171, 175-177, 185, 191, 198-201, 203-204, 206-207, 266
bulk properties 112

C

carbon nanotubes 30, 49, 65, 68-74, 76-78, 81, 85, 107, 113, 120, 124, 126-127, 131, 135-139, 141, 147, 151, 155, 161-162, 164, 170, 172, 174, 176, 178, 190, 196, 199, 202, 204-205, 218, 228, 230-231, 238-239, 241, 243, 248-249, 251-252, 260, 271, 273-275
Carcinogens 253
case studies 80, 82, 102, 140
Cation Exchange Capacity 150, 178
cement 1-2, 7, 11-14, 28-79, 81-82, 84-85, 89, 91, 96-98, 100, 102, 104-108, 117-124, 127, 131-139, 145-149, 151, 154, 164-166, 169-177, 191-192, 204, 206, 272
characterization 26, 71-72, 77, 80, 82, 87, 91, 93, 95, 98, 101-102, 104, 106-107, 109, 134, 140, 151-152, 157, 169, 176-177, 186-187, 198, 201-202, 206, 208, 219, 242-243, 249, 256-258
civil engineering 1-2, 22, 27, 51, 111-112, 116, 128, 131-132, 146-147, 168, 172, 180, 183-185, 198-201, 206, 251, 253, 260
clay particles 5, 65, 89, 149-152, 167-168
compression strengths 10, 16, 20, 26, 30, 33, 36-41, 43-44, 48-50, 53, 56-59, 66, 68-69, 73-77, 81, 116-117, 124-125, 139, 146-148, 161-166, 171, 174, 191, 203
construction 2-4, 6-14, 16-22, 24-27, 29, 48, 64-66, 68-70, 72-73, 75-82, 87, 100, 102, 104, 106-107, 110-113, 115, 128-132, 134-139, 141, 147-148, 165, 167-168, 171, 173, 175-177, 180-183, 185-186, 189, 195, 197-201, 203-204, 206-207, 253, 260, 273
cylinder structure 147, 162

D

Dowel 148, 178

Index

durability 3, 6, 12, 16, 22, 24, 26, 28-34, 38, 40-41, 44, 47-48, 50, 52-54, 57-59, 61-62, 75-77, 81, 100, 108, 111, 117-119, 127, 138, 141, 147-148, 150-151, 160, 165-167, 169, 175, 180, 185-186, 192-194, 197, 200, 253
dyes 209, 230, 241, 245, 247-248

E

empirical relations 142, 151, 157
environment 6, 16, 22, 29-30, 40, 54, 62-64, 66, 69, 82, 119, 132, 134, 167, 170, 181, 183, 191, 193-198, 209, 238-239, 243, 245, 251-254, 257-258, 260-263, 266, 268-269, 273, 275
Equiaxed 144, 178
expansive soils 150, 171, 178

F

fly ash 13, 30, 33, 35-36, 38-39, 41-42, 47, 57-58, 60, 68-72, 75-79, 84, 104, 119-120, 134, 138, 147, 272
FTIR 151, 156-157, 221-223
Fullerenes 144, 178, 260, 274
Functional Properties 62

G

geotechnical engineering 141-143, 148-149, 151, 155, 160, 167, 172, 174, 185
glass 1-2, 14, 17-21, 59, 120-121, 125, 127, 130, 137, 147, 185

H

hazards 141, 152, 167, 244, 253, 258-259, 262, 265, 271, 274-275
heavy metals 209, 230, 246, 262
hexagonal honeycomb 147, 162
high energy 28-29, 187, 196, 211-212
human health 194-195, 204, 209, 238-239, 244, 253, 258-259, 266, 268, 271

hydration 28-31, 33-35, 39-40, 43, 46-47, 49, 57, 59-61, 64, 66, 72, 77, 81, 100, 107, 117-124, 132, 134-135, 146-147, 154, 164-165

I

industrial revolution 128, 130, 174, 181
inter-particle interactions 149, 151, 168

K

Kinetics 233, 236-237, 240-243, 248, 250-251

L

low-temperature cracking 188, 197
LSCM 80, 99-100

M

Magnetotactic Bacteria 143, 178
MD simulation 158-160
MD simulations 157
mechanical properties 28, 30-31, 33-34, 36-38, 40-46, 48-49, 52-55, 57-58, 60-61, 65, 68-69, 71-74, 77, 81-82, 84, 104, 119-121, 124-125, 127, 130-131, 133-136, 138, 146, 149, 161, 163, 165-166, 168, 172-173, 191
MEMS 151-152, 154-155, 168
metamorphic rock 3, 6
micro fabrication 81
microscopic sciences 142
modern structures 2, 18
molecular dynamics 69, 141, 157, 159, 168, 175-176
Monte Carlo 157-159
montmorillonite 89-90, 150-151, 161, 168, 188-189, 199, 206, 245
multidisciplinary research 81

N

nano-alumina 66, 71, 136, 141, 166-167, 169, 173

nanoclay 78, 89, 169, 173, 178, 187-189, 191, 197, 200-201, 206
nanomaterials 28, 31, 34, 38, 59-64, 69-70, 75, 80-82, 87, 90, 102, 107, 111, 113-114, 119, 122, 131, 137, 139, 141-146, 149, 159, 161-163, 169, 173, 175, 180, 183, 186, 188, 194-199, 202, 204, 208, 217-218, 226-227, 229, 238-239, 242-246, 251, 254-255, 261-262, 265-270, 272-274, 276
nano-scale properties 145
Nanoscopy 80
nano-sensors 141-142, 151-152, 155, 185
nano-silica 30, 34, 66, 70, 75-76, 79, 119-120, 135-136, 138, 141, 146, 163, 165-167, 169, 176, 187
nanotechnology 1, 17, 28-31, 33, 66, 68, 70, 76-77, 80-82, 87, 104-105, 107-117, 119, 128-139, 141-144, 146-152, 157-158, 160-161, 165, 167-168, 171-178, 180-188, 190, 192-208, 210-211, 239-240, 248-249, 253-256, 258, 261-264, 266-267, 269-271, 273-276
NeaSNOM 156
NEMS 155, 168
non-crystalline material 142

P

paints 21-23, 26, 81-82, 260-261, 269
particle size 31, 34, 40-43, 55-56, 58, 69, 82, 87, 134, 149, 168, 196, 211-214, 216-218, 221
Pavement Durability 141, 151
pesticides 209
plastics 1-2, 16-18, 112, 178, 260
pollutant species 230, 233
pore structure 30, 47, 75, 79, 88, 192, 207
prevention 148, 253, 256, 269, 274

Q

quantum effects 142, 145

R

rectangular block 7
rheological characteristics 151-152, 159, 164, 169, 206

S

sheet shaped 147, 162
soil composition 148, 167
soil stabilization 141, 147, 157, 160-162
s-SNOM system 156
STM 80, 92-95, 102-103, 106
storage units 254
STXM 80, 96, 98-99, 109
Subgrade 142, 179
surface area 43, 59, 82, 117-118, 124, 142, 145-146, 150, 166, 190, 210, 216, 221, 223-224, 227-230, 233, 238, 241, 257, 265
sustainable transportation facilities 180
synthesis 53, 72, 82, 106-107, 135-137, 139, 198, 208, 211-218, 225, 238-239, 241-242, 244-247, 249-252, 259, 275

T

task forces 112
technology development 112
TEM 31, 80, 86-90, 94, 102, 105, 113-115, 217, 220-222, 224
toxicity 164, 196, 199, 204, 238-239, 242, 244-245, 247-248, 252, 256-257, 260, 265, 271, 273-276
transportation engineering 180, 185-186, 192-193, 197, 206
transverse stresses 7

U

uses 17, 87, 147, 152, 155-156, 195, 270

V

van der Waals forces 85, 150, 179, 215
various shapes 29

Index

W

waste streams 208-209, 225
wastewater treatment 208-210, 223, 225, 240, 248, 261, 275
water runoff 254
water treatment 195, 203, 208, 210-211, 227-228, 230, 240-241, 243, 248, 251, 261
wet deposition 254

workability 12, 28-29, 33-35, 40, 43, 46, 49, 55, 57, 59, 66, 74, 85, 174

X

X-Ray Diffraction 118, 186, 219

Z

zinc oxide 23, 137, 199-200, 215, 219, 238, 249-250

Become an IRMA Member

Members of the **Information Resources Management Association (IRMA)** understand the importance of community within their field of study. The Information Resources Management Association is an ideal venue through which professionals, students, and academicians can convene and share the latest industry innovations and scholarly research that is changing the field of information science and technology. Become a member today and enjoy the benefits of membership as well as the opportunity to collaborate and network with fellow experts in the field.

IRMA Membership Benefits:

- **One FREE Journal Subscription**
- **30% Off Additional Journal Subscriptions**
- **20% Off Book Purchases**
- Updates on the latest events and research on Information Resources Management through the IRMA-L listserv.
- Updates on new open access and downloadable content added to Research IRM.
- A copy of the Information Technology Management Newsletter twice a year.
- A certificate of membership.

IRMA Membership $195

Scan code to visit irma-international.org and begin by selecting your free journal subscription.

Membership is good for one full year.

www.irma-international.org

CPSIA information can be obtained at www.ICGtesting.com
Printed in the USA
BVOW09*0828170516

447879BV00011B/30/P